Geospatial Intelligence for Precision Agriculture

Scrivener Publishing
100 Cummings Center, Suite 541J
Beverly, MA 01915-6106

Publishers at Scrivener
Martin Scrivener (martin@scrivenerpublishing.com)
Phillip Carmical (pcarmical@scrivenerpublishing.com)

Geospatial Intelligence for Precision Agriculture

Edited by

Gurwinder Singh

Vishakha Sood

Narayan Vyas

Surendra Yadav

and

Dankan Gowda V.

Scrivener Publishing

WILEY

This edition first published 2026 by John Wiley & Sons, Inc., 111 River Street, Hoboken, NJ 07030, USA and Scrivener Publishing LLC, 100 Cummings Center, Suite 541J, Beverly, MA 01915, USA

For more information about Scrivener publications please visit www.scrivenerpublishing.com.

Wiley Global Headquarters
111 River Street, Hoboken, NJ 07030, USA

For details of our global editorial offices, customer services, and more information about Wiley products visit us at www.wiley.com.

The manufacturer's authorized representative according to the EU General Product Safety Regulation is Wiley-VCH GmbH, Boschstr. 12, 69469 Weinheim, Germany, e-mail: Product_Safety@wiley.com.

Library of Congress Cataloging-in-Publication Data

ISBN 978-1-394-34660-8

Cover image: Generated with AI using Adobe Firefly
Cover design by Russell Richardson

Set in size of 11pt and Minion Pro by Manila Typesetting Company, Makati, Philippines

Printed in the USA

10 9 8 7 6 5 4 3 2 1

Contents

Preface

With a growing population, rising food demand and climate change, geospatial intelligence heralds a new era of flourishing opportunities. First, it sets the stage for this emerging field by providing an overview of precision agriculture. Precision farming is a method that is partially based on the use of location-based data and smart technologies to improve the agricultural production process and reduce resource wastage. It shows how geospatial tools enhance decision-making throughout the crop cycle, from planting to harvest, by providing higher-resolution and integrated datasets compared with conventional agricultural monitoring approaches. The conversation shifts to geospatial intelligence in the context of agricultural innovations as we delve into the specifics of its applications. Readers will understand how satellite images, drone-based remote sensing, and other technologies enable the detection of problems, resource management, and improved productivity through remote sensing platforms. However, the costs may vary depending on the technology and scale of application.

Integrating geospatial data with Internet of Things (IoT) technologies enhances real-time decision-making in precision agriculture. Different chapters explore the real-time tracking of farm activities through IoT devices, such as smart soil probes and automated irrigation controllers, as well as combinations of global positioning systems (GPSs) and satellite maps. These technologies enable rapid and informed decision-making based on quantitative data with minimal human intervention and conserve time, money, and resources. The book elaborates on recent developments through which such integrated systems are revolutionizing real-time crop management. When conditions change, such as a drop in temperature or declining soil moisture levels, IoT and geospatial technologies enable rapid responses, as demonstrated by the real-time monitoring systems. Precision mapping and geographic information system (GIS) tools enable targeted irrigation and nutrient management, allowing for the application of water and fertilizer specifically in required locations. The book describes the integration of synthetic aperture radar (SAR) and multispectral satellite imagery, utilizing techniques such as HSV-based fusion, to achieve high-accuracy land cover classifications. This enables the differentiation of crop types, identification of crop stress, and monitoring of unauthorized land-use changes with improved resolution.

Once these fused data were obtained, they were combined with machine learning and deep learning algorithms to perform detailed land use land cover (LULC) classification. These methodologies automate and scale previously labor-intensive mapping tasks, providing greater insights into crop types, land degradation, and seasonal changes. The next part of the book, following this chapter, explores the methods used for land planning based on remote sensing and GIS-based sustainable land planning. Through these tools, environmental zones can be fine-tuned for sustainable management in the long run by minimizing harm to the local ecosystem while maximizing production. This will be especially beneficial for governments and policy planners working with large landscapes. Another part focuses on pest and disease management using image classification. Artificial intelligence (AI)-based systems analyze the spectral signatures of crops and detect early signs of pest or disease stress, enabling farmers to take action before substantial damage occurs. This could lead to a significant reduction in pesticide use and crop damage.

The book explores the environmental context through a spatiotemporal analysis of air pollution (NO_2) using Sentinel-5P data, highlighting its impact on agricultural productivity in specific regions. This section illustrates how air quality impacts agriculture and how satellite data can facilitate the tracking and mitigation of these effects through policies and practices. Beyond daily farming, the book explores how tools such as the Google Earth Engine (GEE) are used for natural hazard monitoring, tracking floods, droughts, and other disasters that can severely impact agriculture. These insights are critical for risk mitigation and disaster preparedness planning. As the book approaches its final chapters, it addresses business and economic perspectives. It outlines innovative business models that leverage geospatial technologies, such as agri-tech platforms, data-as-a-service models, and satellite-driven crop insurance schemes, making these tools more accessible to farmers, especially to smallholders.

Eventually, the economics angle expands to evaluate the macroeconomic impacts of these technologies on sustainable farming practices. Geospatial intelligence is not merely a technological advancement for improving agricultural yields; it is a fundamental component of sustainable farming that promotes efficient resource utilization, environmental stewardship, and reduced input costs. The last few chapters look ahead, discussing the Indian policy environment and how geospatial technologies can be best utilized to achieve a sustainable agricultural future. The book concludes by exploring future trends, such as AI-driven analytics, cloud-based data management, and multisensor fusion, demonstrating their potential to transform precision agriculture over the next few decades. We extend our sincere thanks to all the authors, contributors, and reviewers, whose tireless efforts, knowledge, and diverse perspectives significantly contributed to the quality and content of this book. We would especially like to thank our publisher for their continuous support and efforts in bringing this volume to completion.

Dr. Gurwinder Singh
School of Advance Computing, CGC University, Mohali, Punjab, India
Advance Centre of Research & Innovation, CGC University, Mohali, Punjab, India
Dr. Vishakha Sood
Centre of Excellence, Socio-Environmental Sustainability for River Sand Mining (SEnSRS), Indian Institute of Technology Ropar, Rupnagar, Punjab, India
Narayan Vyas
Department of Computer Science and Engineering, Vivekananda Global University, Jaipur, Rajasthan, India
Centre of Excellence in Geospatial Intelligence for Climate Studies, Vivekananda Global University, Jaipur, Rajasthan, India
Dr. Surendra Yadav
Department of Computer Science and Engineering, Vivekananda Global University, Jaipur, Rajasthan, India
Dr. Dankan Gowda V.
Department of Electronics and Communication Engineering, BMS Institute of Technology and Management, Bangalore, India

1

Introduction to Geospatial Intelligence and Precision Agriculture

Neelam Dahiya[1]*, Amanpreet Singh[1] and Sartajvir Singh[2]

[1]Chitkara University Institute of Engineering and Technology, Chitkara University, Rajpura, Punjab, India
[2]Centre of Excellence in Socio-Environmental Sustainability for River Sand Mining (SEnSRS), Indian Institute of Technology Ropar, Rupnagar, Punjab, India

Abstract
Geospatial intelligence (GI) supports precision agriculture by providing data-driven solutions to challenges such as climate change, resource scarcity, and population growth, using data from tools such as remote sensing and analyzed *via* platforms such as Google Earth Engine (GEE) for sustainable farming practices. This chapter describes the adoption of GI technologies, such as remote sensing, global positioning systems (GPS), geographic information systems (GIS), and spatial analytics, by farms to enable better, more efficient, and sustainable agricultural decision making. These tools provide farmers with up-to-date, localized information, allowing them to make informed decisions regarding crops, soil, irrigation, pests, and crop yields. Different applications, such as Normalized Difference Vegetation Index (NDVI)-based crop health monitoring, soil mapping, variable-rate input application, and irrigation scheduling, are elaborated, benefiting from the advantages afforded by GI, including reduced input costs and enhanced productivity, thereby strengthening environmental sustainability. It also highlights several challenges, including high costs, a lack of digital literacy, infrastructure gaps, and data privacy issues, which hinder widespread adoption, particularly in developing regions. This chapter identifies the challenges and suggests potential solutions, such as government assistance, skills development programs, public–private partnerships, and climate-smart agricultural strategies, which can, to some extent, mitigate such barriers. This chapter introduces GI as a transformative approach in agriculture with the potential to drive intelligent, climate-smart, and sustainable agriculture. With appropriate policy support, training, and

**Corresponding author*: neelam.rani@chitkara.edu.in

Gurwinder Singh, Vishakha Sood, Narayan Vyas, Surendra Yadav and Dankan Gowda V. (eds.)
Geospatial Intelligence for Precision Agriculture, (1–18) © 2026 Scrivener Publishing LLC

partnerships, GI can provide smallholder farmers with the tools needed to thrive in a rapidly changing world, as it has done for large agribusinesses.

Keywords: Geospatial intelligence (GI), precision agriculture, remote sensing (RS), smart farming

1.1 Introduction

Agriculture faces numerous challenges, including climate variability, population growth, and resource scarcity. Precision agriculture addresses these challenges by leveraging technology to refine the farming practices. Effective agricultural management is critical to the long-term prosperity of any region [1, 2]. Geospatial technology is a more cost-effective solution that requires a smaller workforce than ground-based surveys, making it ideal for various agricultural applications. It is defined as the analysis and visualization of geospatial data to inform decision-making and plays a pivotal role in precision agriculture [3]. It serves as a game-changing resource, offering valuable insights that support accurate and timely decision-making throughout the farming process. It utilizes tools such as satellite imagery, geographic information systems (GIS), unmanned aerial vehicles (UAVs), and ground-based sensors to generate information about the physical space. They also play a vital role in environmental monitoring, urban planning, disaster response, and precision agriculture [4–6].

Precision agriculture is defined as a method of applying inputs and interventions based on real-time data specific to small areas of land rather than treating entire fields uniformly. The integration of geospatial intelligence (GI) and precision agriculture is highly beneficial for modern farming [7, 8]. It helps farmers by providing a detailed understanding of spatial variability, identifying what constitutes a nutrient deficit in one portion of a field, or whether it is becoming a high-risk region owing to satellite-derived weather patterns. For example, technologies such as the Normalized Difference Vegetation Index (NDVI) enable farmers to assess crop vigor remotely using satellite imagery. Global Positioning System (GPS) sensors guide equipment for precise seed and fertilizer application, improving efficiency and reducing waste [9]. In addition to productivity, it contributes to global sustainability. It benefits the environment by reducing water and chemical use, resulting in less runoff and lower greenhouse gas emissions. It also enhances climate resilience by enabling farmers to respond to changing climate conditions using data-driven strategies, such as planting drought-resistant crops and adjusting irrigation patterns based on current soil moisture levels [10, 11].

GI is an innovative concept, and this chapter discusses its basic principles and role in precision agriculture. It analyzes the primary pillars of this technology, namely remote sensing and GIS, and their applications in agriculture. It addresses current challenges (e.g., cost and technical complexity) and opportunities (e.g., AI-based innovation and policy support) to evaluate the potential of this domain. Based on this foundation, this chapter provides a thorough overview of how GI is transforming agriculture.

1.1.1 Components of GI

GI in precision agriculture encompasses several tools for collecting, analyzing, and visualizing spatial data to help farmers make informed decisions regarding crop management. The main components are described in Figure 1.1.

1.1.2 Remote Sensing

Remote sensing involves collecting data about the Earth's surface without physical contact using sensors on platforms such as satellites, aircraft, or drones [11]. These sensors are used to capture images from various parts of the electromagnetic spectrum. Remote sensing has been widely used

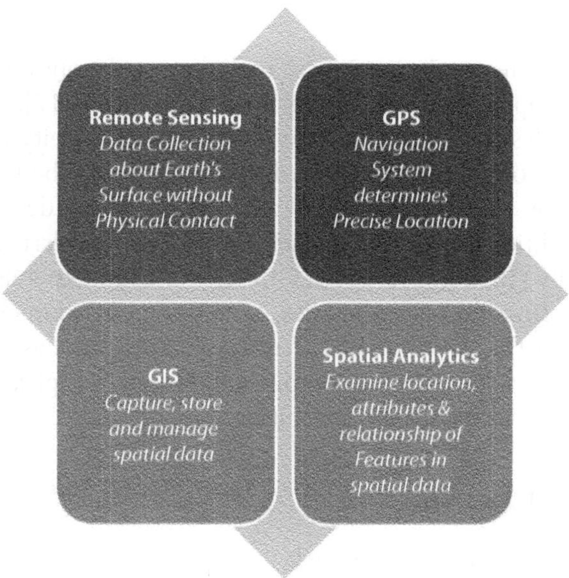

Figure 1.1 Components of GI.

in agriculture for monitoring crop health, tracking seasonal variations in vegetation, and detecting pest infestations. For example, vegetation indices, such as the NDVI, analyze crop reflectance to assess plant vigor and stress levels. This information enables farmers to identify problems early, allowing timely adjustments to prevent production losses. Remote sensing is a low-cost tool for large-scale monitoring and decision-making because of its noninvasive nature and extensive coverage [12, 13].

1.1.3 GPS

GPS contributes to precision agriculture by providing accurate and reliable geolocation information. This is important for mapping the boundaries of fields, directing farm vehicles, and collecting georeferenced information during field surveys [14]. GPS technology is also used in automated systems for tractors, harvesters, and sprayers to ensure that planting, spraying, and harvesting are performed efficiently. Newer forms of GPS, such as Real-Time Kinematic (RTK) positioning, offer centimeter accuracy, which is particularly helpful in row cropping and reducing overlaps and gaps in input applications. Consequently, GPS enhances productivity, reduces the costs of input provisions, and improves accuracy at the field level in farming activities [15, 16].

1.1.4 GIS

The GIS integrates various factors, including soil conditions, altitude, rainfall, and crop history, into a single interactive map [17]. The stratified approach enables the visualization of spatial heterogeneity across a field and the formation of more informed decisions. It can be used in precision agriculture, where management zones in a field are defined based on changing conditions, with inputs such as fertilizer and irrigation applied site-specifically [18–20]. Moreover, GIS can contribute to long-term planning by recording land-use activities, examining environmental impacts, and facilitating the development of infrastructure such as drainage and irrigation systems. Its ability to compile and display complex spatial information is crucial for precision agriculture.

1.1.5 Spatial Analytics

It describes the methods and models used to investigate the spatial distributions and correlations of agricultural data. Spatial analytics is a statistical approach that uses statistical methods and computational algorithms to identify trends, anomalies, and areas of interest in a field. They can identify stressed or low-yielding regions, predict future crop yields based on environmental data, and pinpoint areas susceptible to disease [21, 22]. Spatial interpolation (e.g., kriging), aggregation, and time-series methods help improve our knowledge of field variability. When integrated with machine learning, spatial analytics gains even more capability, enabling predictive modeling and decision support systems that are suited to specific farm conditions.

1.2 Applications of GI in Precision Agriculture

The primary applications of GI are illustrated in Figure 1.2.

1.2.1 Crop Health Monitoring

One of the most critical applications of GI is crop health monitoring in precision agriculture. It involves consistent and systematic monitoring of crop conditions over time and space, utilizing advanced remote sensing technologies, such as satellite imaging, drone-based sensors, and other geospatial platforms. This technology also enables the early detection of biotic and abiotic stresses that harm plant health, such as pests, diseases,

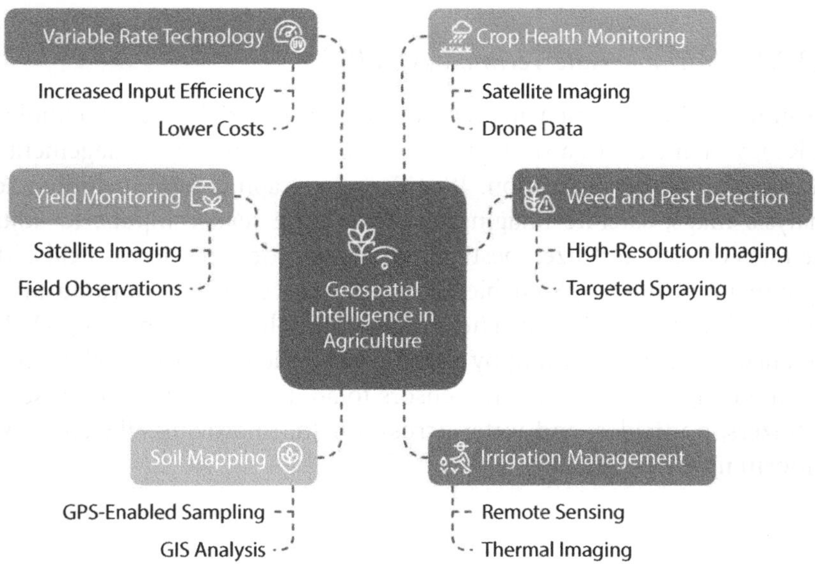

Figure 1.2 Applications of geospatial intelligence.

nutritional deficits, water stress, salt, and physical damage caused by extreme weather events [9, 23]. Farmers and agronomists can employ high-resolution multispectral and hyperspectral imaging to analyze plant vigor, chlorophyll content, leaf area index (LAI), and canopy temperatures. Commonly used vegetation indices, such as the NDVI, Soil-Adjusted Vegetation Index (SAVI), and Enhanced Vegetation Index (EVI), provide a quantitative assessment of crop vitality and growth stages. These indices help detect anomalies or underperforming areas within a field long before visual signs become apparent to the naked eye.

1.2.2 Soil Mapping

It is one of the most basic zonation applications of GI in precision agriculture, offering spatial information on soil parameter variation within a given field. Through GIS and GPS, farmers can sample soils and develop high-resolution maps that illustrate small- and large-scale changes in texture, structure, pH, organic matter, and nutrients. The availability of these spatial data layers enables location-specific practices, allowing the application of fertilizer and soil amendments only where they are needed. It also helps to provide information on irrigation scheduling based on soil water-holding capacity [24]. It is beneficial for farmers to utilize soil maps to select crop varieties that are zone-specific, which helps in improving overall production and resilience.

1.2.2.1 Variable Rate Technology (VRT)

A significant improvement in precision farming is Variable Rate Technology (VRT), which enables a strategic shift from uniform field management to site-specific input application. It utilizes geographic data, including soil analysis maps, satellite imaging, and real-time sensor inputs, to adjust the rates of seed, fertilizer, pesticide, and water delivery across fields. The equipment includes GPS-enabled sprayers and seeders that can adjust their output while moving between field zones. Variable Rate Technology (VRT) enhances precision farming by using geographic data from soil analysis, satellite imagery, and real-time sensors to adjust the application of seeds, fertilizers, pesticides, and water across a field, improving efficiency over uniform methods [25, 26].

1.2.2.2 *Irrigation Management*

The primary agricultural objective is to reduce water wastage, particularly in arid regions with limited water supplies. This issue can be resolved using GI, which integrates data from remote sensing, thermal imaging, and *in situ* soil moisture sensors. Satellites and drones can effectively detect water stress indicators by measuring the canopy temperature. The sensor used for soil moisture provides ground truth data that are used to validate remote measurements. These techniques enable real-time monitoring of moisture across various field zones, allowing farmers to make informed decisions about where and when to irrigate their fields [27].

1.2.2.3 *Yield Monitoring*

Yield monitoring throughout the growing season can be performed more precisely using GI and satellite imagery. Harvesters provide a yield map by combining GI, demonstrating the performance of various parts of the field. Farmers use this information to evaluate earlier management techniques, adjust future input perspectives, and make logistics decisions [28, 29]. When combined with additional geographical datasets, such as soil quality, irrigation patterns, and insect incidence, these maps help identify the elements that influence crop performance. In-season monitoring using drones and satellites also enables real-time assessments, allowing corrective actions to be taken before harvest. Previous season yield data inform future planting management decisions, such as adjustments to seeding rates, crop variety selection, and redesigning irrigation infrastructure.

1.2.2.4 *Weed and Pest Detection*

The primary issues in agriculture are weeds and pests, and early diagnosis is crucial for adequate control. This can be accomplished with the assistance of GI through high-resolution satellite imagery. Vegetation indices, spectral signatures, and specific machine learning algorithms are functional in anomaly detection, which indicates the presence of weeds or pests. Spatial distribution mapping, which includes threats such as pests and weeds, enables highly focused interventions and utilizes site spraying with the aid of VRT technology. This accurate prediction method effectively reduces the amount of pesticides required, leads to lower input costs, and minimizes the risk of chemical runoff to the environment. Additionally, a practical treatment approach prevents the development of resistance among pests and invasive weed species [30].

1.3 Current Trends in the Adoption of GI in Agriculture

The integration of GI into agriculture has made significant progress in recent years, reflecting both technological innovation and the pressing global demand for more efficient, resilient, and sustainable agricultural practices. Spatially referenced data are employed using input sources such as satellites, drones, GPS systems, and GIS, which are being adopted internationally within agricultural systems. This progress is complemented by studies evaluating classifier performance on fused Sentinel datasets to improve agricultural land classification and decision-making through enhanced geospatial insights [31]. It also facilitates the capacity to support site-specific decisions with great precision, which can offer significant enhancements in terms of productivity ratios, input expenditure costs, and environmental conservation. In this section, the researcher examines the essential themes that play a crucial role in the adoption and utilization of geospatial information in contemporary farming, connecting these developments to advancements in technological improvements, usability, and the overall socioeconomic and ecological context, as outlined in Table 1.1. The unification of artificial intelligence (AI) and machine learning (ML) with geospatial data is the most influential trend in the industry. While traditional remote sensing methods require manual interpretation, modern approaches integrated with AI and machine learning enable the automated analysis of large datasets. ML and AI are now utilized to facilitate the automated processing of large datasets captured by satellites and drones, allowing for more rapid, accurate, and direct conclusions and actions.

For example, AI-based picture recognition programs can accurately identify stress in crops, nutrient deficiencies, and weed invasions with high precision. The models also enable the forecasting of yields based on current and historical data from a field, providing an indicative value for supply chain planning and market strategy. As the accuracy and affordability of AI tools continue to improve, they will become accessible to average-sized farmers, expanding the use of technology based on GI to areas beyond large-scale commercial operations.

Another notable trend is the increasing use of UAVs [32]. Drones equipped with multispectral and hyperspectral cameras, thermal sensors, and LiDAR technologies can be used to capture high-resolution images and topographic information. Unlike satellites, which can either limit time resolution or cloud coverage, drones offer an opportunity for flexible and on-demand data collection tailored to specific situations within a farm.

Table 1.1 Current trends in the adoption of GI in agriculture.

Current trends	Description	Key technologies	Benefits
Integration of ML and GI	It automates the analysis of huge satellite and drone datasets. They provide quick and precise agricultural decision-making.	Detecting crop stress, identifying pest and nutritional deficiencies, and developing yield prediction models.	➢ Quick Analysis ➢ Need less manual interpretation
Drones Use	Drones equipped with multispectral, hyperspectral, thermal, and LiDAR sensors capture high-resolution imagery and topography data on demand.	Monitor crop growth, insect emergence, irrigation, flowering, and harvest readiness.	➢ Better resolution ➢ Suitable in small areas ➢ Cost-effective than satellites.
Cloud-based GIS Systems and Mobile Apps	Cloud GIS platforms and mobile apps make spatial analysis tools available on internet-connected devices.	Google Earth Engine (GEE), ArcGIS, QGIS, mobile apps for irrigation predictions.	➢ Easy to access ➢ Empower the typical farmer.
IoT and Geospatial Integration	IoT sensors collect real-time field data and integrate it with GIS to aid in spatial decision making.	Soil sensors, smart irrigation systems, sensor-based GIS.	➢ Stratified agriculture ➢ Resource utilization reduction ➢ Provides possible solutions.

Images of farms provided by drones help farmers monitor plant emergence and flowering, identify pests, manage irrigation, and prepare for harvest. Drone deployment is particularly applicable in small and fragmented areas of land where images obtained through satellite photography may not provide sufficient detail. Drones used in business are now less costly, and regulatory systems in most countries are improving to facilitate safe agricultural applications [32, 33].

Another key trend driving the use of GIS in agriculture is the development of cloud-based GIS systems and mobile applications. Historically, GIS software has required extensive computing resources and specialized knowledge. Modern cloud technologies such as GEE and ArcGIS, as well as open-source tools such as QGIS, have made geospatial analytics more accessible. These platforms enable users to visualize spatial data, conduct analyses, and create actionable maps using any Internet-connected device. Many agritech businesses and public sector efforts have developed smartphone applications that provide farmers with simplified, location-specific analytics [33]. These include real-time weather forecasts, disease alerts, irrigation scheduling, and input recommendations based on local soil and crop conditions. Such digital tools enable farmers to make informed decisions without the need to learn complex GIS software, effectively bridging the knowledge gap between cutting-edge technology and grassroots implementation.

The development of the Internet of Things (IoT) and its integration with geospatial data have altered traditional farming methods. IoT sensors are increasingly utilized in the field to continuously monitor soil moisture, temperature, salinity, and nutrient content. These sensors transmit real-time data that are geotagged and linked to GIS systems. This combination enables exact treatments, such as site-specific irrigation and fertilization, resulting in substantial reductions in water and pesticide use [34].

1.4 Challenges in Adopting GI in Agriculture

Although the advantages of GI in agricultural practices are profound and radical, many obstacles must be overcome before GI can become mainstream, especially in underdeveloped and low-resource countries. These challenges encompass financial, technical, infrastructural, and sociocultural aspects, necessitating a multi-stakeholder approach to ensure that GI implementation is inclusive and effective in the agricultural sector.

1.4.1 Low Resources and High Starting Costs

The significant barriers to the adoption of geospatial technologies in agriculture include exorbitant start-up costs associated with acquiring and installing required tools and infrastructure. Adopting GI may involve tools such as GPS-enabled equipment, UAVs, remote sensing imagery, IoT devices, or GIS software; however, the costs can be prohibitive for smallholder farmers. Nevertheless, affordable alternatives, such as the GEE, exist [35].

Expenditures for maintenance, updates, and technical assistance contribute to the financial burden. Furthermore, proprietary systems and high-resolution commercial satellite imagery often require subscription fees, making them inaccessible to low-income users.

1.4.2 Technical Expertise and Digital Literacy

GI demands a technical understanding and digital literacy for effective use. Farmers must be able to comprehend maps, use mobile applications, manage spatial data layers, and understand the consequences of geographical variability. Such expertise is inadequate particularly rural areas. Extension officers and agricultural consultants may require additional training to assist farmers in adopting and effectively using GI instruments [8, 36]. Even the most advanced platforms risk underutilization if their technical capacity is not sufficiently developed.

1.4.3 Data Integration and Standardization Challenges

Agricultural decision-making often requires the integration of data from multiple sources, including satellites, drones, weather stations, sensors, field surveys, and historical records. However, these data streams commonly differ in terms of their format, resolution, update frequency, and structure. The lack of standards makes it difficult to align datasets with a coherent decision-making framework [37]. Furthermore, restricted interoperability across different GI platforms and software tools impedes seamless data interchange and analysis, preventing the creation of complete and interoperable farm management systems.

1.4.4 Inadequate Infrastructure and Connectivity

Many farming communities, particularly in isolated or developing areas, face fundamental infrastructure issues, including consistent electricity supply, Internet connectivity, and mobile network coverage. These infrastructure limitations severely limit the real-time functioning of GI tools, particularly cloud-based systems and mobile applications that rely on online data. Poor connectivity can cause vital information to be delayed or lost, such as weather alerts, disease outbreak warnings, and irrigation schedules [38].

1.4.5 Data Privacy, Security, and Ownership Concerns

As geospatial systems capture more detailed and continuous data from farm fields, concerns regarding data privacy, abuse, and ownership are growing. Farmers frequently have questions about who owns the data generated by drones or sensors and how third-party service providers, agribusinesses, or government agencies may utilize it. There is also the possibility of commercial exploitation or loss of data sovereignty. Such concerns may result in mistrust and an unwillingness among farmers to use geospatial systems, particularly in cases where the benefits of sharing data and equitable distribution are not effectively communicated [39].

1.4.6 Cultural and Behavioral Resistance

Along with the challenge of resistance to change, cultural traditions and cynicism towards emerging technologies limit the prevalence of GI in agriculture. Conventional agricultural communities often depend on the family's knowledge and experience to determine the best course of action. Placing high-tech equipment without prior adaptation to local practices or clear demonstration of its benefits results in low acceptance rates. Gender inequality in the availability of digital instruments and training can hinder sufficient awareness of the benefits of GI.

1.4.7 Environmental and Physical Constraints

The physical conditions that are unfavorable to satellite-based remote sensing and drone-based surveillance may include dense forest cover, prevailing cloud cover, or rugged rocky terrain. Such inherent constraints render the data obtained from aerial analysis less reliable, potentially necessitating the use of additional technologies or more frequent in-field assessments, which complicate operations [40].

1.5 Opportunities for Enhancing GI Adoption in Agriculture

Although GI faces several obstacles on its path to widespread implementation in agriculture, there are numerous opportunities to expand its use and impact. Such opportunities include technical, legislative, building capacity, and multi-stakeholder partnerships. These possibilities could be leveraged to create a more sustainable, efficient, and resilient agricultural system.

1.5.1 Supportive Government

Governments can contribute to promoting GI in the agricultural sector by implementing incentive policies. This includes removing barriers through enabling legislation, offering incentives, and providing access to government-funded geospatial data [25]. Governments can promote the adoption of GI and PA by funding national digital agriculture initiatives and providing financial support for tools, such as GPS, sensors, and drones. Moreover, policy frameworks that promote open data sharing, ethical spatial data coverage, and safe data management can help build trust and participation among the parties involved.

1.5.2 Capacity Building and Skill Development

Training initiatives for farmers, extension workers, agricultural technicians, and policymakers can help close knowledge and skill gaps. These programs may focus on acquiring skills in remote sensing, GIS software, spatial data interpretation, and digital farming platforms. Educational institutions can incorporate GIS courses into agriculture and rural development curricula [41].

1.5.3 Public–Private Partnerships (PPPs)

Collaboration among official agencies, business enterprises, academic institutions, and non-governmental organizations can accelerate the adoption of geospatial technologies. PPPs can help co-create specialized solutions, share the financial burden of R&D, and expand outreach efforts to disadvantaged areas. For example, satellite imagery providers can collaborate with agricultural cooperatives to offer tailored crop monitoring services. Simultaneously, drone entrepreneurs can partner with extension agencies to provide low-cost mapping services to smallholders [41].

1.5.4 Integration with Climate-Smart Agriculture

GI is an effective tool for addressing climate change concerns in agriculture. By combining GI applications with climate-smart agriculture (CSA) methods, stakeholders can create more resilient farming systems. GI enables real-time monitoring of climate variables, early warning systems for extreme events, and dynamic risk assessments of crop and livestock systems. These technologies help farms implement adaptive practices, reduce emissions, and sustainably manage natural resources. Furthermore, governments and development organizations can utilize GI to design, execute, and monitor CSA policies and interventions, thereby increasing accountability and impact [2, 41].

1.5.5 High-Value Crops

GI technologies have tremendous potential for managing high-value crops such as fruits, vegetables, nuts, and plantation crops. These crops require careful input control, quality assurance, and traceability to meet export market requirements. GI solutions can help achieve international compliance, certification processes, and supply chain optimization. Precision harvesting, pest scouting, and phenological monitoring using geospatial data can improve product quality and reduce post-harvest losses, enhancing profitability for farmers in specialized agriculture [9, 28].

1.6 Conclusion

In the current study, GI integration with precision agriculture has evolved as a critical component of agriculture, providing tools and strategies to transform farming into a more efficient, sustainable, and data-driven form. Farmers can utilize spatial data to gain a better understanding of field variability, use inputs more prudently, monitor crop health in real time, and plan for environmental threats. This study demonstrates how GI technologies, including remote sensing, GPS, GIS, and spatial analytics, enable a deeper understanding of spatial variability in fields, leading to more accurate, productive, and ecologically friendly farming techniques. These methods serve essential objectives, including resource conservation, climate adaptability, and food security, in addition to increasing crop yield and reducing input costs.

AI disclosure statement: The author acknowledges the use of AI-based tools for minor language editing or grammatical correction. However, all intellectual contributions, analyses, and conclusions presented in this chapter are entirely the author's own.

References

1. Wang, J., Wang, Y., Li, G., Qi, Z., Integration of remote sensing and machine learning for precision agriculture: a comprehensive perspective on applications. *Agronomy*, 14, 9, 1975, 2024.
2. Vetrivel, S.C., Arun, V.P., others, Smart Farming and Precision Agriculture Using AI Technologies, in: *Real-World Applications of AI Innovation*, IGI Global Scientific Publishing, pp. 85–106, 2025.
3. Sassu, A., Gambella, F., Ghiani, L., Mercenaro, L., Caria, M., Pazzona, A.L., Advances in unmanned aerial system remote sensing for precision viticulture. *Sensors*, 21, 3, 956, 2021.
4. Bhatti, M.A., *et al.*, Advanced plant disease segmentation in precision agriculture using optimal dimensionality reduction with fuzzy c-means clustering and deep learning. *IEEE J. Sel. Top. Appl. Earth Obs. Remote Sens.*, 2024.
5. Martos, V., Ahmad, A., Cartujo, P., Ordoñez, J., Ensuring agricultural sustainability through remote sensing in the era of agriculture 5.0. *Appl. Sci.*, 11, 13, 5911, 2021.
6. Kok, Z.H., Shariff, A.R.M., Alfatni, M.S.M., Khairunniza-Bejo, S., Support vector machine in precision agriculture: A review. *Comput. Electron. Agric.*, 191, 106546, 2021.
7. Upadhyay, A., *et al.*, Deep learning and computer vision in plant disease detection: a comprehensive review of techniques, models, and trends in precision agriculture. *Artif. Intell. Rev.*, 58, 3, 1–64, 2025.
8. Delavarpour, N., Koparan, C., Nowatzki, J., Bajwa, S., Sun, X., A technical study on UAV characteristics for precision agriculture applications and associated practical challenges. *Remote Sens. (basel)*, 13, 6, 1204, 2021.
9. Sharma, K. and Shivandu, S.K., Integrating artificial intelligence and internet of things (IoT) for enhanced crop monitoring and management in precision agriculture. *Sens. Int.*, 100292, 2024.
10. Liu, J., Xiang, J., Jin, Y., Liu, R., Yan, J., Wang, L., Boost precision agriculture with unmanned aerial vehicle remote sensing and edge intelligence: A survey. *Remote Sens. (basel)*, 13, 21, 4387, 2021.
11. Sangeetha, C., *et al.*, Remote sensing and geographic information systems for precision agriculture: A Review. *Int. J. Environ. Clim. Change*, 14, 2, 287–309, 2024.
12. Pande, C.B. and Moharir, K.N., *Application of Hyperspectral Remote Sensing Role in Precision Farming and Sustainable Agriculture Under Climate Change: A Review*, pp. 503–520, Springer Climate, Springer International Publishing, 2023. doi: 10.1007/978-3-031-19059-9_21.
13. Fuentes, S. and Chang, J., Methodologies used in remote sensing data analysis and remote sensors for precision agriculture. *Sensors*, 22, 20, 7898, Oct. 2022, doi: 10.3390/s22207898.

14. Congalton, R.G., A review of assessing the accuracy of classifications of remotely sensed data. *Remote Sens. Environ.*, 37, 1, 35–46, 1991, doi:10.1016/0034-4257(91)90048-B.
15. Amen, A.R.M., *et al.*, Mapping of flood-prone areas utilizing GIS techniques and remote sensing: a case study of Duhok, Kurdistan Region of Iraq. *Remote Sens. (basel)*, 15, 4, 1102, 2023.
16. Thenkabail, P.S. and Lyon, J.G., *Hyperspectral Remote Sensing of Vegetation*, CRC Press, 2016, doi: 10.1201/b11222.
17. Hamlat, A., Meharzi, S., Guidoum, A., Sekkoum, M., Mokhtari, Y., Kadri, C.B., GIS-based multi-criteria analysis for flood hazard areas mapping of M'zab wadi basin (Ghardaia, North-Central Algeria). *Arid Land Res. Manage.*, 38, 1, 1–25, 2024.
18. Chang, Z., *et al.*, Landslide susceptibility prediction based on remote sensing images and GIS: Comparisons of supervised and unsupervised machine learning models. *Remote Sens. (basel)*, 12, 3, 2020, doi:10.3390/rs12030502.
19. Srinivasan, A. and Richards, J.A., Analysis of GIS spatial data using knowledge-based methods. *Int. J. Geogr. Inf. Syst.*, 7, 6, 479–500, 1993, doi:10.1080/02693799308901978.
20. Thakkar, A.K., Desai, V.R., Patel, A., Potdar, M.B., Post-classification corrections in improving the classification of Land Use/Land Cover of arid region using RS and GIS: The case of Arjuni watershed, Gujarat, India. *Egypt. J. Remote Sens. Space. Sci.*, 20, 1, 79–89, 2017, doi:10.1016/j.ejrs.2016.11.006.
21. Mookambiga, A. and Gomathi, V., Comprehensive review on fusion techniques for spatial information enhancement in hyperspectral imagery. *Multidimension. Syst. Signal Process.*, 27, 4, 863–889, Oct. 2016, doi:10.1007/s11045-016-0415-2.
22. McCaffrey, K.J.W., *et al.*, Unlocking the spatial dimension: Digital technologies and the future of geoscience fieldwork. *J. Geol. Soc. London*, 162, 6, 927–938, Nov. 2005, doi:10.1144/0016-764905-017.
23. Nyéki, A. and Neményi, M., Crop yield prediction in precision agriculture. *Agronomy*, 12, 10, 2460, Oct. 2022, doi: 10.3390/agronomy12102460.
24. Alexopoulos, A., *et al.*, Complementary use of ground-based proximal sensing and airborne/spaceborne remote sensing techniques in precision agriculture: A systematic review. *Agronomy*, 13, 7, 1942, 2023.
25. Saleem, S.R., Zaman, Q.U., Schumann, A.W., Naqvi, S.M.Z.A., Variable rate technologies: development, adaptation, and opportunities in agriculture, in: *Precision Agriculture*, pp. 103–122, Elsevier, 2023.
26. He, L., Variable rate technologies for precision agriculture, in: *Encyclopedia of Smart Agriculture Technologies*, pp. 1–9, Springer, 2022.
27. Vallejo-Gomez, D., Osorio, M., Hincapie, C.A., Smart irrigation systems in agriculture: A systematic review. *Agronomy*, 13, 2, 342, 2023.
28. Ali, A.M., *et al.*, Crop yield prediction using multi sensors remote sensing. *Egypt. J. Remote Sens. Space. Sci.*, 25, 3, 711–716, 2022.

29. Osman, Y., Dennis, R., Elgazzar, K., Yield estimation and visualization solution for precision agriculture. *Sensors*, 21, 19, 6657, 2021.
30. Murad, N.Y., Mahmood, T., Forkan, A.R.M., Morshed, A., Jayaraman, P.P., Siddiqui, M.S., Weed detection using deep learning: A systematic literature review. *Sensors*, 23, 7, 3670, 2023.
31. Vyas, N., Sethi, G.K., Singh, S., Comparative Analysis of Different Classifiers on Fusion of Sentinel Datasets for Agriculture Land, in: 2024 4th International Conference on Technological Advancements in Computational Sciences (ICTACS), pp. 1173–1177, IEEE, Nov. 13, 2024, doi:10.1109/ictacs62700.2024.10840995.
32. Barramou, F., El Brirchi, E.H., Mansouri, K., Dehbi, Y., *Geospatial Intelligence*, Springer International Publishing, 2022. doi: 10.1007/978-3-030-80458-9.
33. Hashi, A.O., Abdirahman, A.A., Elmi, M.A., Hashi, S.Z.M., Rodriguez, O.E.R., A real-time flood detection system based on machine learning algorithms with emphasis on deep learning. *Int. J. Eng. Trends Technol.*, 69, 5, 249–256, May 2021, doi:10.14445/22315381/IJETT-V69I5P232.
34. Bhanumathi, V. and Kalaivanan, K., The role of geospatial technology with IoT for precision agriculture, in: *Cloud Computing for Geospatial Big Data Analytics: Intelligent Edge, Fog and Mist Computing*, pp. 225–250, Springer, 2018.
35. Cesco, S., Sambo, P., Borin, M., Basso, B., Orzes, G., Mazzetto, F., Smart agriculture and digital twins: Applications and challenges in a vision of sustainability. *Eur. J. Agron.*, 146, 126809, 2023.
36. Mehedi, I.M., Hanif, M.S., Bilal, M., Vellingiri, M.T., Palaniswamy, T., Remote sensing and decision support system applications in precision agriculture: Challenges and possibilities. *IEEE Access*, 2024.
37. Singh, A., Review Article: Digital change detection techniques using remotely-sensed data. *Int. J. Remote Sens.*, 10, 6, 989–1003, 1989, doi:10.1080/01431168908903939.
38. Fussell, J., Rundquist, D., Harrington, J.A., On defining remote sensing. *Photogramm. Eng. Remote Sens.*, 52, 9, 1507–1511, 1986.
39. Ghaffarian, S., Kerle, N., Filatova, T., Remote sensing-based proxies for urban disaster risk management and resilience: A review. *Rem. Sens.*, 10, 11, 1760, Nov. 2018, doi: 10.3390/rs10111760.
40. Schmitt, M. and Zhu, X.X., Data Fusion and Remote Sensing: An ever-growing relationship. *IEEE Geosci. Remote Sens. Mag.*, 4, 4, 6–23, 2016, doi:10.1109/MGRS.2016.2561021.
41. Karunathilake, E., Le, A.T., Heo, S., Chung, Y.S., Mansoor, S.,The path to smart farming: Innovations and opportunities in precision agriculture. *Agriculture*, 13, 8, 1593, 2023.

2

The Role of Geospatial Intelligence in Agricultural Innovations

Umesh Kumar Singh[1], Santosh Korav[1]* and Sujatha H. T.[2]

[1]Department of Agronomy, School of Agriculture, Lovely Professional University, Jalandhar, Punjab, India
[2]Department of Agronomy, Keladi Shivappa Nayaka University of Agricultural & Horticultural Sciences, Shivamogga, Karnataka, India

Abstract
Geospatial intelligence has transformed agriculture by enabling innovative farming practices using satellites, drones, and global positioning systems (GPS). Precision farming, enabled by geospatial technologies, optimizes the use of water, fertilizers, and pesticides, thereby reducing environmental impact. This technology enables enhanced monitoring of crop growth and yield prediction, allowing farmers to track their fields and make data-driven decisions throughout the growing season. Geospatial technology supports carbon sequestration by enabling the precise monitoring of soil carbon levels and guiding sustainable practices. In resource management, it assists in groundwater mapping and land use optimization to minimize agricultural activities carried out in the wrong locations. Geospatial intelligence analyzes historical and real-time data to predict environmental risks, enabling farmers to mitigate crop failures and natural disasters. Geospatial data support sustainable agriculture by informing crop choices and reducing pest issues through precision farming and soil management. Additionally, geospatial intelligence enhances the functional performance and flexibility of agricultural systems by providing a comprehensive overview of farm activities and facilitating rapid responses to changes in conditions.

Keywords: Crop monitoring, land use mapping, precision farming, yield monitoring, risk management, sustainable development

**Corresponding author*: santoshkorav@gmail.com

Gurwinder Singh, Vishakha Sood, Narayan Vyas, Surendra Yadav and Dankan Gowda V. (eds.) Geospatial Intelligence for Precision Agriculture, (19–42) © 2026 Scrivener Publishing LLC

2.1 Introduction to Geospatial Intelligence

Geospatial intelligence involves analyzing spatial data and imagery to guide agricultural decision-making and planning. Geospatial technologies, including satellite imagery and geographical information systems (GISs), support comprehensive agricultural development. These technologies support informed decision-making for crop management, soil viability, irrigation, and pest management by providing actionable, spatial data. By analyzing spatial data on crop diseases and stress, farmers can optimize resources and boost yields, making geospatial tools highly effective. Aerial perspectives can also help identify potential problems and support crop yield estimation through image-based analysis. Geospatial intelligence is applicable in creating innovative farming solutions and helping supply food for a growing population. Historical geographical data originating from ancient maps and cartographers were used for transportation and military purposes. The invention of satellite and aerial photography has taken a great leap forward in the way society gathers and analyzes geographic information. Geospatial intelligence has advanced modern agriculture, enabling farmers to assess crop viability and monitor soil moisture using drones and sensors.

2.2 Current Challenges in Agriculture

2.2.1 Global Food Security

A growing population increases food demand; however, unsustainable farming and limited resources jeopardize global food security [3]. Despite increased food availability over the past 25 years, hunger and malnutrition persist due to challenges in market and distribution systems, which are exacerbated by climate unpredictability. The Global Food Security Index assesses the level of food security in 105 countries worldwide. The index looks at three critical elements to measure food security: availability, access and utilization (quality and sustainable consumption of nutrients, income level, health) and factors influenced by Government, NGOs and the Private Sector.

2.2.2 Climate Change and Environmental Pressures

Climate change, including extreme weather events such as droughts, flooding, and extreme temperatures, also affects the agricultural sector. Natural and socioeconomic disasters negatively impact agricultural production, income, and global food security [4]. However, the situation can be improved in areas such as farm reporting and modeling to mitigate these adverse effects. Farmers can use satellite imagery and analysis to monitor climate-related factors such as temperature and precipitation and adopt practices to meet environmental sustainability standards. The integration of geospatial AI and machine learning (ML) helps farmers make informed decisions regarding when to plant crops, apply irrigation, and harvest, thereby conserving resource use while ensuring the highest resource productivity [5]. Early trend analysis allows for initial assessment of current climate and aids in decision making for crop selection, monitoring and management of water and other resources over a short term period.

In agriculture, technology for monitoring and prediction utilizes climate-resilient systems. Extreme weather events disrupt the optimal agricultural conditions. Droughts, flooding, extreme heat, and wet weather, along with numerous other extreme weather events, significantly impact farming in adverse ways. To reduce negative impacts, stronger agricultural monitoring and forecasting systems are required. Farmers can now monitor and distinguish the climate change impacts that are occurring using satellite photography, weather forecasting, and data analysis tools. Farmers can also use these technologies to improve their decision-making regarding planting, irrigation, and harvesting to maximize agricultural yield and resource utilization [5]. Agriculture is embracing artificial intelligence (AI) and ML.

2.2.3 Resource Management

People have to eat, and at the same time, resources must not be depleted; this is at the heart of resource management in food and agriculture. There may be inefficiencies in land, water, and crop management, which can lead to adverse effects such as land and water degradation and low agricultural productivity [6]. These challenges require innovative solutions. Using drones and sensors, farmers can precisely manage resources, such as water and fertilizer, to optimize application and minimize waste. This can be achieved through controlled irrigation, fertilization, and pest control, which minimize the use of chemicals and water while maximizing crop yields. This is achieved by integrating agroecological practices such as crop rotation, cover cropping, and agroforestry, which improve soil quality, control soil erosion, and promote biodiversity [7]. Further development and research on drought-resistant crops, efficient irrigation, and various developmental and organic farming approaches will foster agricultural advancement and resilience.

2.2.4 Sustainability Concerns

Sustainable farms will be the key to the future of food systems. Data-driven precision farming has been implemented to promote sustainability. Data-driven technologies enable farmers to optimize resources, minimize waste, and reduce their environmental impact [8]. For a sustainable and prosperous future, efficient work is necessary. Aerial imaging and soil sensor technology provided by drones open numerous opportunities for sustainable agriculture.

2.3 Application of Geospatial Intelligence in Agriculture

2.3.1 Precision Agriculture

Geospatial technologies improve farm efficiency through precision agriculture. In precision agriculture, geospatial intelligence aids soil monitoring, water management, crop health assessment, and pest control [8]. Farmers can access valuable insights into their fields using satellite imagery, drones, and other geospatial technologies [2]. The integration of geospatial

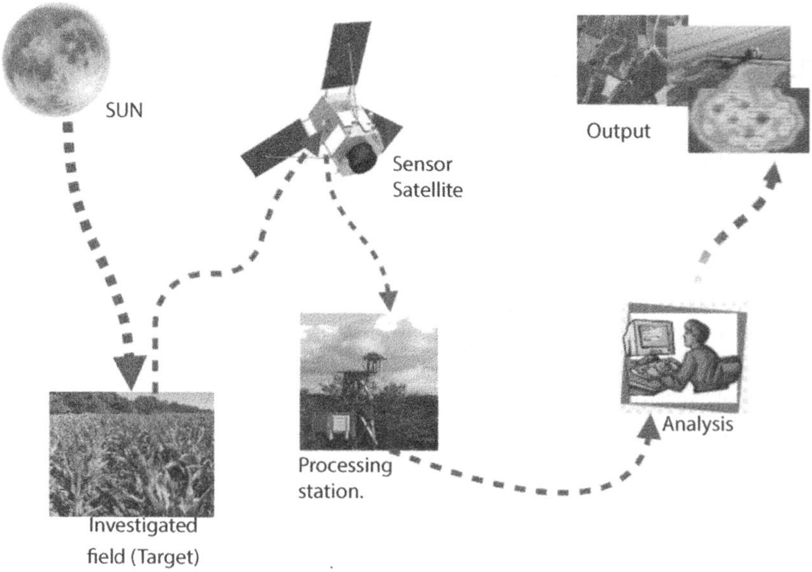

Figure 2.1 Use of geospatial intelligence in agriculture.

data into precision farming enables resource optimization, high yield, and reduced environmental impact. Advanced technologies have maximized crop growth monitoring. Figure 2.1 illustrates the application of technology in agriculture. Satellite sensors capture reflected light from fields and transmit the data to receivers for processing and real-time analysis.

2.3.2 Irrigation Management

In this context, water is used appropriately, meaning that its usage is relatively optimized, and more advanced irrigation technologies are adopted in agriculture. Sensors, drones, and satellite data are used for comprehensive assessments of water stress and evapotranspiration in farming operations. Advanced technologies offer improved irrigation management solutions. Wireless Sensor Networks can theoretically save U.S. agriculture billions of dollars and up to 34% of water in irrigation [9], demonstrating the feasibility of real-time monitoring of temperature, humidity, and soil moisture. More precisely, drones can obtain Water Stress Indices, which help determine the optimal time to water crops, thereby preventing drought. Additionally, the amount of water required for irrigation can be measured using thermal cameras [9]. It optimizes water use to ensure crop health and increase the harvest yield. With an optimal set of rotations, water-use efficiency (WUE) reached 1.07 kg/m^3, resulting in a more significant crop yield [10]. Intelligent irrigation systems help predict water needs and adjust to different crop types for optimal growth. Unfortunately, in resource-limited configurations, exposure to these technologies can inhibit the efficiency improvements that can be achieved through these innovations.

2.3.3 Pest and Disease Monitoring

Geospatial intelligence plays a vital role in pest surveillance and disease control in agriculture. Farmers can target pest and disease control interventions by relying on up-to-date data on crop conditions and distribution patterns, utilizing satellite images and GIS mapping [7]. Moreover, data analytics can be used to predict, prevent, and enable early disease detection and control [2]. The application of remote sensing and GIS approaches can improve the risk assessment of geographical areas (Figure 2.2). With geoinformatics, farmers can apply the proper amounts of pesticides to control diseases and increase production while promoting the concept of sustainable agriculture. As observed by [2], geospatial intelligence in agriculture is only practical in the presence of challenges, such as staff training and data integration across systems.

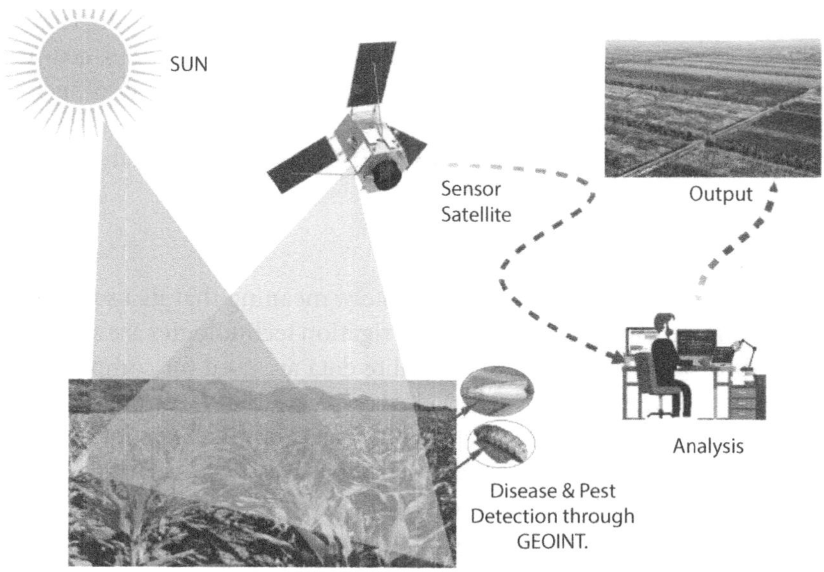

Figure 2.2 Geospatial intelligence-based disease and pest detection.

2.3.4 Land Use Planning and Farm Management

Geospatial intelligence enables advanced farming methodologies to assess which crops to grow and how the land can be utilized. Farmers use remote sensing and GIS to analyze environmental conditions and determine optimal crop-planting locations. Using GIS technology, farmers can create a three dimensional terrain model of their farm and use this together with weather and soil attribute data to plan optimally, which acts as a centralized spatial database management system for spatial data through its core. Such geospatial intelligence facilities enable the preparation of soil maps with high accuracy characteristics for crops, correlated with elevation, slope inclination, soil type, and other relevant factors. Furthermore, the more complex method of hyperspectral imaging contributes to maintaining crop health and soil management monitoring for the proper utilization of resources. This study aimed to investigate the efficiency and environmental impact of this process. Hence, geospatial data present a better alternative for farmers to effectively select crops and manage their land with minimal harm to the environment. The geospatial intelligence system in supporting precision agriculture would help increase watering efficiency, reduce the volume of chemicals used, and promote suitable habitats for farming [11]. However, farmers often lack sufficient funding and access to these technologies.

2.3.5 Disaster and Risk Management

Geospatial data enhance the effectiveness of early warning systems for climate-related disasters such as droughts and floods. Stakeholders can precisely track conditions using satellite imagery and identify potential risks using GIS. Geospatial data enable timely analysis, allowing interventions within critical time frames to protect crops and food sources. Geospatial technologies significantly improve the response and recovery capabilities of disaster management. Remote sensing technologies that utilize satellite imagery provide real-time data on environmental changes, enabling authorities to assess and implement preventive measures [12]. GIS can be used to analyze historical and present vegetation health and water availability data to predict potential disasters. It is a collaborative data delivery system. Strong collaboration between data providers and end users is essential for effective disaster risk management, as it provides accessible and actionable geospatial information [13]. With new cartographic visualizations, decision-makers can now interact with the data based on their specific needs, driving better response strategies. Despite the challenges of cumbersome data standardization and integration, geospatial data are integral to disaster preparedness. Solving these problems is essential for realizing the full potential of geospatial technologies in disaster management.

2.4 Technological Progress in Geospatial Intelligence Applications in Agriculture

2.4.1 Satellite Remote Sensing

The high-resolution and real-time satellite data acquisition available through advancements in satellite technology has led to a shift in agricultural practices in recent years. Precision agriculture can be transformed by monitoring crops and optimizing the resource inputs. Commercial satellites, such as IKONOS and QuickBird, enable the delivery of images with a ground sample distance of 1 m or lower, facilitating the detection of plant stress and diseases through the imaging of agricultural land use. Remote sensing technologies, such as those using Sentinel-1 and Sentinel-2 data, enable the precise detection of agricultural variations, thereby supporting crop management and planning [14].

2.4.2 Drones and Unmanned Aerial Vehicles

Technological advancements have made it possible to conduct critical low-altitude aerial imaging and data gathering facilitated by UAVs, which has advanced agriculture in numerous ways. The Internet of Things (IoT) enables real-time monitoring and management in agriculture. UAVs provide high-quality imagery during key growth stages, surpassing satellite data [7]. IoT technology has been applied to lightweight drone systems, which have been utilized for the remote monitoring of crop health, heat stress evaluation, water stress evaluation, and pest management evaluation [15]. The advantages and challenges of UAVs are very real, including institutional constraints and the need for skilled pilots. Thus, they focus on addressing the most pressing drone issues to meet the agricultural needs.

2.4.3 AI and ML

With the application of AI and ML techniques, there is currently a revolution in agriculture, which incorporates the analysis of geospatial data to facilitate predictive modeling. Additionally, owing to these significant advancements in technology, farmers are now equipped to incorporate more scientific principles into their farming methods, enabling them to increase output while maintaining environmental sustainability. A study has shown that neural networks and ensemble approaches have been employed to predict crop yield using ecological parameters, including soil, precipitation, and temperature [16]. The application of remote sensing data fusion can result in precision farming-grade digital soil map. Advanced models, such as graph-based neural networks (GNN) and recurrent neural networks (RNN), enhance the predictive accuracy of geospatial and temporal data analyses. Combining remote sensing images with convolutional neural networks (CNNs) outperform existing methods that utilises single-source satellite imagery or other data sources, such as the use of previous images obtained from satellites or translating the data, such as numerical values from satellite images [16]. Agricultural predictive modeling has improved with the integration of AI and NN; however, challenges still exist in data synthesis and model development for heterogeneous climatic and geographical settings.

2.4.4 IoT Sensors

Utilizing geospatial intelligence to integrate ground-based sensors with IoT technology enables real-time monitoring and informed decision-making under dynamic environmental conditions. To manage resources efficiently

Table 2.1 Different IoT-based sensors are used in the geospatial intelligence system.

Different kinds of sensors	Utility	Operational procedures	Source
Audio-visual based sensors	Monitoring pest infestation and fruit gathering/ harvesting.	Assesses changes in noise levels in agricultural fields.	[18]
Electromagnetic sensors	Electromagnetic responses will be recorded; residual nitrate and organic matter of the soil will be determined; transpiration rate will be measured on a real-time basis.	Employs an electrical circuit to record the conduction or the accumulation of biological electricity within the soil.	[18]
Light detection and ranging (LIDAR)	Among the agricultural land-based utilities are 3D-draft, monitoring of the erosion of land, mapping of agricultural lands or soils as well as soil type detection.	This sensor sends out light wave pulses towards the object that has been targeted. The range light waves that were previously aimed at the target objects also return to the sensor after physical contact of the objects with it. For evaluation, the time taken for light waves to return back to the sensors post contact with the target item is analyzed.	[18]

(*Continued*)

Table 2.1 Different IoT-based sensors are used in the geospatial intelligence system. (*Continued*)

Different kinds of sensors	Utility	Operational procedures	Source
Sensors with optical capabilities	Serves for determining soil characteristics such as texture, mineral content, clay content, moisture, soil color, etc. The use of optical sensors that function based on fluorescence allows one to assess the ripening stage of the fruit. Optical and microwave scattering techniques are integrated for the monitoring of the orchard canopy.	This applies to the transition of light as it approaches an object.	[19]
Pneumatic sensors	Soil rigidity and particle compaction measures.	Sensors measure the pressure decided on by load cells or strain gauges.	[20]
Massive flow sensors.	Estimate the grain flow through a combine harvester to evaluate yield.	Within the mass flow of grain of several modules, including, moisture content detecting sensor, data storage, and an internal system, there are sensors that monitor grain.	[20]

(*Continued*)

Table 2.1 Different IoT-based sensors are used in the geospatial intelligence system. (*Continued*)

Different kinds of sensors	Utility	Operational procedures	Source
FPGA (Field programmable gate array) sensors	Measures the irrigation, humidity, moisture content and transpiration rate in plants.	Utilizes digital circuits around silicon-based processors and logic blocks.	[19]
Electrochemical sensors	Aids in the measurement of soil phosphorus status and pH.	In agricultural soil, individual sensors monitor the electrochemical gradients.	[20]
Ultrasonic sensors	Aids in the detection of pests, monitoring the crop canopy, as well as identifying weeds.	An object is located by an ultrasonic sensor that sends ultrasonic pulses and receives them back.	[19]
Adaptable and mobile sensors	And it helps in sensing the form, size, and extent of development of different plant parts along with the temperature.	Sensors are made from flexible polymers that can be mounted to different parts of the plants.	[20]

that you have carefully sourced and put in place, data collection, analysis, and dissemination should run smoothly across all collaborative content creation processes. Using pressure sensors, groundwater monitoring systems are wirelessly linked to provide real-time data for managing water resources. IoT sensors can facilitate continuous monitoring and real-time data collection, making air and soil quality data available and ensuring that accessibility issues are addressed [17]. IoT sensors, combined with geospatial intelligence, improve urban environment monitoring by providing real-time data; however, challenges such as complex data management and optimized communication protocols remain for agricultural sensors (Table 2.1).

2.5 Case Studies and Practical Applications

2.5.1 Examples of Implemented Agricultural Innovations

GPS and remote sensing technologies help farmers optimize land use and their reduce costs. Case studies, such as Integrated Crop-Livestock Systems (ICLS) in Brazil and France, demonstrate that geospatial innovations enhance productivity and sustainability in agriculture. Farmers in southern Brazil, the Northern Great Plains, and southern France have successfully established thriving farming enterprises through the adoption of Integrated Crop-Livestock Systems (ICLS). These systems of crop-livestock integration in conservation agriculture, combined with precision farming and automated irrigation systems, enhance both soil and financial health, despite overcoming the initial challenges of changing the way farming is conducted. Use of GIS-Based Applications in Agriculture: GIS applications in agriculture have been reported to yield better crop yields and resource savings, although factors such as cost and expertise level have been identified as constraints to broader adoption. Farmers can enhance their outputs while minimizing costs by using GPS and remote sensing. Technology is becoming increasingly essential in modern farming. While these improvements sound great, many implementers, mainly small-scale farmers, face installations that seem complicated because of a lack of resources and education. Overcoming these constraints will make it possible to achieve global implementation.

2.5.2 Government and Policy Integration

Geospatial intelligence is crucial for developing both national agricultural policies and initiatives related to resource management and food security. They demonstrated how the government could utilize geospatial intelligence to refine decision-making processes, thereby enhancing market outreach and price equalization in the agricultural sector, as seen in India following specific policy adjustments. Remote sensing and image processing have enabled the accurate identification of land use and crop types, which is crucial for monitoring agricultural production and planning the use of water resources. This is supported by spatial and thematic databases containing economic and environmental data for research on resource-management strategies. Sub-Saharan regions require resources to address inland investment and alleviate food insecurity. Geospatial intelligence in agriculture offers food security and management benefits, which are best illustrated through visual data.

2.5.3 International Collaborations

International agricultural Research and Development is mostly enhanced by geospatial intelligence through international collaborations, particularly in developing countries. Collaboration in geospatial technology enhances the scientific potential, resource management, and knowledge exchange in agriculture. Such a worldwide environment is evolving with empirical research on the systemic features of science in developing nations, including the growth of dual scientific capacity through collaborative research. Data sharing, joint research, and the adoption of standard operating procedures, including risk assessment, are key components of most collaborative strategies in agriculture worldwide. Conducting evaluations of crop varieties and agricultural methods is relatively inexpensive when conducted through organized networks of researchers from different countries. Information management and remote sensing-based land cover/land use change monitoring for sustainable agriculture are two essential tasks that require satellite geospatial datasets to be performed. These data support the ARDA in promoting food security preparedness and optimizing resource use [1]. However, certain regions face political and financial barriers that slow the development of cross-national agricultural cooperation.

2.6 Advantages of Geospatial Intelligence in Agriculture

2.6.1 Increase in Productivity and Efficiency

Geospatial intelligence enhances agricultural decision-making, boosting returns and reducing costs (Figure 2.3). These also aid in estimating crop yields before the harvesting season. GIS, AI, and wireless sensor networks can assist in accelerating this transformation [4].

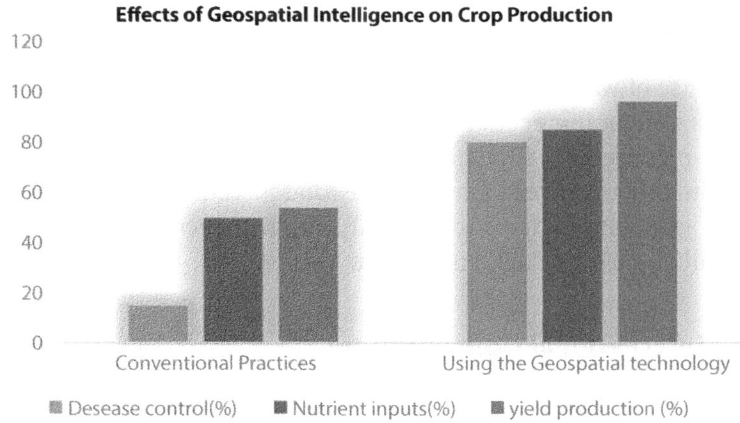

Figure 2.3 Geospatial intelligence influences crop production.

2.6.2 Enhanced Crop Management

Farmers can receive real-time information through decision support systems and wireless sensor networks to prevent pests and diseases and to make informed use of resources. The use of AI in soil irrigation, weather, and disease prediction, among other factors, contributes significantly to improved agricultural productivity. Using IoT, intelligent Spatial Decision Support Systems in the agro-industry sector have been further developed, enabling better spatial data handling and decision-making [21]. Computer optimization algorithms help in crop selection balance to maximize profits from a particular crop's demand and supply [22].

2.6.3 Sustainability

Geospatial intelligence optimizes resource utilization in environmentally friendly agricultural practices (Figure 2.3). With the introduction of geoinformatics technology, farmers will also be able to effectively manage water, fertilizers, and pest control to achieve maximum sustainability in the future. Geospatial data support sustainable farming practices by reducing pollution while maintaining yields. Utilizing geospatial tools reveals how modern agriculture addresses plant infection issues with minimal use of toxic substances, such as pesticides, and employs soil organic matter and electrical conductivity as effectiveness indices for agricultural disease management [2]. While substantial improvements in resource optimization have been achieved using geospatial intelligence, the utilization of these technologies in various agrarian regions remains a challenge and requires ongoing research and collaboration among stakeholders in the field.

2.6.4 Carbon Sequestration

Geospatial technologies improve carbon sequestration through precise data acquisition and timely interventions. Promotion of climate-smart agriculture will ensure that if agricultural practices are changed to enhance productivity and improve livelihoods, it will not lead to environmental degradation [23]. The reduction of soil disturbance through conservation agriculture methods, such as residues with cover crops, will significantly reduce carbon sequestration and soil organic carbon [24]. Precision agriculture utilizes geospatial tools to target specific crops, thereby enhancing soil health and increasing the amount of carbon it contains [2]. Carbon credits are managed through a GPS-GIS system, which verifies emissions *via* temporal data management systems. Geospatial technologies help achieve a greater level of transparency and accountability while maximizing land-use morphologies for carbon removal [1].

2.6.5 Climate Change Adaptation Plan

Agroecological adaptation to climate change can be enhanced using early warning systems and geospatial intelligence adaptive planning tools. Additionally, satellite images in conjunction with AI can provide solutions for changing weather conditions, enabling farmers to adjust their practices. Geospatial technology facilitates mapping efforts and accelerates the response to extreme weather conditions. COSPARIN and GOBEYOND, on the other hand, also utilize satellite images to counter agricultural natural hazards [25]. The tools are designed to assist the agricultural sector's policymakers in planning for climate change by integrating weather and climate information into their planning processes, including historical climate years, in which AI in GIS predicts optimum agricultural conditions based on decision-making processes [26]. GeoAI frameworks have enabled the modernization of cropping system management through strategies such as crop yield estimation and farm management under changing climatic conditions [2, 26].

However, hurdles still prevail in the distribution of resources to poor farmers. It is also important to fill the gaps that limit the use of geospatial intelligence in agriculture. Adaptive Planning implements adaptive paradigms against scenarios based on the identification of historic climate

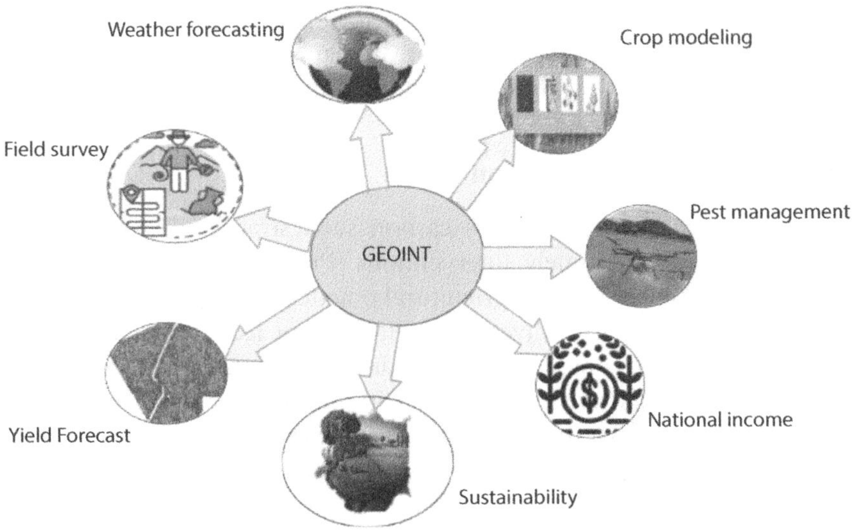

Figure 2.4 Advantages of geospatial intelligence in agriculture.

patterns and agricultural outcome probabilities, where AI with GeoAI technology is incorporated to significantly aid decision-making [26]. GeoAI frameworks have made it possible to improve the efficiency of appraising cradle yield and adopt farm management solutions, irrespective of weather changes [2]. However, dependency limitations among poor farmers remain a threat to equity. Finally, to maximize the advantages of using geospatial intelligence in agriculture, the unexpected and existing insufficiencies being reported should be addressed (Figure 2.4).

2.7 Challenges and Limitations of Geospatial Intelligence in Agriculture

2.7.1 Data Accessibility

Geospatial Intelligence has tremendous potential but is being held back by several challenges such as lack of relevant data, high costs and technology infrastructure deficiency. Satellite images and AI enable farmers to adapt to variable temperatures and effectively manage their farming practices. Early Warning Systems: The mapping and response times to extreme weather disasters improve with geospatial data [25]. Satellite data supports COSPARIN and GOBEYOND in assessing hydro-meteorological risks for early warning systems. Early warning systems and adaptive planning tools for climate change are made available by geospatial intelligence to improve agricultural resilience. Pairing satellite data with AI enables farmers to easily identify and mitigate issues while positioning them to adjust their farming practices more rapidly in response to extreme weather and other conditions. Resource-limited farmers often lack access to advanced agricultural tools [23, 25]. An equitable distribution of resources is required to maximize the benefits of geospatial intelligence in agriculture.

2.7.2 Cost and Infrastructure Barriers

High costs and complexity limit the adoption of geospatial technology across disciplines. There are also economic difficulties, spatial or infrastructural requirements, and the absorption and dissemination of expertise. Technologies associated with geospatial information have high start-up costs, including hardware and software, as well as extensive maintenance, which makes it difficult for local governments in developing regions to adopt them [27, 28]. Thus, satisfactory infrastructure, which is largely dependent on the capacity to store and manipulate data, is not guaranteed in all locations [29].

2.7.3 Privacy and Ethical Considerations

There is neither protection of personal data nor consent given before such data are shared, and this absence of consent mechanisms permits ethical dilemmas concerning data ownership and privacy in IoT agriculture technology. The vulnerability of these devices to hacking and unauthorized access also leads to confidentiality vulnerabilities in agricultural information [30, 31]. Addressing these issues is essential to gaining farmers' trust and ensuring safe and ethical use of their data. Devices used in agriculture are vulnerable to hacking and unauthorized data access, potentially compromising the confidentiality of agricultural information [30]. If there is no legitimate consent, misused data present ethical challenges related to ownership and privacy rights [31]. Informed consent, facilitated through a user-centric form, is thus required to empower individuals to manage their data. Algorithmic biases should be addressed through ethical frameworks, and the transparency of procedures that handle data should be encouraged to maintain accountability [32, 33]. Robust security protocols that employ encryption and device authentication are required to maintain data integrity and privacy in agricultural data management, which must be discussed and adapted over time as new technologies evolve.

2.8 Future Trends and Opportunities

2.8.1 Advancements in Satellite and Sensor Technology

Compact, low-cost sensors enable rapid deployment and energy-efficient tracking in agricultural applications. Compact and inexpensive, the bases of nanosatellites are perfectly suited for the speedy establishment of sensor networks, leading to enhanced data acquisition in remote locations [34]. Once ML and AI are embedded in the sphere of data processing and network management, they will become dynamic, responsive, and active systems [35]. Advanced imaging spectroscopy enhances the monitoring of agriculture. Hyperspectral Imaging is a technique that collects both Image and Spectral data for every single pixel and is an integral part of agriculture, environmental monitoring, and other necessary applications [36]. The fusion of IoT, AI, and advanced sensors enables the automation of industrial processes based on data [27]. Especially for the most thrilling advances in these domains, the so far encountered hurdles of security, interoperability, and power efficiency must be addressed for any chance of realizing their full potential.

2.8.2 AI Integration and Predictive Analytics

The integration of AI and predictive analytics in agriculture is fundamentally and rapidly changing the way we predict crop yields, track market trends, and how the world produces food. Advanced AI frameworks can be used to refine agricultural methodologies and improve decision-making processes. Of importance to agricultural planning, the accuracy of crop yield forecasting dramatically improves from ML and AI methodologies [37]. Climate variability and resource constraints [38] are well addressed using hybrid AI models that integrate various techniques. Using AI-integrated predictive analytics, agricultural demand forecasting and supply chain optimization have been revolutionized to achieve unparalleled operational efficiency [39]. Real-time data analytics and big data techniques have been combined to facilitate agricultural supply chain management and risk reduction.

2.8.3 Market Trends and Global Food Dynamics

AI's application is not limited to identifying industry trends in the food market and helping stakeholders adapt to changes in the global food chain.

Developing robust data systems and skilled personnel is essential to maximize AI's potential in agriculture [39]. Despite all these general and specific key benefits that can be gleaned from integrating AI into agriculture, the factors of standardizing data and investing in infrastructure need to be addressed.

2.8.4 Cross-Sector Collaborations

Interdisciplinary benefits arise from partnerships between government agencies and private businesses in geospatial intelligence incorporation efforts, with academic research institutions in the agricultural domain. Agricultural productivity and sustainability can be significantly enhanced through the joint utilization of data and resource sharing. Cooperative projects with corporations and governments leverage geospatial technology investment synergies in underdeveloped regions, as well as interdisciplinary partnerships with corporations that utilize corporate data for problem-solving. For example, drones and their associated data analytics can improve yields while minimizing environmental impacts [40–42]. Open data enable more reactive and spatially precise analyses, facilitating advanced decision-making in the agricultural sector [43]. However, pursuing Sustainable Development Goals through integrative collaboration with marginalized agricultural producers and improving food security can be modified [44].

2.8.5 Conclusion

This study explored how geospatial intelligence improves agricultural decision-making and policy formulation. In agriculture, geospatial intelligence has a significant impact on precision farming, which utilizes advanced geospatial technologies. These technologies enable farmers to make informed decisions based on the geographical information of the soil, crops, and environment. Through the application of geospatial intelligence-based decision support systems, farmers can increase productivity, thereby reducing waste and demonstrating effective resource management. GIS and statistical analyses create maps and charts to support decision-making. This approach improves the ecological context of farming activities and the economic efficiency of agricultural production. Geospatial intelligence offers practical solutions to help farmers address modern agricultural challenges and enhance productivity and sustainability.

AI disclosure statement: The author acknowledges the use of AI-based tools for minor language editing or grammatical correction. However, all intellectual contributions, analyses, and conclusions presented in this chapter are entirely the author's own.

References

1. Skakun, S., Kussul, N., Kussul, O., Shelestov, A., Quantitative estimation of drought risk in Ukraine using satellite data; in: *2014 IEEE Geoscience and Remote Sensing Symposium*, pp. 5091–5094, Jul. 2014, Accessed: Oct. 21, 2024 [Online] Available: http://dx.doi.org/10.1109/igarss.2014.6947642.
2. Roberts, D.P., Short Jr, N.M., Sill, J., Lakshman, K.D., Hu, X., Buser, M., Precision agriculture and geospatial techniques for sustainable disease control. *Indian Phytopathol.*, 74, 2, 287–305, Feb. 2021, [Online] Available: http://dx.doi.org/10.1007/s42360-021-00334-2.
3. Zahra, S.M., Shahid, M.A., Maqbool, Z., Sabir, R.M., Safdar, M., Majeed, M.D., Sarwar, A., Application of Geospatial Techniques in Agricultural Resource Management in Agricultural Sciences. IntechOpen, London, UK, 2024, Accessed: Oct. 21, 2024 [Online] Available: http://dx.doi.org/10.5772/intechopen.112222.
4. Lemi, T., Effects of Climate Change Variability on Agricultural Productivity. *IJESNR*, 17, 1, Feb. 2019, [Online] Available: http://dx.doi.org/10.19080/ijesnr.2019.17.555953.
5. Berhane, A., Climate Change and Variability Impacts on Agricultural Productivity and Food Security. *Journal of Climatology and Weather Forecasting*, 06, 03, 2018. [Online]. Available: http://dx.doi.org/10.4172/2332-2594.1000240.
6. Srivastav, L.A., Dhyani, R., Ranjan, M., Madhav, S., Sillanpaa, M., Climate-resilient strategies for sustainable management of water resources and agriculture. *Environ. Sci. Pollut. Res.*, 28, 31, 41576–41595, Jun. 2021, [Online] Available: http://dx.doi.org/10.1007/s11356-021-14332-4.
7. Zhang, P., Guo, Z., Ullah, S., Melagraki, G., Afantitis, A., Lynch, I., Nanotechnology and artificial intelligence to enable sustainable and precision agriculture. *Nat. Plants*, 7, 7, 864–876, Jun. 2021, [Online] Available: http://dx.doi.org/10.1038/s41477-021-00946-6.
8. Dubey, K.P., Singh, G.S., Abhilash, P.C., Resource Conserving and Innovative Practices for Agricultural Sustainability, in: *Springer Briefs in Environmental Science*, pp. 63–92, Springer International Publishing, Cham, 2019, Accessed: Oct. 21, 2024 [Online] Available: http://dx.doi.org/10.1007/978-3-030-15519-3_4.
9. Alexandris, S., Psomiadis, E., Proutsos, N., Philippopoulos, P., Charalampopoulos, I., Kakaletris, G., Paraskevopoulos, A., Integrating

Drone Technology into an Innovative Agrometeorological Methodology for the Precise and Real-Time Estimation of Crop Water Requirements. *Hydrology*, 8, 3, 131, Sep. 2021, [Online] Available: http://dx.doi.org/10.3390/hydrology8030131.
10. El-Shirbeny, M.A., Abdelraouf, M., Savin, I., Poddubskiy, A., Dokukin, P., Agricultural Water Monitoring for Water Management Under Pivot Irrigation System Using Spatial Techniques. *Earth Syst. Environ.*, 5, 2, 341–351, Jul. 2020, [Online] Available: http://dx.doi.org/10.1007/s41748-020-00164-8.
11. Thorp, K.R., Hunsaker, D.J., French, A.N., Bautista, E., Bronson, K.F., Integrating geospatial data and cropping system simulation within a geographic information system to analyze spatial seed cotton yield, water use, and irrigation requirements. *Precis. Agric.*, 16, 5, 532–557, Apr. 2015, http://dx.doi.org/10.1007/s11119-015-9393-x.
12. Gahler, M., Remote Sensing for Natural or Man-made Disasters and Environmental Changes, in: *Environmental Applications of Remote Sensing.* InTech, InTech – Rijeka, Croatia, 2016, Accessed: Oct. 21, 2024 [Online] Available: http://dx.doi.org/10.5772/62183.
13. Miyazaki, H., Nagai, M., Shibasaki, R., Reviews of geospatial information technology and collaborative data delivery for disaster risk management. *ISPRS Int. J. Geo-inf.*, 4, 4, 1936–1964, Sep. 2015, http://dx.doi.org/10.3390/ijgi4041936.
14. Vyas, N., Singh, S., Sethi, G.K., An efficient posterior probability-based image fusion change detection model for the estimation of seasonal agricultural changes using microwave and optical datasets. *Comput. Electron. Agric.*, 237, 110683, Oct. 2025, DOI:10.1016/j.compag.2025.110683.
15. Yaqot, M. and Menezes, B.C., Unmanned Aerial Vehicle [UAV] in Precision Agriculture: Business Information Technology Towards Farming as a Service, in: *Proceedings of the 2021 1st International Conference on Emerging Smart Technologies and Applications [eSmarTA]*, pp. 1–5, IEEE, Aug. 2021, DOI:10.1109/esmarta52612.2021.9515736.
16. Rayaguru, N.K., Gowrishankar, J., Deshpande, B.G., Grover, M., Vekariya, D., Emerging Trends: Nano-Scale Wireless Sensor Networks and Applications. *J. Intell. Syst. Internet Things*, 13, 2, 25–34, 2024, DOI:10.54216/jisiot.130202.
17. Abraham, M.T., Satyam, N., Pradhan, B., Alamri, A.M., IoT-based Geotechnical Monitoring of Unstable Slopes for Landslide Early Warning in the Darjeeling Himalayas. *Sensors*, 20, 9, 2611, 2020, DOI:10.3390/s20092611.
18. Schuster, J.N., Darr, M.J., McNaull, R.P., Performance Benchmark of Yield Monitors for Mechanical and Environmental Influences, in: *2017 ASABE*, Spokane, Washington, pp. 1–5, 2017. DOI:10.13031/aim.201700881.
19. Dvorak, J.S., Stone, M.L., Self, K.P., Object Detection for Agricultural and Construction Environments Using an Ultrasonic Sensor. *J. Agric. Saf. Health*, 22, 2, 107–119, 2016, DOI:10.13031/jash.22.11260.

20. Yew, T.K., Yusoff, Y., Sieng, L.K., Lah, H.C., Majid, H., Shelida, N., An Electrochemical Sensor ASIC for Agriculture Applications, in: *Proceedings of the 37th International Convention on Information and Communication Technology, Electronics and Microelectronics [MIPRO]*, pp. 227–231, IEEE, May 2014, DOI:10.1109/mipro.2014.6859538.
21. Yusianto, R. and Hardjomidjojo, H., Spatial Analysis for Crop Land Suitability Evaluation: A Case Study of Potatoes Cultivation in Wonosobo, Indonesia, in: *Proceedings of the 2020 International Seminar on Application for Technology of Information and Communication [iSemantic]*, pp. 313–319, IEEE, Sep. 2020, DOI:10.1109/isemantic50169.2020.9234284.
22. Nde, R.K., Fendji, J.L.E.K., Yenke, B.O., Schöning, J., Crop selection. *Smart Agric. Technol.*, 9, 100602, Dec. 2024, DOI:10.1016/j.atech.2024.100602.
23. Yu, L., *et al.*, Enhancing global agricultural monitoring system for climate-smart agriculture. *Clim. Smart Agric.*, 2, 1, 100037, Feb. 2025, DOI:10.1016/j.csag.2024.100037.
24. Srinivasarao, C., Lal, R., Kundu, S., Thakur, P.B., Conservation Agriculture and Soil Carbon Sequestration, in: *Conservation Agriculture*, pp. 479–524, Springer International Publishing, Cham, Switzerland, 2014, DOI:10.1007/978-3-319-11620-4_19.
25. Reznik, L., *Intelligent Security Systems*. Wiley, Hoboken, NJ, USA, Oct. 19, 2021, DOI:10.1002/9781119771579.
26. Majidi, B., *et al.*, Geo-Spatiotemporal Intelligence for Smart Agricultural and Environmental Eco-Cyber-Physical Systems, in: *Studies in Computational Intelligence*, pp. 471–491, Springer International Publishing, Cham, Switzerland, 2020, DOI:10.1007/978-3-030-52067-0_21.
27. Rajora, R., Rajora, A., Sharma, B., Aggarwal, P., Thapliyal, S., Sensing the Future: Challenges and Trends in IoT Sensor Technology, in: *Proceedings of the 2024 4th International Conference on Innovative Practices in Technology and Management [ICIPTM]*, pp. 1–5, IEEE, 2024, DOI:10.1109/iciptm59628.2024.10563962.
28. Schwartz-Belkin, I. and Portman, M.E., Exploring Barriers to the Implementation of Geospatial Technologies in Marine Spatial Planning: Reports from Practitioners. *Mar. Policy*, 159, 105891, 2024, DOI:10.1016/j.marpol.2023.105891.
29. Liang, S.H.L., Huang, C.-Y., Geospatial Cyberinfrastructure for Addressing the Big Data Challenges on the Worldwide Sensor Web, in: *Big Data*, pp. 204–218, CRC Press, Boca Raton, FL, USA, 2024, DOI:10.1201/9781003406969-13.
30. Ray, S., Jarquin, D., Howard, R., Comparing Artificial-Intelligence Techniques with State-of-the-Art Parametric Prediction Models for Predicting Soybean Traits. *Plant Genome*, 16, 1, 2022, DOI:10.1002/tpg2.20263.

31. Phattanaviroj, T., Moslehpour, M., Walawalkar, A.M., Data Ethics and Privacy, in: *Advances in Computational Intelligence and Robotics*, pp. 321–353, IGI Global, Hershey, PA, USA, 2024, DOI:10.4018/979-8-3693-3860-5.ch010.
32. Narayanan Valli, L., Sujatha, N., Mech, M., V. S., L., Ethical Considerations in Data Science: Balancing Privacy and Utility. *Int. J. Sci. Res. Arch.*, 11, 1, 011–022, 2024, DOI:10.30574/ijsra.2024.11.1.1098.
33. Golyan, A., Panchal, S., Vaghasiya, D., Parekh, H., Data Ethics and Privacy, in: *Advances in Systems Analysis, Software Engineering, and High-Performance Computing*, pp. 259–268, IGI Global, Hershey, PA, USA, 2024. DOI:10.4018/979-8-3693-3609-0.ch011.
34. Rajak, P., Ganguly, A., Adhikary, S., Bhattacharya, S., Internet of Things and smart sensors in agriculture: Scopes and challenges. *J. Agric. Food Res.*, 14, 100776, Dec. 2023, DOI:10.1016/j.jafr.2023.100776.
35. El Khediri, S., *et al.*, Integration of Artificial Intelligence [AI] with Sensor Networks: Trends, Challenges, and Future Directions. *J. King Saud Univ. Comput. Inf. Sci.*, 36, 1, 101892, 2024, DOI:10.1016/j.jksuci.2023.101892.
36. MacKinnon, J., *et al.*, Deep Neural Networks for Evaluating Future Satellite-Based Hyperspectral Microwave Sensor Designs, in: *IGARSS 2023 IEEE International Geoscience and Remote Sensing Symposium*, pp. 5210–5213, IEEE, Jul. 2023, DOI:10.1109/igarss52108.2023.10282220.
37. Shahriar, S., Corradini, M.G., Sharif, S., Moussa, M., Dara, R., The role of generative artificial intelligence in digital agri-food. *J. Agric. Food Res.*, 20, 101787, Apr. 2025, DOI:10.1016/j.jafr.2025.101787.
38. Choudhary, R., Mantri, S.P., Barse, V.A., Chitnis, S., Leveraging AI in Smart Agro-Informatics: A Review of Data Science Applications. *Int. Res. J. Adv. Eng. Manag.*, 2, 06, 1964–1975, 2024, DOI:10.47392/irjaem.2024.0291.
39. Elufioye, O.A., Ike, C.U., Odeyemi, O., Usman, F.O., Mhlongo, N.Z., AI-Driven Predictive Analytics in Agricultural Supply Chains: A Review: Assessing the Benefits and Challenges of AI in Forecasting Demand and Optimizing Supply in Agriculture. *Comput. Sci. IT Res. J.*, 5, 2, 473–497, 2024, DOI:10.51594/csitrj.v5i2.817.
40. Tonneau, J.-P., *et al.*, Geospatial Information for African Agriculture: A Key Investment for Agricultural Policies. *Perspective*, 51, 1–4, 2019, DOI:10.19182/agritrop/00068.
41. Susha, I., Janssen, M., Verhulst, S., Data Collaboratives as a New Frontier of Cross-Sector Partnerships in the Age of Open Data: Taxonomy Development, in: *Proceedings of the Annual Hawaii International Conference on System Sciences*; Hawaii International Conference on System Sciences, 2017, DOI:10.24251/hicss.2017.325.

42. Basso, B., Dobrowolski, J., McKay, C., From the Dust Bowl to Drones to Big Data: The Next Revolution in Agriculture. *Georgetown J. Int. Aff.*, 18, 3, 158–165, 2017, DOI:10.1353/gia.2017.0048.
43. Wisnubhadra, I., Adithama, S.P., Baharin, S.S.K., Herman, N.S., Agriculture Spatiotemporal Business Intelligence Using Open Data Integration, in: *Proceedings of the 2019 International Seminar on Research of Information Technology and Intelligent Systems [ISRITI]*, pp. 534–539, IEEE, Dec. 2019, DOI:10.1109/isriti48646.2019.9034635.
44. Florini, A., Pauli, M., Collaborative Governance for the Sustainable Development Goals. *Asia Pac. Policy Stud.*, 5, 3, 583–598, 2018, DOI:10.1002/app5.252.

3

Data Acquisition and Sensing Technologies in Geospatial Intelligence

Vishvajeet D. Jadhav[1], Santosh Korav[1]*, Sandeep Menon[1] and Prakash Kumar Jha[2]

[1]*Department of Agronomy, School of Agriculture, Lovely Professional University, Jalandhar, Punjab, India*
[2]*Department of Plant and Soil Sciences, Mississippi State University, Starkville, MS, USA*

Abstract
The field of geospatial intelligence (GEOINT) has come a long way since the first oblique aerial photograph was taken from a balloon over Paris in 1858. GEOINT is a fundamental element in monitoring and achieving Sustainable Development Goals (SDGs), relying on remote sensing and geospatial technologies to track numerous indicators effectively. GEOINT involves the collection, management, analysis, and dissemination of imagery and geospatial data to characterize, analyze, and visualize the Earth's physical features and human activities. This multidisciplinary field relies on various types of data, including primary methods such as field surveying, global navigation satellite system (GNSS), photogrammetry, LiDAR, remote sensing, and meteorological sensors, as well as secondary methods such as digitizing existing maps and converting paper files to digital formats. The collected data include vector, raster, and metadata types gathered using both active and passive sensors. GEOINT supports environmental monitoring, urban development, and disaster management through precise spatial analysis but faces challenges in maintaining data quality, managing large datasets, and adapting to rapid technological advancements. Despite these hurdles, GEOINT remains a valuable means of understanding spatial dynamics more effectively and continues to play a crucial role in ecological sustainability by addressing the resource balance. GEOINT will play an increasingly important role in leveraging emerging technologies to enhance our capacity to monitor and respond to global problems

**Corresponding author*: santoshkorav@gmail.com

Gurwinder Singh, Vishakha Sood, Narayan Vyas, Surendra Yadav and Dankan Gowda V. (eds.) Geospatial Intelligence for Precision Agriculture, (43–62) © 2026 Scrivener Publishing LLC

in ways that enable sustainable development and informed decision-making across sectors.

Keywords: Sustainable development, global challenges, dimensions, technology, disaster management, environmental monitoring

3.1 Introduction

With the evolution of geographic information systems, geospatial intelligence (GEOINT) has emerged as one of the most revolutionary developments in modern intelligence and spatial analysis. Fundamentally, GEOINT involves the systematic collection, analysis, and presentation of geospatial data associated with the physical features of the Earth. This discipline originated from traditional cartography and remote sensing practices. New technologies have catalyzed a remarkable transformation in this field. Modern GEOINT operations integrate a wide range of sources, including high-resolution satellites, autonomous drones, and IoT devices, to provide comprehensive spatial data for analysis [1]. GEOINT has a wide range of practical applications, demonstrating its usefulness in modern society. GEOINT is indispensable for forecasting growth patterns, monitoring infrastructure development, and mapping susceptibility to prospective security threats in urban planning and security contexts. GEOINT delivers significant value in environmental monitoring and urban resilience applications where it can be used to assess the affects of climate change, monitor natural hazards and proactively determine optimal management of resources to meet upcoming challenges [2].

With GEOINT and IoT networks, we can achieve a new form of environmental surveillance that allows near-real-time monitoring of multiple parameters, including air quality, temperature changes, and soil moisture levels [3]. Geographic Information Systems (GIS) have become transformative tools in contemporary public health, offering capabilities for real-time disease surveillance, spatial analysis, and integration of diverse epidemiological data. GIS supports the detection and monitoring of disease outbreaks, facilitates efficient allocation of health resources, and strengthens data-driven public health decision-making through advanced geospatial visualization. The utility of GIS was particularly evident during the COVID-19 pandemic, where it played a vital role in delivering timely and accurate information, thereby enhancing the responsiveness of public health authorities to rapidly changing epidemiological conditions [4].

Table 3.1 GOINT-based satellites.

Launching agency	Name	Date of launch	Type	Purpose	Source
	NVS-01 (IRNSS-1J)	29/05/23	Geo Synchronous	Navigation satellite	[7]
ISRO	IRNSS-1I	12/04/18	Geo Synchronous	To deliver precise real-time positioning and timing services across India.	[8]
	EOS-4 (RISAT- 1B)	14/02/22	Sun Synchronous	Agriculture, forestry, and plantations.	[9]
	OceanSat-3	26/11/22	Sun Synchronous	Surface winds and ocean surface strata.	[10]
NASA	Landsat 9	27/09/21	Sun Synchronous	Repeat global observations for monitoring, understanding, and managing Earth's natural resources	[11]
	OCO-3 on ISS	04/05/19	Other	To investigate the distribution of carbon dioxide on Earth	[12]
	ICESat-2	15/09/18	Near polar	Measure the elevation of ice sheets, sea ice, and forests	[13]
	GRACE-FO	22/05/18	Near polar	Track water movement and surface mass changes across the planet	[14]
Student satellites	Jugnu	12/10/11	low Earth orbit	To image the earth for agriculture and disaster monitoring purpose.	[15]
	SATAYABAMA SAT	22/06/16	Sun Synchronous Polar Orbit	The primary mission is pollution monitoring and data communication to the earth station.	[16]

The future of GEOINT is promising, with advanced frameworks, such as spatial, analytical, and multidimensional meta-models, enhancing data organization and analysis. Complex geospatial data can be organized and analyzed using advanced frameworks such as GIS-based spatial models and multidimensional data analytics. This is instrumental in making GEOINT operations efficient in environments where multiple variables are considered simultaneously. As the impacts of climate change become more prevalent and most countries face rapid urbanization, there is a growing need for comprehensive environmental health risk management in cities worldwide, which GEOINT can address [5]. It is effective, and indeed, GEOINT can significantly enhance what emergency management is about in terms of situational awareness for emergency response coordination and optimizing resource utilization. As such development continues in this area, it will be an exciting time for more predictive capabilities, real-time analysis, and environmental monitoring systems. Its ability to combine a wide range of data and provide actionable insights in almost real time makes it indispensable in future spatial analyses and decision-making [6]. The recently launched GEOINT-based satellites are listed in Table 3.1.

3.2 Data Acquisition Techniques

A data acquisition system is an integrated setup consisting of sensors, measurement devices, and computers. It is used to gather and process

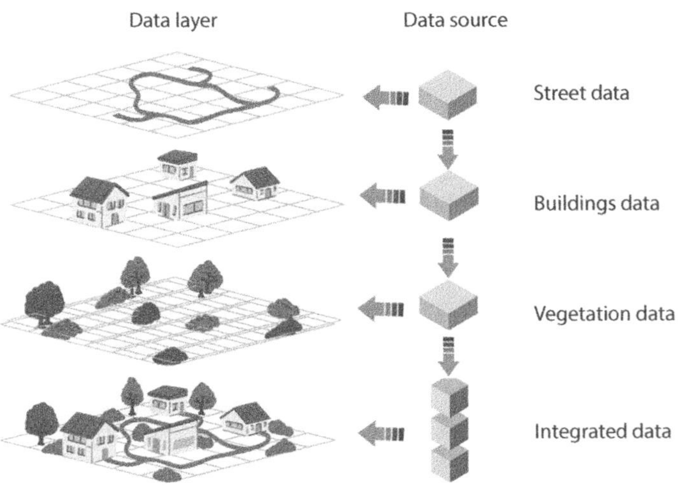

Figure 3.1 Data acquisition, analysis, storage, and integration techniques.

data to analyze electrical or physical phenomena, enabling a comprehensive understanding of the collected information. Data acquisition systems or DAQ systems provide solution for measuring and recording electrical signals and physical attributes (Figure 3.1). The acquired organised data stream can then be used for analysis, discrimination between different forms of data and decision making in many disciplines within the scientific and engineering industries [17]. These systems capture data from real-world systems and store them in a format that is readily accessible for further engineering or scientific analysis [17].

3.2.1 Types of Data Acquisition

Geospatial data has become available through a myriad of sophisticated acquisition methods that serve different mapping needs, ranging from publication of static products such as orthorectified imagery through real-time monitoring and sensing. This comprehensive overview examines the primary techniques used in modern digital forensic investigations, highlighting their applications, advantages, and potential challenges.

Static Acquisition (Non-Volatile Memory): Non-volatile memory acquisition forms the cornerstone of computer forensic investigations, focusing on storage media that retain data after power loss. With the support of tools such as FTK Imager, Pro Discover, EnCase, and X-Ways Forensics, forensic analysts can accomplish this task.

Logical Acquisition: This approach captures only active data, which is helpful in some instances. Logical acquisition is a practical course of action used primarily for dealing with large storage systems, such as Redundant Array of Independent Disk (RAID) configurations, or when time is of the essence in an investigation. RAID is a standardized architecture for configuring redundant multidisk storage systems. It is designed to achieve two primary objectives: enhancing system performance by increasing the data throughput (bandwidth) and improving fault tolerance through the implementation of data redundancy mechanisms [18].

Sparse Acquisition: Sparse acquisition captures active and deleted data, including fragments, making it suitable for efficient data collection in complex systems such as RAID. This is a suitable acquisition methodology because, in general, sparse acquisition is appropriate in RAID systems (or when suspects have limited proficiency in data obfuscation techniques). The trade-off lies between the breadth of data collection and efficiency in resource utilization [19].

Network Forensics: These are among the newest and most challenging topics in the digital investigation field. Investigators face several key obstacles:

- Managing large volumes of data, particularly in RAID system investigations
- Requiring diverse technical expertise for analyzing evidence across different device types
- Navigating organizational constraints regarding service disruption
- Addressing multi-jurisdictional challenges and varying privacy regulations across different regions [20].

Cloud Data Acquisition: The changes in IT service delivery and management brought about by cloud computing necessitate adapted forensic approaches. Traditional digital forensic practices often prove inadequate in cloud environments for the following reasons:

- The dynamic nature of cloud architecture
- Heavy reliance on virtualization technology
- Geographical distribution of cloud components
- Complex legal and technical challenges across multiple jurisdictions

Physical Acquisition: Physical acquisition manifests in two distinct forms: *bit-stream disk-to-image files*. The widely adopted method produces exact bit-by-bit replications (images) of the source drives. The principal advantage of its use is that it is possible to create multiple copies with the integrity of the original media remaining intact. Therefore, it is usually the preferred tool for use in most investigations because of the type of evidence it records.

Bit-Stream Disk-to-Disk: Disk imaging, also referred to as disk cloning or bit-stream imaging, involves creating an exact bit-for-bit replica of the source storage medium, encompassing all active data as well as free and slack space. In a disk-to-disk copy, data are directly transferred from the original medium to the target medium. It is essential to preserve the integrity of the original storage device throughout the backup or imaging process. To prevent any modification of data on the source media, forensic investigators must employ a write-blocking device during the acquisition procedure [21].

3.3 GEOINT and Sensing Technologies

In the field of GEOINT, sensors are the foundation of modern technology, providing cost-effective and sophisticated devices that can detect and distinguish various types of environmental inputs. These devices convert physical phenomena, such as light, heat, motion, moisture, and pressure, into quantifiable signals. These signals are then converted into a human-readable form or transmitted electronically for further processing [22].

Active Sensing Systems in GEOINT: Active sensing systems, such as radar and LiDAR, emit their own energy to illuminate targets, thereby enabling precise data collection under various conditions. Major active sensing systems include radar, LiDAR, sounders, laser range finders, and scatterers. Compared to optical active sensors, wind radar systems have the advantage that they are less sensitive to atmospheric conditions since they operate at a higher frequency within the microwave range.

Passive Sensing Technology: Most remote sensing systems operate in passive mode, utilizing naturally available solar energy as their primary source of radiation. The different types of sensors that make up this system work efficiently across the visible, infrared, thermal infrared, and microwave regions of the electromagnetic spectrum, utilizing technologies such as accelerometers, hyperspectral radiometers, imaging radiometers, and spectrometers.

Imaging Sensor Technologies and Applications: Imaging sensors have revolutionized the collection of spatial data, transforming it into a science of creating detailed digital representations of whatever falls within their field of view. By employing various sophisticated technologies, including CCD cameras, they record the spatial distribution of signal strength across the observed area. This category includes optical imaging sensors operating in the visible and reflective IR ranges, thermal IR imaging sensors functioning between 9 μm and 14 μm wavelengths, and radar imaging sensors working within the 1 mm–1 m electromagnetic spectrum range. Each of these types has different uses, from making detailed surface maps to detecting thermal anomalies and measuring surface characteristics.

Non-Imaging Sensor Systems and Their Role: However, non-imaging sensors are directed at the signal intensity across the entire field of view and ignore spatial variations. These systems, including radiometers, altimeters, spectrometers, and spectroradiometers, perform excellently in measuring specific environmental parameters, such as height, temperature, and wind speed. Among these, a spectroradiometer provides very high spectral resolution measurements across several wavelength bands, whereas a radiometer offers quantitative measurements of electromagnetic radiation intensity across specific spectral regions [23].

Advanced Scanning Methodologies: Remote sensing systems are effective in the transition between whisk-broom and push-broom approaches. Whisk-broom sensors employ a rectangular scanning pattern with varying angles, suitable for either airborne (90°–120°) or satellite-based platforms (10°–20°). Whisk-broom sensors are suitable for the detailed mapping and monitoring of features. Push-broom sensors, on the other hand, operate in a linear pattern and are highly effective for large-scale mapping and detecting temporal changes.

Integration of Sensor Technologies: Modern GEOINT applications benefit from the integration of multiple sensor types, each of which contributes unique capabilities to the overall system. Optical sensors provide detailed surface imaging in the visible and near-infrared spectra, whereas radar sensors offer all-weather observation capabilities [24]. LIDAR systems enable precise 3D mapping and modeling, whereas infrared sensors specialize in measuring thermal properties. Microwave sensors complete this technological ecosystem by providing crucial data on precipitation patterns and sea surface temperatures.

Sensors Used in Daily Life: In our modern world, various sensors play crucial roles in multiple industries and everyday applications. Level sensors are fundamental tools for measuring fluid quantities in open and closed systems. While commonly associated with fuel-level monitoring, these sensors are extensively utilized in liquid-handling industries, including recycling facilities and beverage production plants, where precise liquid asset tracking is essential (Figure 3.2). Temperature sensors have evolved beyond their traditional roles in climate control systems. Today, they are integral to manufacturing processes, where maintaining specific ambient conditions is crucial for the operation of equipment. In agriculture, these sensors are used to monitor soil temperature to ensure optimal plant growth and maximize crop yields. The healthcare sector has embraced temperature-sensing technology for various medical applications.

Proximity sensors are ubiquitous in retail and automotive applications. They play a significant role in vehicles by assisting with parking and obstacle detection [25]. Pressure sensors are among the most critical components of any liquid- or gas-based system. These sensors enable IoT-integrated monitoring of pressure-driven equipment, allowing system administrators to identify and rectify anomalies at the earliest opportunity. They are used

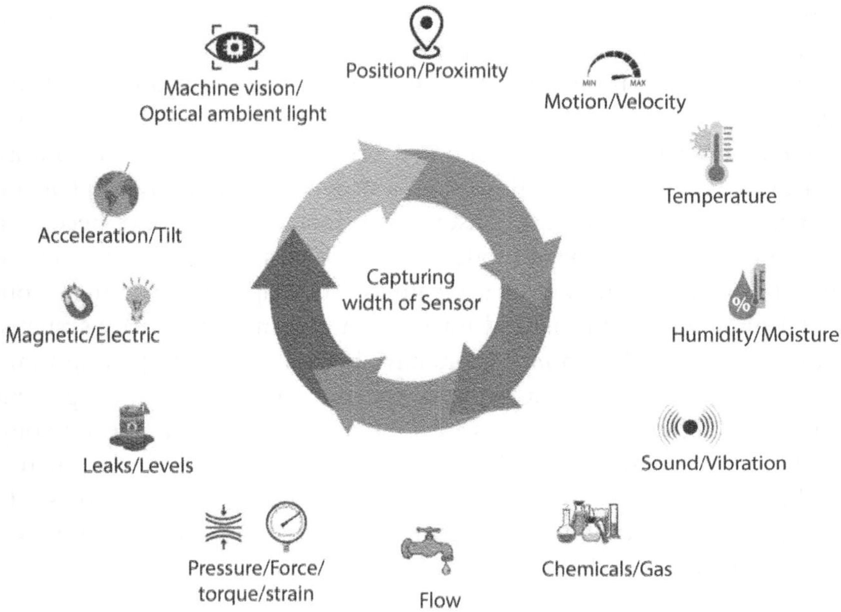

Figure 3.2 Sensors digitally capture the reality of physical objects.

to maintain water distribution networks and heating systems, as fluctuations in pressure can indicate potential issues. Water quality sensors play a crucial role in monitoring water systems in various industries. They are essential to water distribution systems, where they can detect multiple sources of contamination, such as cross-connections with non-potable water, pollution from compromised pipes, and microbial growth within the distribution system [26].

It also encompasses the use of chemical sensors, which have become integral to monitoring functions in urban and industrial environments, as well as gas sensors. In turn, chemical sensors detect modifications in the liquid composition or air quality. In contrast, gaseous sensors control the general state of the atmosphere to trace the presence of specific gases. However, such sensors are essential in industrial settings, laboratories, and urban environments for their application in environmental monitoring and safety [27]. The two most common smoke detectors are photoelectric and ionization smoke detectors. Ionization smoke detectors are good at detecting flaming fires, and this type of smoke detector can sense small, invisible particles; therefore, it provides a better response toward flaming fires, whereas photoelectric smoke detectors are better at detecting smoldering fires, which are better at detecting slow-developing fires that produce more smoke instead of flames [28].

Infrared sensors have applications in healthcare and consumer electronics. These devices can measure radiated heat and have general applications in blood flow, smart devices, and other forms of non-contact temperature measurement. Their applications span a wide variety of home appliances, automotive systems, and art conservation. Image sensors have become a fundamental component of modern technology, serving as converters of optical information into electrical signals. Their applications range from digital photography to medical imaging and biometric security systems. Similarly, motion detection sensors have become increasingly important in security applications, automated systems, and energy management solutions. Accelerometer sensors are another class of deployed technology, including mobile devices and vehicle monitoring. These devices detect mechanical motion by converting acceleration due to inertial forces into electrical signals, enabling applications ranging from simple vibration detection to smart pedometers [29].

3.4 Aerial Photography—A Vital Part of GEOINT

Aerial photography, a significant component of GEOINT, relies on several key elements for the visual interpretation of objects. The size is essential to identify by comparing the relative scale within the image, while the pattern shows the spatial arrangement and repetition of features [30]. High-resolution unmanned aerial vehicle (UAV) imagery enhances precision agriculture by enabling accurate crop disease detection and monitoring, thereby complementing advanced deep learning models such as YOLOv8 [31].

3.4.1 Data Fusion: Advanced Integration Techniques for Multi-Sensor Environments

Data fusion is a sophisticated method used to integrate and analyze information from multiple sources, providing a more accurate and detailed inference than that attainable with data from a single source. Traditionally used only in sensor networks, this technology spans the spectrum of text processing and decision-making systems.

3.4.2 Classification Framework

Several key frameworks guide the classification of data fusion techniques, each addressing specific aspects of the fusion process. These provide a useful categorization that helps explain the ways to combine and process data effectively. Forms links to the actual data type and source relation.

3.4.3 Data Fusion Techniques Can Be Categorized Based on the Relationships Between Data Sources

Different parts of a scene are complementary (distinct from each other), redundant (providing overlapping or repeated data from multiple sources), and cooperative (the union creates a complex understanding).

3.4.4 Abstraction and Architecture

The abstraction levels progress from signal-level processing through pixel, characteristic, and symbol levels, each serving a specific analytical purpose. Data fusion systems can be implemented in various configurations:

- Centralized: utilizing a single fusion node [32].
- Decentralized: employing a network of processing nodes
- Distributed: processing data independently before fusion [33].
- Hierarchical: combining different architectural approaches [34].

3.4.5 System Components

A comprehensive data fusion system typically incorporates three essential components:

- Source management for input data handling
- Human-computer interaction interfaces for operator engagement
- Database management systems for information storage and retrieval

3.4.6 Implementation Significance

The implementation of data fusion techniques offers significant advantages in multiple applications:

- Enhanced detection accuracy and reliability
- Improved decision-making capabilities
- More comprehensive environmental understanding
- Efficient resource utilization [35].

3.5 Remote Sensing Resolution

Remote sensing systems are characterized by four fundamental types of resolution that collectively determine the quality and utility of the collected data. The effectiveness of a remote sensing system in a particular application is determined by its spatial, spectral, radiometric, and temporal resolutions.

Spatial Resolution: Spatial resolution refers to the smallest spatial feature that can be detected in an image, which is primarily determined by the instantaneous field of view (IFOV) of the sensor. Landsat 1, 2, and 3 multispectral scanner sensors had a spatial resolution of approximately 80 m. Lower platforms provide better detail over smaller areas, whereas higher platforms cover more area with less detail [36].
Spectral Resolution: The spectral resolution of an imaging sensor is defined as the number of spectral bands or the number of spectral bandwidths that can be resolved by the sensor across the electromagnetic spectrum. This capability is required to distinguish between features based on their spectral response patterns. Application areas that demand sensors capable of detecting nuances, such as water bodies and vegetation, are best served by wide-wavelength-range sensors. However, those that require more subtle distinctions need sensors with finer spectral resolution capabilities.
Radiometric Resolution: The sensor's ability to detect slight differences in radiation flux is called radiometric resolution, which is commonly characterized in binary digits (bits). For instance, an 8-bit system can recognize 256 different brightness (2^8) levels, yielding significant gradations of image information content [23].
Temporal Resolution: The temporal resolution (or revisit period) refers to the frequency at which a sensor can capture imagery at an exact location. The temporal resolution varies by system; for example, IRS-1A had a revisit

period of 22 days. According to the applications of regular monitoring of surface changes, the actual revisit period may not be equal to the orbital cycle duration [37].

3.6 Applications of Big Data in Agriculture

Precision agriculture has been revolutionized by the application of big data technology and the comprehensive analysis of multi-source data to facilitate more precise and intelligent farming. The agricultural big data framework encompasses diverse data sources, including real-time sensor network data from IoT devices that measure soil conditions, historical farming records documenting various agricultural activities, and spatial data from satellite remote sensing and mapping systems. The success of agricultural big data applications relies heavily on effective data integration and quality management [38]. Through Extract, Transform, Load (ETL) processes, data from heterogeneous sources are consolidated into centralized databases, ensuring data consistency and accessibility across all systems. The system utilizes advanced decision-support models that employ deep learning, artificial neural networks, and machine learning algorithms to provide dynamic agricultural solutions. The challenge lies in developing flexible algorithms to address temporal and event-driven changes in agricultural production, as static models cannot effectively adapt to evolving needs (Figure 3.3) [39].

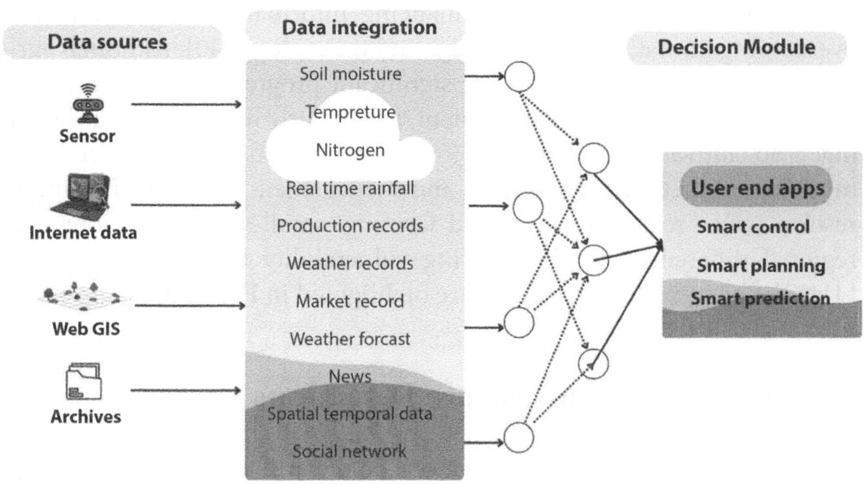

Figure 3.3 Big data analytics framework for precision agriculture.

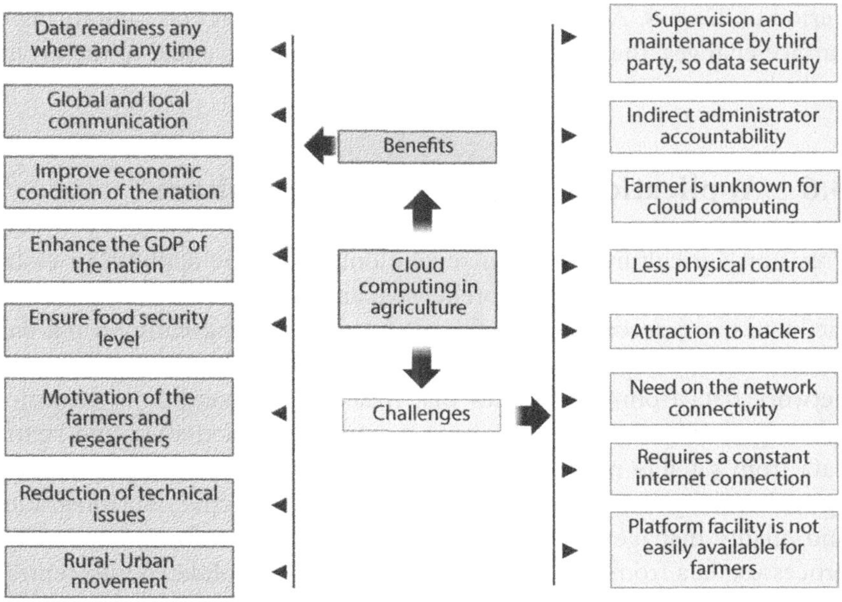

Figure 3.4 Benefits and challenges of cloud computing in agriculture.

3.7 Cloud Computing in Agriculture

Cloud computing has emerged as a transformative technology in agriculture, offering hosted services over the Internet through three main categories: IaaS, PaaS, and SaaS. This technology addresses key challenges in agricultural modernization by enhancing information accessibility and overcoming limitations in knowledge and technology [40]. Although agricultural mechanization has made significant progress, challenges persist in technical support and management, particularly in areas such as breeding, crop cultivation, and weather forecasting. Cloud computing facilitates modern production organization and management methods, helping to resolve issues related to dispersed farming, small-scale operations, and regional differences, while reducing vulnerability to climate variations [41]. The benefits and challenges are highlighted in Figure 3.4.

3.8 Challenges and Future Directions

Data quality and ethical considerations are central challenges in the rapidly evolving field of GEOINT. Crowd-sourced data and low-resolution imagery are particularly problematic sources, as they are inconsistent and unreliable, posing significant concerns regarding data quality. Advanced validation and correction techniques are required to address these issues and maintain the reliability and actionability of the GEOINT [42]. Additionally, ethical problems extend well beyond privacy concerns; as AI models in GEOINT become more automated, questions about ethical responsibility shift to ensuring the moral implications of decision-making when there is minimal human oversight. In the case of urban security, GEOINT's contribution requires it to remain effective while ensuring ethical accountability, mainly because such data are often extracted from surveillance sources. Artificial intelligence models and their efficient legal frameworks will take responsibility and ethically manage geospatial data [43].

Computational Capacity and Quantum Computing: The efficient processing of large geospatial datasets requires computational advancements owing to the immense complexity of geospatial data. Distributed computing has been practical; however, with the advent of quantum computing, the potential applications of GEOINT are transforming, enabling the processing of complex data in real time. This has enabled researchers to solve geospatial optimization problems that are central to geospatial analysis, such as logistics route planning or disaster resource allocation. Such integration may provide a significant leap forward for GEOINT capabilities, enabling efficient urban planning and transportation network optimization, particularly in large cities. User-friendly quantum platforms and GEOINT professionals skilled in quantum technologies will be key to the widespread adoption of such analyses [44].

Sustainability and Human-Centered GEOINT: The development of GEOINT is becoming increasingly important for sustaining human needs. GEOINT for sustainable development involves minimizing environmental impacts to support the global Sustainable Development Goals (SDGs) [45]. They provide real-time data for conservation and policy formulation, supporting sustainable development. GEOINT also has the potential for social equity, the ability to provide insights into environmental injustice, and the ability to help inform policies that benefit vulnerable communities [6].

3.8.1 Conclusion

Since its inception, GEOINT has undergone a dramatic transformation into a critical resource for addressing complex global challenges in multiple sectors. Through the integration of sophisticated data acquisition techniques, modern sensing technologies, and computational methods, GEOINT has revolutionized the monitoring, analysis, and understanding of Earth's physical characteristics and human activity. Its strength lies in its ability to extract multidimensional data from various sources, including satellite imagery and ground sensors, and to utilize a combination of both active and passive sensing systems. The integration of aerial photography, remote sensing, and big data analytics with its four key resolutions (spatial, spectral, radiometric, and temporal) has promoted precision agriculture and environmental monitoring. Nonetheless, GEOINT faces many challenges in terms of data quality, computational, and ethical issues. However, with the advent of quantum computing, quantum computers can provide huge speed improvements for complex computations involving large amounts of geospatial data, and cloud computing enables greater accessibility and enhanced data management capabilities in agricultural applications. In the future, GEOINT will move toward sustainability and human-focused approaches and will contribute to the SDGs. Although the essential role that GEOINT plays in fostering informed decision-making and sustainable development will only grow in importance as technology advances, the advancement of technology will depend on developing effective means of managing its ethical implications in responsible data handling and maintaining technological innovation.

AI disclosure statement: The author acknowledges the use of AI-based tools for minor language editing or grammatical correction. However, all intellectual contributions, analyses, and conclusions presented in this chapter are entirely the author's own.

References

1. Li, W. and Hsu, C.Y., Geo-AI for large-scale image analysis and machine vision: recent progress of artificial intelligence in geography. *ISPRS Int. J. Geo-inf.*, 11, 7, 385, 2022.
2. Bacastow, T.S., Definitions of GEOINT (Geospatial Intelligence), in: International Encyclopedia of Geography, pp. 1–6, Wiley, Mar. 31, 2023, doi:10.1002/9781118786352.wbieg2184.
3. Tagliabue, L.C., Re Cecconi, F., Moretti, N., Rinaldi, S., Bellagente, P., Ciribini, A.L.C., Security assessment of urban areas through a GIS-based analysis of lighting data generated by IoT sensors. *Appl. Sci.*, 10, 6, 2174, 2020.

4. Laosupap, K., Wongpituk, K., Butsorn, A., Boonsang, A., Thammaboribal, P., Chankong, W., Pokomnird, C., Advancements in Disease Surveillance: The Role of GIS in Global Health Preparedness. *Int. J. Geoinf.*, 20, 10, 95–108, 2024.
5. Biu, P.W., Oliha, J.S. and Obi, O.C., The evolving role of geospatial intelligence in enhancing urban security: A review of applications and outcomes. *Eng. Sci. Technol. J.*, 5, 2, 483–495, 2024.
6. Kolluru, V., Mungara, S., Chintakunta, A.N., Kolluru, L., Telaganeni, C.S., Geospatial Intelligence Enhancement using Advanced Data Science and Machine Learning: A Systematic Literature Review. *SSRN*, 4929468, 2024.
7. Cui, H. and Zhang, S., Satellite Availability and Service Performance Evaluation for Next-Generation GNSS, RNSS and LEO Augmentation Constellation. *Remote Sens.*, 13, 18, 3698, 2021.
8. Varsha, H.S., Chougule, S., Vighnesam, N.V., Sudha, K.L., IRNSS orbit and clock bias estimation using navic ground receiver data: extended kalman filter. *Int. J. Eng. Adv. Technol.*, 9, 713–7, 2019.
9. Haldar, D., Thulariraman, D., Mitra, M., Singh, R.P., RISAT-1A (EOS-4) Medium and Fine resolution data utilization for field crop and plantation monitoring, 2024.
10. Barale, V., Half a Century of Oceans from Space: Features and Futures. *Remote Sens.*, 15, 16, 4064, 2023.
11. Choate, M.J., Rengarajan, R., Hasan, M.N., Early in mission Landsat 9 geometric performance, in: *Earth Observing Systems XXVII*, vol. 12232, pp. 234–248, SPIE, 2022.
12. Taylor, T.E., Eldering, A., Merrelli, A., Kiel, M., Somkuti, P., Cheng, C., Rosenberg, R., Fisher, B., Crisp, D., Basilio, R., Bennett, M., OCO-3 early mission operations and initial (vEarly) XCO2 and SIF retrievals. *Remote Sens. Environ.*, 251, 112032, 2020.
13. Palm, S.P., Yang, Y., Herzfeld, U., Hancock, D., Hayes, A., Selmer, P., Hart, W., Hlavka, D., ICESat-2 atmospheric channel description, data processing and first results. *Earth Space Sci.*, 8, 8, p.e2020EA001470, 2021.
14. Loomis, B.D., Felikson, D., Sabaka, T.J., Medley, B., High-spatial-resolution mass rates from GRACE and GRACE-FO: Global and ice sheet analyses. *J. Geophys. Res.: Solid Earth*, 126, 12, e2021JB023024, 2021.
15. Murugan, P., Satellite Projects by Indian Students. *Int. J. Eng. Res. Technol. (IJERT)*, 9, 3, Mar. 2020, [Online] Available: https://www.ijert.org/satellite-projects-by-indian-students.
16. Rani, B.S., Santhosh, R.R., Prabhu, L.S., Federick, M., Kumar, V., Santhosh, S., A survey to select microcontroller for Sathyabama satellite's On Board Computer subsystem, in: *Recent Advances in Space Technology Services and Climate Change 2010 (RSTS and CC-2010)*, pp. 134–137, IEEE, 2010.
17. Sarker, I.H., Machine learning: Algorithms, real-world applications and research directions. *SN Comput. Sci.*, 2, 3, 160, 2021.

18. Rodriguez-Olivares, N.A., Gomez-Hernandez, A., Nava-Balanzar, L., Jimenez-Hernandez, H., Soto-Cajiga, J.A., FPGA-based data storage system on NAND flash memory in RAID 6 architecture for in-line pipeline inspection gauges. *IEEE Trans. Comput.*, 67, 7, 1046–1053, 2018.
19. Quach, T.T., Agarwal, S., James, C.D., Marinella, M.J., Aimone, J.B., Sparse Data Acquisition on Emerging Memory Architectures. *IEEE Access*, 7, 1685–1693, 2018.
20. Ng, Q.A., Chiew, Y.S., Wang, X., Tan, C.P., Nor, M.B.M., Damanhuri, N.S., Chase, J.G., Network data acquisition and monitoring system for intensive care mechanical ventilation treatment. *IEEE Access*, 9, 91859–91873, 2021.
21. Mann, H.K. and Chhabra, G.S., Volatile memory forensics: a legal perspective. *Int. J. Comput. Appl.*, 155, 3, 11–15, 2016.
22. Protasowicki, T. and Napiorkowski, J., High resolution imagery data acquisition and analysis in support of geospatial intelligence operations, in: *26th Geographic Information Systems Conference and Exhibition "GIS ODYSSEY 2019"*, p. 217, 2019.
23. Bhatta, B., *Remote sensing and GIS Vol. 2*, Oxford University Press, New Delhi, 2015.
24. Jiao, Z., The Application of Remote Sensing Techniques in Ecological Environment Monitoring. *Highlights Sci. Eng. Technol.*, 81, 449–455, 2024.
25. Javaid, M., Haleem, A., Rab, S., Singh, R.P., Suman, R., Sensors for daily life: A review. *Sens. Int.*, 2, 100121, 2021.
26. Myers, D.N., Innovations in monitoring with water-quality sensors with case studies on floods, hurricanes, and harmful algal blooms, in: *Separation science and technology*, vol. 11, pp. 219–283, Academic Press, 2019.
27. Hooshmand, S., Kassanos, P., Keshavarz, M., Duru, P., Kayalan, C.I., Kale, İ., Bayazit, M.K., Wearable nano-based gas sensors for environmental monitoring and encountered challenges in optimization. *Sensors*, 23, 20, 8648, 2023.
28. Ho, E.X., *IOT-based Smoke Alarm System*, Doctoral dissertation, UTAR, 2019.
29. Faisal, I.A., Purboyo, T.W., Ansori, A.S.R., A review of accelerometer sensor and gyroscope sensor in IMU sensors on motion capture. *J. Eng. Appl. Sci.*, 15, 3, 826–829, 2019.
30. Sabins Jr, F.F. and Ellis, J.M., *Remote sensing: Principles, interpretation, and applications*. Waveland Press, 2020.
31. Vyas, N. and Dutt, V., Advancing Precision Agriculture: Leveraging YOLOv8 for Robust Deep Learning Enabled Crop Diseases Detection, in: *2023 International Conference on Integration of Computational Intelligent System (ICICIS)*, 2023, Pune, India, pp. 1–6, 2023, doi:10.1109/ICICIS56802.2023.10430280.
32. Abu Bakr, M. and Lee, S., Distributed multisensor data fusion under unknown correlation and data inconsistency. *Sensors*, 17, 11, 2472, 2017.
33. He, H., Zhu, Z., Makinen, E., Task-oriented distributed data fusion in autonomous wireless sensor networks. *Soft Comput.*, 19, 2305–2319, 2015.

34. Dautov, R., Distefano, S., Buyya, R., Hierarchical data fusion for innovative healthcare. *J. Big Data*, 6, 1, 1–23, 2019.
35. Casian, T., Nagy, B., Kovács, B., Galata, D.L., Hirsch, E., Farkas, A., Challenges and opportunities of implementing data fusion in process analytical technology-a review. *Molecules*, 27, 15, 4846, 2022.
36. Omali, T.U., Impacts of sensor spatial resolution on remote sensing image classification. *GSJ*, 6, 1, 63, 2018.
37. Pazur, R., Price, B., Atkinson, P.M., Fine temporal resolution satellite sensors with global coverage: an opportunity for landscape ecologists. *Landscape Ecol.*, 36, 2199–2213, 2021.
38. Tantalaki, N., Souravlas, S., Roumeliotis, M., Data-driven decision making in precision agriculture: The rise of big data in agricultural systems. *J. Agric. Food Inf.*, 20, 4, 344–380, 2019.
39. Chen, J., He, S., Li, X., A study of big data application in agriculture, in: *J. Phys.: Conf. Ser.*, IOP Publishing, vol. 1757, no. 1, 012107, 2021.
40. Paramale, P.P. and Patil, P.P., Roll of Cloud Computing In Agriculture. *J. Emerg. Technol. Innov. Res.*, 8, 7, ISSN-2349-5162, 2021.
41. Debauche, O., Mahmoudi, S., Manneback, P., Lebeau, F., Cloud and distributed architectures for data management in agriculture 4.0: Review and future trends. *J. King Saud Univ. Comput. Inf. Sci.*, 34, 9, 7494–7514, 2022.
42. Menthe, L., Goldfeld, D.A., Tingstad, A., LINGEL, S., Geist, E., Brunk, D., Wicker, A., Soliman, S., Gintautas, B., Stickells, A., Cordova, A., *Technology innovation and the future of Air force intelligence analysis. Volume 1*, p. 23, RAND Corporation, Santa Monica, CA, USA, 2021.
43. Dracott, K., Trimble, M., Jollineau, M., Time for tools: a review on geospatial tools and their role in co-management. *Can. Geogr.*, 64, 1, 169–179, 2020.
44. Levin, E., Shults, R., Habibi, R., An, Z., Roland, W., Geospatial virtual reality for cyberlearning in the field of topographic surveying: Moving towards a cost-effective mobile solution. *ISPRS Int. J. Geo-inf.*, 9, 7, 433, 2020.
45. Nwachukwu, M.A., Nwachukwu, J.I., Ulo, C., Anyanwu, J., Okorondu, J., Acholonu, C., Ugochukwu, C., A Model of Sustainability Education and Partnership to Achieve SDGs in sub-Saharan Africa. *J. Sustainability Educ.*, 28, 2023.

4

Integrating IoT and Geospatial Technologies for Real-Time Crop and Soil Monitoring in Precision Agriculture

Aritra Guin[1], Santosh Korav[2*] and Sujatha H.T.[3]

[1]*SIFF, AP Community-Managed Natural Farming, RySS, Gorantla, Andhra Pradesh, India*
[2]*Department of Agronomy, School of Agriculture, Lovely Professional University, Jalandhar, Punjab, India*
[3]*Department of Agronomy, Keladi Shivappa Nayaka University of Agricultural & Horticultural Sciences, Shivamogga, Karnataka, India*

Abstract

The agricultural sector plays a significant role in global and national economies, significantly enhancing the economies of both developed and emerging nations. With such an enormous increase in the need for food that is sustainable for humans, smart farming is becoming increasingly important. This presents a significant opportunity to transform conventional farming through a synergistic combination of geographic technologies and the Internet of Things (IoT). This study proposes a integrated framework that combines IoT and geospatial technologies for near-real-time crop management in agricultural regions. The proposed methodology involves deploying IoT devices equipped with a suite of sensors for the continuous monitoring of essential soil parameters, including temperature, pH, moisture content, and nutrient concentration. IoT devices transmit data wirelessly to a cloud database for processing, analyzing, and storing. Furthermore, the gathered data, utilizing geographic information system (GIS) capabilities, enable spatial mapping, providing a clear picture to farmers and agricultural professionals of the prevailing nutritional status and health of soils across the entire farm or area. Real-time soil health and nutritional state monitoring enables farmers to make informed decisions on the timing and quantity of fertilizer and water applications to crops, as well as other management techniques, for optimal crop growth.

**Corresponding author*: santoshkorav@gmail.com

Gurwinder Singh, Vishakha Sood, Narayan Vyas, Surendra Yadav and Dankan Gowda V. (eds.) Geospatial Intelligence for Precision Agriculture, (63–82) © 2026 Scrivener Publishing LLC

***Keywords*:** Smart farming, geographic technology, Internet of Things (IoT), cloud database, soil monitoring, Geographic Information System (GIS), sensors

4.1 Introduction

It is projected that the global population will reach 10 billion over the next 35 years owing to ongoing population growth. To guarantee sufficient food for the world's population, breakthroughs in agricultural output and disaster management are essential [1]. Sustainable agricultural practices are critical for addressing limited arable land, variable soil quality, and declining groundwater levels in many regions [2]. The use of technology, such as global positioning systems (GPS), geographic information systems (GIS), and remote sensing, has revolutionized agricultural operations through precision agriculture. Crop production techniques have evolved as a consequence of this integration, leading to a significant shift in the way people perceive agricultural automation [3]. Precision crop production is made possible by integrating geospatial technologies and Internet of Things (IoT) sensors. Modern technology has utilized sensors and geospatial technologies to achieve maximum crop growth and yield. Farmers can gather intensive field data and use them to make informed decisions regarding fertilization, irrigation, and pest management. Additionally, a reduction in temporal production variability has led to increased yield stability and improved tolerance to climate variability, becoming the most significant output advantage of precision agriculture [4].

To strategically manage their crop fields, precision farmers require detailed information on the spatial variability of their farmland. Geospatial technologies, including satellite images, remote sensing, geofencing, and GPS, provide access to geographic information from various centers, enabling the IoT to establish global networks. Relevant data can then be retrieved at any time and from anywhere, if the user has the necessary access. For example, the IoT has been identified as one of the most important areas that integrates software, hardware, and other physical objects to facilitate the collection, sharing, and communication of information. These technologies can enhance almost every application for problem-solving and diagnostic analysis [5]. A wireless sensor network (WSN) is one of the most cost-effective, time-efficient, and simple-to-deploy technologies for collecting and monitoring the real spatiotemporal variability of a particular crop field.

A farmhouse or mainframe decision-support system utilizes GIS combined with prior knowledge and data collected from the field to inform the best management options for water irrigation, pesticide application timing, and fertilization scheduling [6]. This creates many new arenas for advancement in research in the field of agriculture and helps farmers in estimating and making use of easily available resources, such as water and manure, in the right amounts and time, as well as diagnosing crops or diseased plants appropriately and in time for preservation of yield as well as increasing the yield. Site-specific data collection enables precise crop management during specific growth stages, thereby reducing resource waste. This method can identify the potential processes and their causes that underlie improved adaptation to climate variability and stress conditions. Therefore, interventions designed to mitigate adverse environmental effects can be developed and employed. It describes the spatial and temporal control of space, which, in turn, allows for a more complex decision support system than those currently available [7, 8]. Below, we provide an overview of the most recent techniques applied for crop management. It carefully examines and demonstrates state-of-the-art methods and technologies that, at this very moment, are revolutionizing agricultural practices. The core of these studies has focused on big data and the use of IoT in research and the agricultural industry. This study elucidates the actual uses of these technologies and how they can improve agricultural yield, resource efficiency, and overall farm management. The challenges that arose while integrating IoT sensors with geographic technology in agriculture were also discussed.

4.2 Overview of IoT Sensors in Agriculture

Sensors are employed throughout the agricultural cycle, from planting the seed to packing the product. Within agriculture, several key processes can be distinguished, including planting, soil management, and nutrient and water control. Sensing technologies can contribute significantly to post-harvest activities, including processing, yield harvesting, and pest and disease management [9, 10]. As construction elements of real agricultural IoT systems, sensors are substantially responsible for the transmission of physical information in a digital form. Technological solutions that leverage trends aligned with environmental and economic sustainability approaches can be identified, as well as data sourced from sensors embedded in soil, plants, and animal systems. This enhances the soil, plant, and animal aspects of precision agriculture [11]. Users, especially IoT system developers, should not only be able to deploy sensors selectively, but also apply these sensors to the task at hand. The advancement of sensor technology has been the primary driver of the exponential increase in IoT usage.

Information transfer compatibility, operating reliability, sensitivity, power consumption, durability, and repeatability are the major attributes that must be considered when selecting sensors to develop IoT systems. In agriculture, various types of sensors are used on plants during different stages of production, from sowing to packaging. Different types of farming activities can be generally categorized into planting, soil cultivation, and nutrient and water management, among others. Pest and disease control in the post-harvest stage of the crop, as well as other activities following the harvesting of the produce, are areas that can benefit from the use of advanced sensor technology [12]. Low-altitude NDVI cameras mounted on Unmanned Aerial Vehicles, Unmanned Ground Vehicles, and Satellites can provide useful proxy estimates of crop yield potential, though accuracy varies with crop type, canopy structure, and sensor resolution. The accuracy of the estimation can be further improved using hyperspectral and normalized difference red edge (NDRE) cameras. The color parameters of crops (fruits) for maturity levels can be analyzed using Principal Component Analysis (PCA) and Artificial Neural Networks (ANN). Various embedded sensors, including pressure, touch, imaging, inertia measurement, position, and radio frequency identification (RFID), are used to sense the gloves worn by workers while classifying fruits [11–16]. These gloves have also been used to identify fungal disease.

Another category of measurements is represented by sensors that are moved vertically and horizontally into the soil or placed stationary in the ground. Advancements in nanotechnology have led to an increased use of smaller, miniature, or matched sensors [17]. Such sensor placement, along with data collection, has the potential to reduce losses for agricultural enterprises while increasing their effectiveness. Collected data provide the best means to establish stringent guidelines for classification, which would fall on the natural characteristics of the crop to minimize the instances of misclassifications and the consequent losses [18]. Some typical uses of chemical sensors include the determination and monitoring of pH, salinity, and nutrient levels in soil, as well as analytes such as conductivity and pH in irrigation water, and gases such as methane (CH4), oxygen (O2), and carbon dioxide (CO2), which are relevant to photosynthesis. The pH of soil and water sources is crucial to determine because it influences the solubility of nutrients in plants during the nutrient development and uptake processes.

Chemical sensors are classified into two broad types: photochemical and electrochemical sensors. To spread it further, electrochemical sensors measure the electrical signatures associated with the chemical activity or presence of a substance. In contrast, photochemical sensors observe the spectral signatures of specific substances or compounds [19]. Sensors

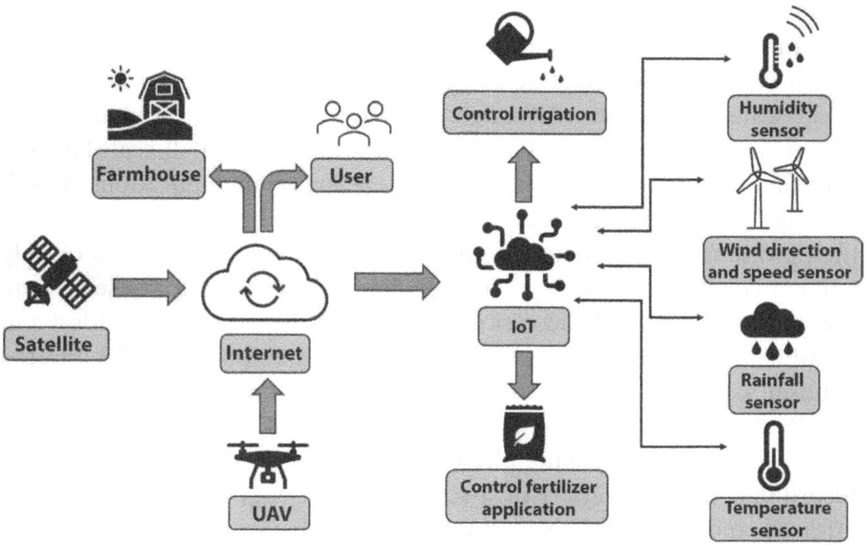

Figure 4.1 IoT-driven efficient communication in farming [23].

measure grain properties, such as protein content, mass density, and moisture, to assess quality and market value post-harvest. Grain yield variation can also be explained using qualitative parameters associated with grain structure [20]. In recent years, sensors have been developed and deployed to measure soil burrowing parameters. These sensors are designed to measure specific parameters, including the temperature, pH value of the soil, moisture, and concentration of nutrients in the soil. A distinctive feature of these sensors is their ability to biodegrade in the soil. In addition, soil-degradable sensors can address the ecological pollution issues associated with traditional sensors, eliminate the need for sensor retrieval, and provide insights into soil behavior without disrupting the environment [21]. Inexpensive and self-contained soil moisture sensors can further revolutionize our understanding of the temporal and spatial density of soil water availability on a field scale, as illustrated in Figure 4.1 [22].

4.3 Geospatial Technologies in Agriculture

Aerial and satellite image development has enhanced the capacity to monitor crop growth and soil moisture content, among other key aspects affecting crop health. Such innovations have revolutionized how farmers manage agriculture and agribusinesses by providing key information on the health and development of crops, thereby enabling them to make more informed decisions [24]. Satellite imagery provides spatial resolutions, which is suitable for large-scale agricultural monitoring, but limited for small-scale applications. Although most of these coarse-resolution products are commercially available, they are unsuitable for small-scale farming applications. The costs associated with these and other applications are still too high for widespread adoption in agriculture and elsewhere, even for the newest satellite systems offering very high spatial resolutions of only a few meters or less. Data specifically devoted to vegetation research are now available on widely accessible satellite platforms, such as Landsat and Sentinel, including the most recent Sentinel-2 mission from ESA and the Landsat 9 mission from NASA (Table 4.1). For effective large-scale crop monitoring, high-resolution readings, short return times, and broad coverage are ideal [25]. Satellite photos are also multispectral, meaning they record data across a variety of spectra and wavelengths, including visible, infrared, and thermal. However, owing to the distance between the satellite and the ground, satellite systems have limitations that affect the quality of the data, resulting in signal attenuation caused by various elements, including aerosols, water vapor, and atmospheric gases [26, 27].

Many researchers have established different mathematical formulas or imaging indices that can help identify crop and non-crop areas [28, 29]. In other words, a thorough understanding of the specific characteristics of different crops, their developmental stages, and the environmental factors that influence their appearance in satellite images is essential. These indices are further vulnerable to adverse conditions, including cloud and atmospheric changes, changes in illumination, or the presence of obstructive structures such as cloud cover, shadows, or atmospheric interference, which may affect the actual performance of identifying crop patches in the field. Satellite systems face challenges such as coarse resolutions for small-scale farming and delays in data delivery, although modern platforms such as Sentinel-2 mitigate these issues. This would also mean that novices or those without specialized training may have limited access to such metrics and technologies because meeting expert knowledge requirements might be quite difficult [30, 31]. However, satellite images are currently being used to gather useful data on agricultural water needs, irrigation effectiveness, and the identification of water stress in plants. They provide better water management for agricultural resources, water availability monitoring, and irrigation-schedule optimization.

Nonetheless, because UAVs offer numerous benefits that enhance productivity, accuracy, and efficiency, their application in agriculture has increased. These include yield assessment, field assessment, livestock

Table 4.1 Satellite platforms that are widely accessible for use in agriculture [33].

Satellite	Cost	Output	Applications in agriculture
MODIS	Free	Image data with low and high resolution	Calculating vegetation indicators, tracking land surface temperatures, and keeping updated on droughts
Sentinel-2	Free	Multispectral high-resolution pictures	Monitoring crops, identifying changes in vegetation, managing land, and detecting fires
Landsat	Free	Multispectral images and elevation information	Crop observation, disease and pest identification, management of water resources, and soil assessment
TerraSAR-X	Paid	SAR pictures with high quality	Monitoring floods, detecting soil deformation, and assessing crop structure
WorldView-3	Paid	Multispectral high-resolution pictures	Identification of vegetation changes, accurate crop monitoring, and thorough field mapping
RADARSAT-2	Paid	SAR pictures with high quality	Monitoring of soil moisture, identification of vegetation changes, and handling of natural disasters

monitoring, and mapping of weeds, among others that apply to agriculture. A general UAV design is typically inspired by specific agricultural needs, including precision spraying, field mapping, and crop monitoring. The various types of UAVs used in agriculture include [32]:

- Fixed-wing unmanned aerial vehicles: These require runways for takeoff and landing. They have a greater range and the ability to carry larger payloads.

- UAVs with rotor wings: higher resolution owing to the speed and altitude of flight. Compared to fixed-wing UAVs, they cover a smaller area.
- Hybrid unmanned aerial vehicles overcome the drawbacks of both types by reaping their respective benefits. They combine the cruise flight of fixed-wing UAVs with the VTOL capabilities of rotary-wing UAVs.

Different types of UAVs used in agricultural management have been discussed. The selection of the type of UAV depends on various criteria, including the extent of the agricultural area, the level of detail required for gathering data, obstacles, and specific activities to be performed using the UAV. Such technology will enable farmers to increase agricultural outputs while enhancing practices with fewer inputs. Aerial and satellite images are essential for tracking agricultural health-related variables, such as soil moisture content and crop development. They provide agribusinesses and farmers with valuable information on health and development to inform decision-making. This enhances agricultural yields and productivity. Applications associated with satellite and aerial photography in agriculture will become increasingly significant as imaging technologies continue to evolve.

4.4 Integration of IoT and Geospatial Technologies

The use of ICT in precision agriculture ensures that sustainable agricultural development is one of the key means of catering to the rapidly growing global population. It has demonstrated the development of advanced techniques that enhance agricultural productivity, production, and management while protecting the environment. Advanced technologies, such as blockchain [34], IoTs [35], and Artificial Intelligence (AI) techniques [36], will be implemented in the agricultural sector to ensure long-term agricultural productivity. The use of these technologies for data-driven agriculture is a promising approach to solving this problem. IoT-based supply chains support the entire agricultural production process through data collection and management [37]. In today's world, the Internet is substantial in many diversified businesses.

In this study, an IoT-based method was proposed for monitoring crop fields in agricultural areas. Based on these, this methodology utilizes sensors employing WSN technology to monitor various parameters in agricultural fields [38]. The term "IoT" refers to the concept of connecting all devices and gadgets, including sensors, to the Internet, allowing them to exchange data. The IoT is a vast network comprising an unimaginably large number of interconnected devices within a global information infrastructure. There is a need to identify an appropriate medium or channel and recognize a distinct sender and recipient for effective communication [39]. The application of IoT technology in agriculture can facilitate transformational changes in agricultural operations, thereby enhancing industrial production and its efficiency. IoT can also be deployed for precision farming methods in agriculture, where sensors and other technologies are used for data collection, enabling timely and accurate adjustments to various aspects of agricultural operations. This will, for instance, allow the proper application of fertilizers and insecticides using GPS-guided equipment, economizing resources and reducing waste. Despite this, farmers face difficulties in utilizing smart farming and IoT technologies, as they require expertise in interfacing with sensors, the Internet cloud, and end-user applications [40].

In the field of digital agriculture, the IoT has various applications, including determining the physiological state of plants, recommending fertilizer replenishment, and assessing water needs [38, 41]. The ability to predict and prepare for possible disruptions that may originate in remote locations, including insect pest invasions and the monitoring of soil water, nutrients, dynamics, and pest management, are just a few of the many advantages that the IoT offers for agriculture and crop sustainability. The capacity to estimate production, costs, and energy savings, as well as the creation of indoor vertical farming systems using data from existing IoT stations, are further benefits of integrating IoT technology into agricultural systems [40]. The Internet of Everything (IoE) is gaining increasing attention because of its ecological benefits [42]. Consequently, expenditures are reduced, and protection is more effective. Connecting distant IoT devices to store in cloud computing capabilities is an example of spatiotemporal use for big data [43, 44].

Each farm attempts to practice a more "intelligent" use of its land with various IoT and geospatial systems. This has led to the practice of "Precision Farming" and "Sustainable Farming," which have generally been more straightforward methods for easing the problems regarding food production by improving both quality and quantity. Precision farming considerably reduces the negative effects on crop yield by investigating and monitoring the geographical variability in water, soil, and ambient variables. Furthermore, farming with such practices requires a comprehensive understanding of the spatial variability of farmland to evaluate the condition of a crop field for informed management decisions [45].

Remote sensing, satellite photography, GPS, geofencing, and other geospatial technologies are used to collect valuable geographic data from various sources. With this information, the IoT creates a global communication channel *via* the Internet [46]. Consequently, stakeholders can access important data at any time and from any location. The IoT [47] is an essential element that combines hardware, computational devices, physical things, and software to create, gather, and share data, and offers wireless infrastructure to realize the full potential of agricultural management. Precision farming is essential in conjunction with geospatial technology.

This includes decision support systems, visualization, storage, processing, analysis, and inferences employing technical components such as remote sensing, GPS, GIS, and WSNs. Geospatial technologies serve as instruments for effectively gathering, evaluating, and modifying a variety of agricultural data *via* various map layers [48]. This has the potential to significantly alter agricultural decision-making and crop production management at the field, regional, national, and international levels. As a result, GIS is beneficial for various agricultural tasks, including crop forecasting, irrigation management, stress mapping, and the control of fertilizers and pesticides [49].

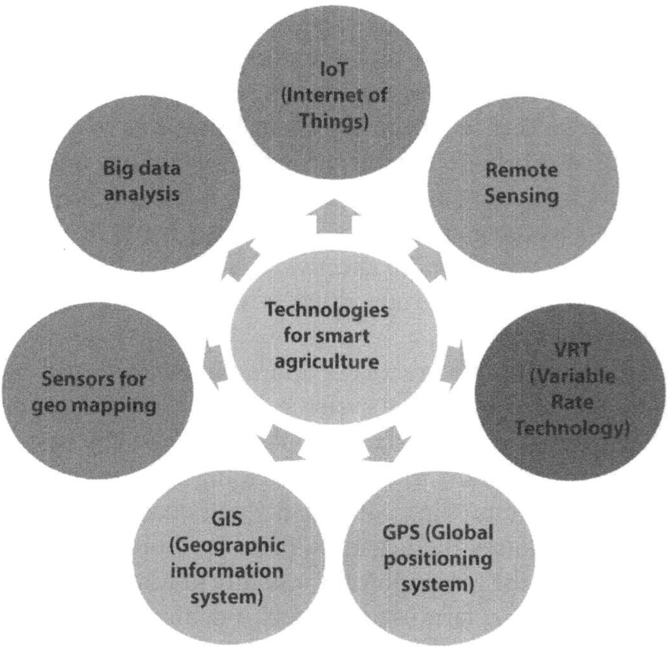

Figure 4.2 Geospatial technology used in smart agriculture [23].

However, the incorporation of IoT and geospatial technology into the agricultural sector still faces several significant challenges in global farming modernization, which disadvantages producers and consumers in existing agricultural product transactions. Challenges in smart agriculture include complex supply chains, food waste, costly intermediaries, delayed data authentication, climate variability, and limited financing options for farmers (Figure 4.2).

4.5 Benefits of Real-Time Crop Management

Real-time crop management systems are transforming agriculture by enabling technologies that place farmers in the key positions they deserve to be in.

4.5.1 More Disease-Free Crops and Higher Yields

Another major advantage is that the health of a crop can be tracked at any time. Sensors are used in the cultivation area to monitor soil moisture, nutrient levels, and pest infestations. You will realize much quicker if something is going wrong. You will know the proper time to water, add fertilizers, or address pest issues without stressing the plants, thereby increasing the yield. Internet of things (IoT) technology has been widely integrated into agricultural management practices to enhance crop productivity and improve water use efficiency through enhanced monitoring and decision-making processes [50].

4.5.2 Resources Used Efficiently

Resources such as water, fertilizers, and pesticides must be used efficiently to reduce costs and environmental impacts. These are expensive and harmful when applied in excess. Real-time crop management ensures the efficient use of resources in a more targeted manner, minimizing wastage. For instance, a precision irrigation system can determine the exact amount of water each part of a field requires, thereby saving water and preventing fertilizer runoff, which is harmful to the environment. In this regard, such systems help conserve water by preventing fertilizer runoff, which is assumed to harm the environment [51].

4.5.3 Smart Decision-Making

With real-time data at your fingertips, making decisions regarding crops becomes much easier. Advanced analytics and machine learning (ML) can analyze these data and provide actionable insights. Find the optimal planting time or determine the best moment for harvest to ensure maximum quality. In any case, these systems eliminate much of the guesswork from farming.

4.5.4 Adaptation to Climate Change

Climate change is associated with unstable and highly destructive weather conditions and events that impact crops negatively. Real-time systems enable adaptation owing to the current availability of updated weather information, which allows for effective and rapid responses to unfavorable conditions. IoT weather stations in your fields can monitor the local climate. This enables the improvement of practices using actual real-time data [52].

4.5.5 Increased Productivity and Profitability

Optimizing crop management will go a long way to increasing your productivity and profitability. Healthier crops yield bigger yields, and efficient resource usage reduces costs. This can be further enhanced with better decision-making and adaptation to climate conditions, making farming operations more stable and predictable.

4.5.6 Sustainability and Environmental Benefits

Real-time crop management will be beneficial not only for the wallet but also for the planet. These systems conserve water and soil quality by reducing the overuse of chemicals. It also encourages other sustainable approaches, such as crop rotation and field cover, which improve soil structure and biodiversity. Thus, making farming more viable with environmental sustainability ensures its long-term sustainability [53].

4.6 Challenges and Limitations

4.6.1 Security Related Concerns

Environmental factors, including humidity, wind speed, temperature, acidity, CO_2, and soil moisture, are closely tied to agricultural production; any changes made to these factors, whether by addition or subtraction, can affect crop growth and yield. For example, reduced application of fertilizer leads to a decrease in agricultural output, whereas overapplication leads to environmental pollution and further degrades the quality of groundwater. Thus, significant efforts must be invested in protecting environmental data from network hackers and eavesdroppers [54].

4.6.2 Information Management

In smart agriculture, multiple sources of data are utilized, including agricultural sensor, satellite, GPS, and weather data. All data sources should be hosted on the cloud. All these processes, namely data collection, diagnosis, analysis, decision-making, and assessment, are often distributed across multiple cloud platforms, making data management complex; thus, it is difficult for agricultural specialists to handle large amounts of data. Therefore, this combination of IoT and SDN is beneficial and requires new physical and software deployment tools as well as effective communication and resource management.

4.6.3 Fault Tolerance

This is one of the key features required by WSNs for efficient smart agriculture. There may be different types of faults in WSNs owing to the following reasons: (i) incorrect value reportage while measuring a certain parameter, (ii) deployment of faulty sensors, (iii) inefficient architecture in the design of networks, (iv) loss in the connectivity of communications, and (v) rapid drainage of battery life [55].

4.6.4 Requirement of Memory

It requires a considerable amount of memory to store massive amounts of data received from multiple deployed sensor nodes, along with other resources. The received data are used to send valid instructions to farmers *via* SMS or mobile agricultural applications and to estimate and draw conclusions about the current condition of the field, crop growth, fertilizer use, and insect presence in the field [56]. The use of cutting-edge technology may completely transform data collection for improved farming methods, ultimately increasing crop yields. Various issues have also been identified, including the successful integration of various technologies, management of colossal amounts of data, lack of standardized connectivity protocols for technical clarity, surpassing budgets against high implementation costs, data privacy, impact assessment on the environment, training farmers to adapt to technology, addressing regulatory challenges, sensor reliability in extreme conditions, and scaling-up of solutions through frequent application.

4.7 Future Trends

The future of real-time crop management looks promising, which can be attributed to the continuous development of IoT sensors and geospatial technologies. These tools enable farmers to observe crop health, soil moisture, and environmental conditions with unprecedented accuracy and in real time. The integration of AI and ML further enhances predictive analytics, enabling timely interventions to prevent outbreaks of crop diseases and optimize resource utilization. Furthermore, high-resolution details about in-field variability, provided through geospatial technologies, connected satellite imagery, and drones, will enable farmers to apply only the necessary treatments to specific parts of the field, thereby minimizing waste and reducing environmental degradation. These developments will make farming more effective, sustainable, and resilient to the challenges posed by climate change.

4.8 Conclusion

The IoT is a key tool in precision agriculture that helps measure soil and water quality, monitor environmental factors, and increase agricultural output. Through the Internet, farmers can utilize IoT technology to remotely monitor crops and make informed decisions about farming operations, such as water irrigation, greenhouse maintenance, and insect monitoring at any time and from any location. This chapter elaborates on the components and technology of IoT, as well as its advantages, disadvantages, and challenges. The need for geospatial data with IoT devices is discussed. The goal of this chapter is to increase awareness of precision farming and the control software application toolkits available on the market, as well as to encourage users to utilize them. This enables farmers to access real-time crop data and management recommendations *via* digital platforms.

AI disclosure statement: The author acknowledges the use of AI-based tools for minor language editing or grammatical correction. However, all intellectual contributions, analyses, and conclusions presented in this chapter are entirely the author's own.

References

1. Galanakis, C.M., The Future of Food. *Foods*, 13, 4, 506, Feb. 2024, doi:10.3390/foods13040506.
2. Karandish, F., Liu, S., de Graaf, I., Global groundwater sustainability: A critical review of strategies and future pathways. *J. Hydrol.*, 657, 133060, Aug. 2025, doi:10.1016/j.jhydrol.2025.133060.
3. Aijaz, N., Lan, H., Raza, T., Yaqub, M., Iqbal, R., Pathan, M.S., Artificial intelligence in agriculture: Advancing crop productivity and sustainability. *J. Agric. Food Res.*, 20, 101762, Apr. 2025, doi:10.1016/j.jafr.2025.101762.
4. Yost, M.A., Kitchen, N.R., Sudduth, K.A., Sadler, E.J., Drummond, S.T., Volkmann, M.R., Long-Term Impact of a Precision Agriculture System on Grain Crop Production. *Precis. Agric.*, 18, 823–842, 2017.
5. Barik, R.K., Dubey, H., Misra, C., Borthakur, D., Constant, N., Sasane, S.A., Mankodiya, K., Fog assisted cloud computing in era of Big Data and Internet-of-Things: systems, architectures, and applications, in: *Cloud Computing for Optimization: Foundations, Applications, and Challenges*, pp. 367–394, Springer, Cham, 2018.
6. Thorp, K.R., Hunsaker, D.J., French, A.N., Bautista, E., Bronson, K.F., Integrating geospatial data and cropping system simulation within a geographic information system to analyze spatial seed cotton yield, water use, and irrigation requirements. *Precis. Agric.*, 16, 5, 532–557, 2015.
7. Nyéki, A., Milics, G.K.A.J., Neményi, M., Improving Yield Advisory Models for Precision Agriculture with Special Regards to Soil Compaction in Maize Production, in: *Precision Agriculture '13*, J.V. Stafford, (Ed.), pp. 443–450, Wageningen Academic Publishers, Wageningen, The Netherlands, 2013.
8. Nyéki, A., Milics, G., Kovács, A.J., Neményi, M., Effects of Soil Compaction on Cereal Yield. *Cereal Res. Commun.*, 45, 1–22, 2017.
9. Kakarla, S., Ampatzidis, Y., Park, S., Adosoglou, G., Pardalos, P., Emerging Sensing Technologies for Precision Agriculture, in: *Information and Communication Technologies for Agriculture—Theme I: Sensors,* D.D. Bochtis, M. Lampridi, G.P. Petropoulos, Y. Ampatzidis, P. Pardalos, (Eds.), pp. 3–16, Springer International Publishing, Cham, Switzerland, 2022, ISBN 978-3-030-84144-7.
10. Bochtis, D.D., Lampridi, M., Petropoulos, G.P., Ampatzidis, Y., Pardalos, P. (Eds.), *Information and Communication Technologies for Agriculture—Theme I: Sensors*, vol. 182, Springer International Publishing, Cham, Switzerland, 2022, ISBN 978-3-030-84143-0.
11. Vohland, M., Ludwig, M., Thiele-Bruhn, S., Ludwig, B., Quantification of Soil Properties with Hyperspectral Data: Selecting Spectral Variables with Different Methods to Improve Accuracies and Analyze Prediction Mechanisms. *Remote Sens.*, 9, 1103, 2017.

12. Jia, S., Li, H., Wang, Y., Tong, R., Li, Q., Hyperspectral Imaging Analysis for the Classification of Soil Types and the Determination of Soil Total Nitrogen. *Sensors*, 17, 2252, 2017.
13. Mouazen, A.M., Alhwaimel, S.A., Kuang, B., Waine, T.W., Fusion of Data from Multiple Soil Sensors for the Delineation of Water Holding Capacity Zones. In Precision Agriculture'13, J.V. Stafford (Ed.), pp. 745–751, Wageningen Academic Publishers, Wageningen, The Netherlands, 2013.
14. Navas, E., Fernández, R., Sepúlveda, D., Armada, M., Gonzalez-de-Santos, P., Soft Grippers for Automatic Crop Harvesting: A Review. *Sensors*, 21, 2689, 2021.
15. Kultongkham, A., Kumnon, S., Thintawornkul, T., Chanthasopeephan, T., The Design of a Force Feedback Soft Gripper for Tomato Harvesting. *J. Agric. Eng.*, 52, 1091, 2021.
16. Vyas, N. and Dutt, V., Advancing Precision Agriculture: Leveraging YOLOv8 for Robust Deep Learning Enabled Crop Diseases Detection, in: *2023 International Conference on Integration of Computational Intelligent System (ICICIS)*, Pune, India, pp. 1–6, 2023, doi:10.1109/ICICIS56802.2023.10430280.
17. Bernd, N., Pollution Prevention and Treatment Using Nanotechnology, in: *Nanotechnology*, pp. 1–15, Wiley, Hoboken, NJ, USA, Jul. 15, 2010, doi:10.1002/9783527628155.nanotech010.
18. Neményi, M., Kovács, A.J., Oláh, J., Popp, J., Erdei, E., Harsányi, E., Ambrus, B., Teschner, G., Nyéki, A., Challenges of Sustainable Agricultural Development with Special Regard to Internet of Things: Survey. *Progress. Agric. Eng. Sci.*, 18, 95–114, 2022.
19. Baranwal, J., Barse, B., Gatto, G., Broncova, G., Kumar, A., Electrochemical Sensors and Their Applications: A Review. *Chemosensors*, 10, 9, 363, Sep. 2022, doi:10.3390/chemosensors10090363.
20. Reyns, P., Missotten, B., Ramon, H., De Baerdemaeker, J., A Review of Combine Sensors for Precision Farming. *Precis. Agric.*, 3, 169–182, 2002.
21. Sui, Y., Atreya, M., Dahal, S., Gopalakrishnan, A., Khosla, R., Whiting, G.L., Controlled Biodegradation of an Additively Fabricated Capacitive Soil Moisture Sensor. *ACS Sustain. Chem. Eng.*, 9, 2486–2495, 2021.
22. Dahal, S., Yilma, W., Sui, Y., Atreya, M., Bryan, S., Davis, V., Whiting, G.L., Khosla, R., Degradability of Biodegradable SoilMoisture Sensor Components and Their Effect on Maize (Zea mays, L.) Growth. *Sensors*, 20, 6154, 2020.
23. Zhao, W., Wang, M., Pham, V.T., Unmanned aerial vehicle and geospatial analysis in smart irrigation and crop monitoring on IoT platform. *Mobile Inf. Syst.*, 2023, 1, 4213645, 2023.
24. Cucho-Padin, G., Loayza, H., Palacios, S., Balcazar, M., Carbajal, M., Quiroz, R., Development of Low-Cost Remote Sensing Tools and Methods for Supporting Smallholder Agriculture. *Appl. Geomat.*, 12, 247–263, 2020.
25. Wulder, M.A., *et al.*, Fifty years of Landsat science and impacts. *Remote Sens. Environ.*, 280, 113195, Oct. 2022, doi:10.1016/j.rse.2022.113195.

26. Nguyen, T.T., Hoang, T.D., Pham, M.T., Vu, T.T., Nguyen, T.H., Huynh, Q.T., Jo, J., Monitoring Agriculture Areas with Satellite Images and Deep Learning. *Appl. Soft Comput. J.*, 95, 106565, 2020.
27. Li, Z., Shen, H., Weng, Q., Zhang, Y., Dou, P., Zhang, L., Cloud and Cloud Shadow Detection for Optical Satellite Imagery: Features, Algorithms, Validation, and Prospects. *ISPRS J. Photogramm. Remote Sens.*, 188, 89–108, 2022.
28. Zhang, M., Lin, H., Wang, G., Sun, H., Fu, J., Mapping Paddy Rice Using a Convolutional Neural Network (CNN) with Landsat 8 Datasets in the Dongting Lake Area, China. *Remote Sens.*, 10, 1840, 2018.
29. Eitel, J.U.H., Basler, D., Braun, S., Buchmann, N., D'Odorico, P., Etzold, S., Gessler, A., Griffin, K.L., Krejza, J., Luo, Y., *et al.*, Towards Monitoring Stem Growth Phenology from Space with High Resolution Satellite Data. *Agric. For. Meteorol.*, 339, 109549, 2023.
30. Matese, A., Toscano, P., Di Gennaro, S.F., Genesio, L., Vaccari, F.P., Primicerio, J., Belli, C., Zaldei, A., Bianconi, R., Gioli, B., Intercomparison of UAV, Aircraft and Satellite Remote Sensing Platforms for Precision Viticulture. *Remote Sens.*, 7, 2971–2990, 2015.
31. Hunt, E.R. and Daughtry, C.S.T., What Good Are Unmanned Aircraft Systems for Agricultural Remote Sensing and Precision Agriculture? *Int. J. Remote Sens.*, 39, 5345–5376, 2018.
32. Delavarpour, N., Koparan, C., Nowatzki, J., Bajwa, S., Sun, X., A Technical Study on UAV Characteristics for Precision Agriculture Applications and Associated Practical Challenges. *Remote Sens.*, 13, 1204, 2021.
33. Fuentes-Peñailillo, F., Gutter, K., Vega, R., Carrasco Silva, G., Transformative technologies in digital agriculture: Leveraging Internet of Things, remote sensing, and artificial intelligence for smart crop management. *J. Sens. Actuator Netw.*, 13, 4, 39, 2024.
34. Kamilaris, A., Fonts, A., Prenafeta-Bold´*u*, F.X., The Rise of Blockchain Technology in Agriculture and Food Supply Chains. *Trends Food Sci. Technol.*, 91, 640–652, 2019.
35. Bhati, N. and Vyas, N., Elevating IoT Sensor Data Management and Security Through Blockchain Solutions, in: *Advances in Systems Analysis, Software Engineering, and High Performance Computing*, pp. 58–77, IGI Global, Hershey, PA, USA, Mar. 18, 2024, doi:10.4018/979-8-3693-0968-1.ch003.
36. Mark, R., Ethics of Using AI and Big Data in Agriculture: The Case of a Large Agriculture Multinational. *ORBIT J.*, 2, 1–27, 2019.
37. Kim, H. and Laskowski, M., Agriculture on the Blockchain: Sustainable Solutions for Food, Farmers, and Financing. *SSRN J.*, 2017, doi:10.2139/ssrn.3028164.
38. Rajeswari, S., Suthendran, K., Rajakumar, K., A Smart Agricultural Model by Integrating IoT, Mobile and Cloud-Based Big Data Analytics, in: *Proceedings of the 2017 International Conference on Intelligent Computing and Control (I2C2)*, Coimbatore, India, pp. 1–5, 23–24 June 2017.

39. Vermesan, O., Harrison, M., Vogt, H., Kalaboukas, K., Tomasella, M., Wouters, K., *Cluster of European Research Projects on the Internet of Things: Vision and Challenges for Realising the Internet of Things*, European Commission, Luxembourg, 2010.
40. Ali, A., Hussain, T., Tantashutikun, N., Hussain, N., Cocetta, G., Application of Smart Techniques, Internet of Things and Data Mining for Resource Use Efficient and Sustainable Crop Production. *Agriculture*, 13, 397, 2023.
41. Chugh, M., Chugh, N., Essentials of Internet of Things in Smart Agriculture, in: *Applications of AI, Digital Twin, and Internet of Things for Sustainable Development*, pp. 186–199, IGI Global, Hershey, PA, USA, 2023.
42. Kumar, R., Narayanan, S., Kaur, G., Future of Internet of Everything (IOE). *Int. Res. J. Comput. Sci.*, 8, 84–92, 2021.
43. Baldassarre, G., Giudice, P.L., Musarella, L., Ursino, D., A Paradigm for the Cooperation of Objects Belonging to Different IoTs, in: *Proceedings of the 22nd International Database Engineering & Applications Symposium on—IDEAS 2018*, Villa San Giovanni, Italy, 18–20 June 2018, ACM Press, New York, NY, USA, pp. 157–164, 2018.
44. Jaiganesh, S., Gunaseelan, K., Ellappan, V., IOT Agriculture to Improve Food and Farming Technology, in: *Proceedings of the 2017 Conference on Emerging Devices and Smart Systems (ICEDSS)*, Mallasamudram, India, pp. 260–266, 3–4 March 2017.
45. Ph. D. Thenkabail Prasad S., *Land Resources Monitoring, Modeling, and Mapping with Remote Sensing*. CRC Press, Oct. 02, 2015. doi: 10.1201/b19322.
46. Wolfert, S., Ge, L., Verdouw, C., Bogaardt, M.J., Big data in smart farming a review. *Agric. Syst.*, 153, 69–80, 2017.
47. Prathibha, S.R., Hongal, A., Jyothi, M.P., IOT based monitoring system in smart agriculture, in: *International Conference on Recent Advances in Electronics and Communication Technology*, IEEE, 2017, https://doi.org/10.1109/ICRAECT.2017.52.
48. Sarmah, K., Deka, C.R., Sharma, U., Sarma, R., Role of GIS based technologies in sustainable agriculture resource planning & management using spatial decision support approach. *Int. J. Innovative Res. Eng. Manag. (IJIREM)*, 5, 2018.
49. Sharmaa, R., Kamblea, S.S., Gunasekaran, A., Big GIS analytics framework for agriculture supply chains: a literature review identifying the current trends and future perspectives. *J. Comput. Electron. Agric.*, 6, 103–120, 2018.
50. Kamilaris, A., Kartakoullis, A., Prenafeta-Boldú, F.X., A review on the practice of big data analysis in agriculture. *Comput. Electron. Agric.*, 143, 23–37, 2018.
51. Sharma, A., Jana, P., Paul, R., Smart agriculture: A journey from manual techniques to smart techniques. *J. Autom. Control Eng.*, 8, 3, 141–149, 2020.
52. Ribeiro, L., Costa, J., Cunha, A., Precision agriculture and the future of farming. *Agric. Syst.*, 180, 102–120, 2020.

53. Bhanumathi, V. and Kalaivanan, K., The Role of Geospatial Technology with IoT for Precision Agriculture. *Studies in Big Data*, pp. 225–250, Springer International Publishing, Cham, Switzerland, Dec. 11, 2018. doi: 10.1007/978-3-030-03359-0_11.
54. Ammar, M., Russello, G., Crispo, B., Internet of Things: a survey on the security of IoT frameworks. *J. Inf. Secur. Appl.*, 38, 8–27, 2018.
55. Kalaivanan, K. and Bhanumathi, V., Reliable location aware and cluster-tap root based data collection protocol for large scale wireless sensor networks. *J. Netw. Comput. Appl.*, 118, 83–101, 2018.
56. Bhanumathi, V. and Kalaivanan, K., Application specific sensor-cloud: Architectural model, in: *Computational Intelligence in Sensor Networks. Studies in Computational Intelligence*, B. Mishra, S. Dehuri, B. Panigrahi, A. Nayak, B. Mishra, H. Das, (eds.), vol. 776, pp. 277–305, Springer, Berlin, Heidelberg, 2019.

5

Real-Time Crop Management with IoT Sensors and Geospatial Technologies

R. Kanthavel[1], R. Adline Freeda[2] and R. Dhaya[1]*

[1]School of Electrical and Communication Engineering, PNG University of Technology, Lae, Morobe Province, Papua New Guinea
[2]Department of Information Technology, KCG College of Technology, Chennai, Tamil Nadu, India

Abstract

Real-time crop management is a high requirement for improving agricultural productivity and its sustainability. The integration of Internet of Things (IoT) sensors and geospatial technologies enables farmers and agricultural stakeholders to collect and analyze data efficiently, facilitating precise real-time decision-making for crop and resource monitoring. IoT sensors installed throughout farms continue to measure environmental factors such as soil moisture, temperature, humidity, and light intensity. This data, aided by geospatial technologies such as Global Positioning Systems (GPS), remote sensing, and Geographic Information Systems (GIS), enables mapping and spatial analysis of patterns related to crop health, growth, and environmental conditions. Geospatial technologies and IoT sensors enable farmers to identify variations in soil and crop conditions across different farm sections using data analysis and visualization. Moreover, the cloud-computing mobile platform enables remote monitoring and control, providing real-time access to data and recommendations that support farmers' decision-making. This chapter discusses the role of the IoT and geospatial technologies in crop management, including various applications, advantages, and challenges. It also opens avenues for improving agricultural efficiency, sustainability, and profitability based on real-time data-driven solutions.

**Corresponding author*: dhayavel2005@gmail.com
R. Kanthavel: ORCID: 0000-0001-5537-2571
R. Adline Freeda: ORCID: 0009-0002-3335-0907

Gurwinder Singh, Vishakha Sood, Narayan Vyas, Surendra Yadav and Dankan Gowda V. (eds.) Geospatial Intelligence for Precision Agriculture, (83–108) © 2026 Scrivener Publishing LLC

Keywords: Internet of Things, crop management, agriculture, geospatial, sensors, agricultural efficiency, sustainability

5.1 Introduction

The agricultural sector has reached a revolutionary point in the use of technology, with the Internet of Things (IoT) and geospatial technologies already playing a significant role in modernizing crop management. With the rise in global food demand and intensification of environmental challenges (e.g., climate change and soil degradation), precision agriculture offers a promising solution to enhance efficiency, productivity, and sustainability.

5.1.1 Introduction of IoT and Overview of Geospatial Technologies in Agriculture

IoT devices continuously record massive amounts of data, which are further processed and analyzed for actionable information. IoT technologies can be applied to real-time data collection and analytics to make better decisions by optimizing resource utilization and timely interventions in crop management. Geospatial technologies, including Global Positioning Systems (GPS), Geographic Information Systems (GIS), and remote sensing, are major constituents of precision farming. These geospatial technologies capture spatial data about the farm field itself and over wider areas for viewing by farmers regarding the conditions of the crops, soil variability, and other environmental factors. GPS provides location information, which is correct and required by farmers for the location of a certain piece of equipment, monitoring of the field, and variable-rate applications of inputs, such as water, fertilizers, and pesticides. In GIS, spatial data are stored, analyzed, and visualized to describe field variability better and implement appropriate interventions. Satellites and drones equipped with remote sensing technology capture images and data to monitor crop health, soil moisture, and potential pest or disease indicators. Together, these tools provide farmers with basic geospatial information to optimize resource use, improve crop yields, and reduce environmental impacts.

5.1.2 Convergence of IoT and Geospatial Technologies

Central to all these technologies, the integration of IoT and geospatial technologies is the backbone of precision agriculture. Such integration enables the creation of highly comprehensive and sophisticated crop management systems that allow real-time monitoring, decision support, and automation capabilities. For example, IoT sensors may record soil moisture changes, and the differences between zones of a field can be highlighted in GIS maps. Data integrated into this form will trigger automated irrigation systems to ensure that the right amount of water is provided at the right place and time. While remote sensing data would be useful for monitoring the health of crops over time, IoT sensors can detect early signs of pest infestations and disease development in the environment.

5.1.3 Impact on Contemporary Agriculture

Crop management has improved significantly owing to the introduction of IoT and geospatial technologies. These facilitate easier decision-making by allowing agricultural practitioners to achieve more accuracy, timeliness, and efficiency in their interventions without wasting resources and overutilization of resources. The interventions assured higher crop yields with lower input costs and environmental degradation. It further offers real-time crop management through IoT and geospatial tools that would enable farmers to evolve according to climatic change and market demand. This is followed by sections explaining, in detail, the major constituents of IoT and geospatial technologies, their applications in crop management, and the challenges and opportunities for the future of agriculture. The following section is related to "Real Time Crop Management with IoT Sensors and Geospatial Technologies."

5.2 Real-Time Crop Management with IoT Sensors and Geospatial Technologies

Real-time crop management has emerged as an important approach for enhancing agricultural productivity and efficiency, while maintaining crop sustainability. All of this is made possible using IoT sensors and geospatial technologies. Therefore, this initiates proactive and optimized interventions, leading to a significant improvement in resource efficiency and crop yields.

5.2.1 IoT Sensors for Real-Time Data Collection

IoT sensors have become vital for data collection in farm settings. The potential of IoT sensors to be spread over the field provides an avenue for the constant monitoring of factors that impede or enhance crop development and health. Common types of IoT sensors used in crop management are illustrated in Figure 5.1.

Soil moisture sensors provide in-soil water content to crops. In real time, it enables farmers to fine-tune irrigation schedules and avoid over- and under-watering [6].

Temperature and Humidity Sensors: These sensors are used to measure the surrounding environment of the crops. Such information is useful for predicting and managing crop stress, disease outbreaks, and growth rates. This allows farmers to act in time on the different changes presented by various environmental factors.

Light and Radiation Sensors: They measure the amount of solar radiation reaching the plants, which is important for photosynthesis. Tracing light intensity allows farmers to control shading or supplemental lighting inside controlled environments, thereby optimizing plant growth.

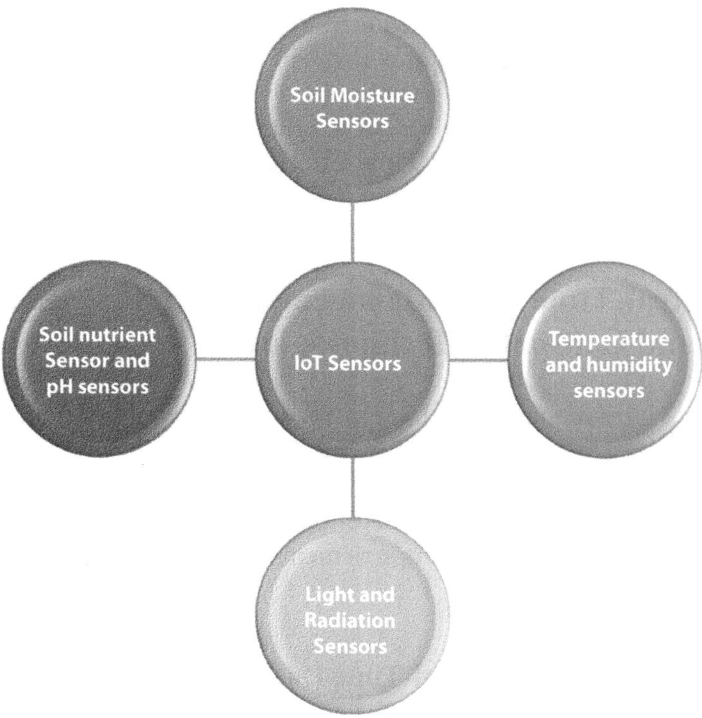

Figure 5.1 IoT sensors for real-time data collection.

Soil Nutrient Sensors and pH Sensors: These sensors monitor the nutrient levels and pH of the soil for suitable amounts of nutrient absorption by the crops. In turn, real-time nutrient concentration levels, can be used for the targeted application of fertilizers. The data captured from the sensors are wirelessly transmitted to cloud-based platforms, where they can be accessed, analyzed, and used to make real-time decisions [7].

5.2.2 Geospatial Technologies for Spatial Data Analysis

Geospatial technologies, such as GPS, GIS, and remote sensing, provide vital spatial insights into IoT sensor data. This can help farmers view the spatial variability occurring in their fields and find more accurate ways of managing their crops, as shown in Figure 5.2.

Global Positioning System: Applications of GPS technology can be put in place for tracking and mapping fields, machinery, and crops with high accuracy. It uses this technology in mapping fields, conducting planting, spraying, and harvesting with precision, and recording and analyzing spatial data across the farm.
Geographic Information System: It is an accepted powerful tool for managing, analyzing, and visualizing spatial data. GIS helps farmers map conditions in the fields, soil variability, crop performance, and other variations

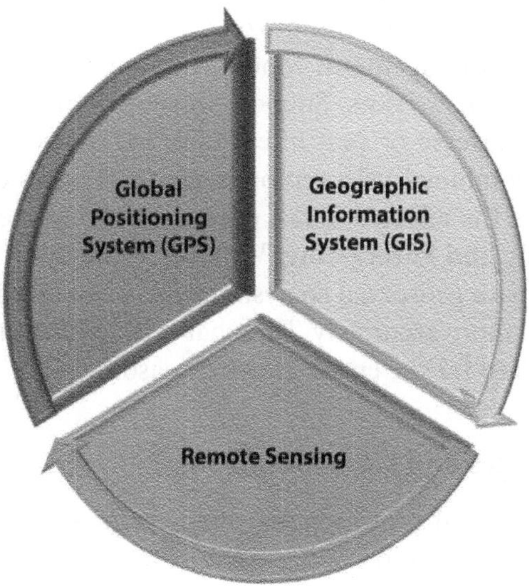

Figure 5.2 Geospatial technologies for spatial data analysis.

across different zones. It also creates management zones through site-specific variability in the use of resources, such as water, fertilizers, and pesticides.

Remote Sensing: Satellite image acquisition, together with the imaging and data acquisition capabilities of drones or any other airborne platforms, can acquire pictures and data of crops and soil. These images can sense changes in crop health, water stress, or infestations by pests or diseases that often appear visibly even before naked-eye observation. Such data need to be supplemented with data generated from IoT sensors. This will allow farmers to make better decisions regarding when and where to intervene in crop management.

Real-Time Crop Management: The combination of IoT sensors and geospatial technologies offers a potent system for real-time crop management. While IoT sensors provide continuous, ground-level data concerning the environment and crop conditions, geospatial technologies provide a wider spatial overview of the farm.

Precision Irrigation: Because the current usage of IoT soil moisture sensors can handle real-time data and GIS maps show the variability in the soil profile, farmers must install precision irrigation systems. Such a system then applies the exact amount of water needed per zone to use less water with minimal waste.

Variable-Rate Fertilization and Pesticide Application: IoT nutrient sensors and remote sensing data can guide the variable rate application of fertilizers and pesticides, sending them only where they are needed, saving inputs, and minimizing impacts on the environment.

Disease and Pest Management: Monitoring temperature, humidity, and crop health using IoT sensors. Suppose a farmer's area has a predisposition to a certain disease or pest infestation. IoT sensors can alert farmers. Using data combined with remote sensing, farmers can now quickly and easily isolate the affected areas, even before it worsens for the crop.

Yield Monitoring and Forecasting: With real-time information from IoT sensors merged with geospatial analysis, farmers can monitor crop growth and predict yield more accurately. Thus, harvesting schedules and resource use can be prepared for the entire growing season.

5.2.3 Challenges and Considerations

While the merits of real-time crop management are evident, the challenges associated with the use of IoT and geospatial technology are as follows:

Connectivity Issues: There are many areas where the Internet is still patchy and unable to deliver reliable connectivity, thus hindering the transmission and accessibility of real-time data.
Data Management: The massive amount of data generated by IoT sensors and geospatial tools requires storage, processing, and analysis systems that can effectively manage such data.
Technical Expertise: The implementation and maintenance of IoT and geospatial systems require technical know-how that can be acquired through training or hiring a competent workforce.

5.3 Sensor Deployment in The Field

Sensor deployment concerns the strategic placement of various IoT devices to monitor environmental and crop-specific parameters. Proper calibration, maintenance, and placement of sensors are necessary to ensure smooth data collection for high-quality and sustainable crop management. This section discusses the types of sensors suitable for agriculture, best practice deployment, and challenges and solutions of sensor deployment.

5.3.1 Best Practices for Sensor Deployment

Real-time crop management will only be successful if the sensors are strategically located in the field to capture accurate and representative data. The following best practices highlight some of the key factors that must be considered when placing sensors:

Site Selection: The appropriate selection of locations for sensor placement is crucial to ensure representative data collection in the field. The following factors should be considered:
Field Variability: Sensors must be in areas that represent the various zones of the filed. The zones can be differentiated according to soil type, topography, and microclimate.
Crop Stage: Another factor influencing the sensor location is the crop growth stage. For instance, during critical growth stages, soil moisture sensors should be positioned near the root zone.
Field Size: For larger fields, more sensors are required to capture spatial variability, whereas for smaller fields, fewer sensors are needed to monitor effectively.

Sensor Density: The Density of field sensors is dependent on the precision of the data needed and the rate at which the field conditions may

change. Highly dense sensor networks can capture very fine details, but they increase the expense and complexity. A low-density network may be cheaper and may not reflect vital local variations in field conditions.
Zonal Placement: Sensors may be placed in representative zones of the field for highly heterogeneous soils to capture differences in conditions.
Grid or Randomized Deployment: In uniform areas, a grid or randomized sensor deployment strategy may suffice to collect representative data.

Depth of Deployment: The depth of the sensor determines the monitored parameters. Soil moisture sensors must be deployed at depths representative of the root zone depths of crops, whereas temperature sensors may be placed on the soil surface and in the air to obtain ambient temperatures.
Shallow Depths: This is suitable for crops with shallow root systems, where early seedling or surface soil moisture can be monitored.
Deeper Depths: For crops with deep root systems or long-term moisture and temperature trends.

Wireless Connectivity and Power Supply: Real-time monitoring has no scope for loss of communication between the sensor and data management platform. WSNs are typically used to transmit data from the field to cloud-based platforms [1].
Mesh Networks: Sensors are set up in a mesh network. Each sensor acts as a node that may forward data to other nodes to ensure that the data can be transmitted over long distances, even in large fields.
Power Supply: Most IoT sensors rely on either batteries or solar panels for power supply [2]. It is highly important to regularly check the battery level or ensure that the solar panels receive enough sunlight, as interruptions in data result.

Calibration and Maintenance: The correct reading of sensors depends on proper calibration before and during the deployment process. Sensors also require regular maintenance to continue functioning correctly.
Calibration: Sensors can be calibrated according to specific soil types, crops, and environmental conditions. In this regard, soil moisture sensors may be calibrated differently for varying soil textures, such as sand, clay, or silt.
Cleaning and Replacing: Dust, dirt, and vegetation debris may interfere with the sensor readings. Cleaning needs to be performed periodically, and sensors should be replaced from time to time to ensure better data quality.

5.3.2 Issues in Sensor Deployment

However, IoT sensors placed in agricultural fields face challenges [9]. Knowledge of and mitigation strategies for such challenges can ensure optimal sensor performance and effective data acquisition.

Exposure to Environment: Agricultural fields are not gentle to IoT sensors with extreme environmental conditions, including rain, wind, extremes in temperature, and soil erosion, which may or may not be very helpful in their longevity and performance [10].
Waterproof and robust sensors: The sensors must be waterproof and rugged to accommodate environmental circumstances, such as heavy downpours or dust storms.
Protection from Animals or Insects: Some animals or insects can damage a sensor if it is left open. Data loss can be expected in such situations. Encasements or positioning that avoids danger may be a solution.

Connectivity Issues: Internet or cellular connectivity is not available or is unreliable in rural or other remote settings. Such a connection challenge may hinder the transmission of real-time data from field sensors to computing sites in the cloud [8].
Local Data Storage: Some sensors are deployed with local data storage to store data locally until it becomes possible to send data to the central server.
LoRa and mesh networks: In such regions, which often experience difficulties in data transfer, wide area and mesh networks are preferred for good communication.

Innovative Solutions for Sensor Deployment: Innovative sensor deployment solutions designed by researchers and technological developers addressed the challenges of sensor deployment through improvements in sensor performance and simplification of deployment. Drone-assisted sensor deployment contributes to the complexity of deployment through their availability with sensors for data collection over vast areas without fixed installations. This tends to reduce labor input and improve the flexibility of crop monitoring conditions. *Automated Calibration Systems:* One of the advancements of AI and machine learning within various sensors is that they can now automatically calibrate using real-time raw data, eliminating manual calibration with precision [20].

With real-time crop management using IoT and geospatial technologies, the proper deployment of sensors is crucial [11]. Strategically deployed and calibrated sensors ensure that accurate and quality data are collected to inform agricultural decisions. However, environmental exposure, connectivity issues, and costs represent limiting factors, although innovative solutions are emerging, making sensor deployment more efficient and feasible for farmers worldwide.

5.4 Data Collection and Aggregation

Hence, data collection and aggregation are essential for real-time crop management. Farmers can then make strategic decisions based on a comprehensive understanding of the conditions in their fields. IoT sensors and geospatial technologies generate immense amounts of data regarding several environmental and crop-specific parameters [12]. This section explains the data collection methods and technologies that exist for collecting data in the field, the aggregation of data from several sources, and the associated issues with data management.

5.4.1 Data Collection from IoT Sensors

Field-deployed IoT sensors continuously monitor and collect data on crop health, environmental conditions, and resource availability [13]. Some important aspects of data collection using IoT are as follows:

Types of Data Collected: IoT sensors collect a variety of data types that are essential for crop management, as shown in Figure 5.3:
Soil Moisture Data: It measures the availability of moisture in the soil. This information helps optimize irrigation scheduling and water application levels.
Temperature and Humidity Data: Tracks atmospheric conditions. Data drive the understanding of the stress factors experienced by plants and disease outbreaks.
Light and Solar Radiation Data: The intensity of light entering the crops is recorded. This is essential because it ensures that photosynthesis and the growth cycles of crops are well managed.
Nutrient Levels and pH Data: Record keeping of the chemical composition of soil. The measurement procedure guides the processes of fertilization and soil health management.

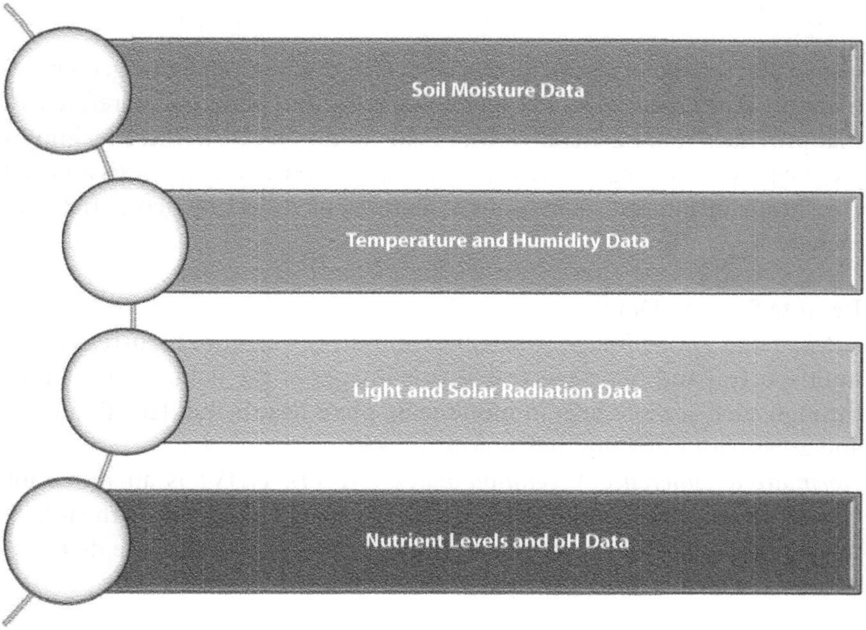

Figure 5.3 Data types for crop management.

Real-Time Data Transmission

The vast majority of IoT sensors are equipped with wireless communication modules, such as Wi-Fi, LoRa, and Zigbee. This enables them to forward data from the sensor in real time to a central server or cloud. In some cases, considering the type of sensor and the requirements for processing the collected data, data transmission is performed wirelessly over short intervals or continuously.

Wireless Sensor Networks (WSN): WSNs are generated by multiple IoT sensors in the field, with data transmission to a central gateway [16]. The central gateway collects all data from its scattered sensors and forwards them to the more processed cloud-based platforms.

Edge Computing: At times, the sensors perform preliminary processing of the data at the edge, closer to where the data are being collected. This reduces the data transmitted to the cloud and enhances the response time in decision-making by checking the data locally on the edge device.

5.4.2 Data Collection from Geospatial Technologies

Geospatial technologies, including GPS, GIS, and remote sensing, complement the data captured by IoT technologies by providing spatial references that provide a more general view of field conditions [17]. These technologies help monitor large areas and detect spatial variability in crops and soil conditions of interest. Some salient features of collecting geospatial data include:

Remote Sensing Data
Satellite and Drone Imagery: Satellites and drones capture high-resolution multispectral and hyperspectral sensor images of the field. Images taken through such sensors help in monitoring plant health and detecting pests and areas with water and nutrient stresses [5].
Normalized Difference Vegetation Index (NDVI): NDVI is an index for remote sensing that computes the health of a plant by measuring the reflectance of light from the vegetation. Thus, the index facilitates the determination of which areas of the field must be monitored immediately.

GPS Data
Field Mapping: Using GPS technology, field boundary maps are created with accuracy and the movement of machinery. It also assists with location data to automate tasks such as planting, spraying, and harvesting.
Spatial Data Layers: GIS enables the user to create spatial data layers, including soil type, topography, and crop health maps. The layers help distinguish what is different from one area to another and how many variables affect crops in different areas of the field.

5.4.3 Data Aggregation

The data collected from the various sensors of IoT and geospatial technologies are then aggregated to provide a comprehensive outlook on the field conditions [21]. The following steps are important for efficient data aggregation.

Centralized Data Platforms
Aggregated data are stored and processed on centralized cloud-based platforms or farm management systems. These data-compiled platforms aggregate data from various IoT sensors, remote sensing tools, and GPS systems, which farmers can view together through a single interface.

Cloud-Based Platforms: Sensor and geospatial data are stored in large quantities on cloud computing platforms [3]. These platforms offer scalable storage and power processing, providing an opportunity for farmers to analyze data in real time from anywhere.
Farm Management Information Systems (FMIS): An FMIS is an integrated data-collecting tool that integrates sensor data, weather, past crop performance, and other relevant information. These systems are designed to aid decision-making by providing an easy interface through which data visualization and analysis can be performed.

Data Fusion
Data fusion integrates information from multiple sensors and sources to produce more meaningful insights. For example, data from IoT sensors (soil moisture, temperature, and pH) are fused with geospatial data, such as satellite imagery or NDVI. This enables farmers to have a clearer view of field heterogeneity and make management decisions accordingly.
Temporal Data Fusion: This offers the fusion of information from different time points with the objective of monitoring changes in field conditions, such as fluctuations in soil moisture or plant stages of development.
Spatial Data Fusion: This focuses on how data emanating from different parts of a field can be combined to identify patterns and relationships, such as variations in soil types or nutrient levels within the field.

Data Standardization
It must also be standardized for compatibility and accuracy before any data can be accumulated from different sources. Sensors collect data in numerous formats, units, and intervals. Therefore, standardizing it involves converting all data into one format for analytical purposes.
Normalization: Multi-sensor data are normalized to make the data easily comparable and combinable. For instance, soil moisture data gathered from different parts of the field may be transformed into a standard unit of measurement before aggregation.

5.4.4 Data Processing and Analysis

The data were collected and combined into a dataset, after which processing and analysis were performed to extract meaningful insights from the data. This has been an integral process since the raw data are manipulated into usable information for the farmer. Data processing is performed to clean and structure the data by removing errors or inconsistencies. For instance, malfunctioning sensors or transmission problems can produce missing or invalid data values that must be detected and corrected before data analysis begins. Data processing techniques may involve the following:

Outlier Elimination: This involves eliminating data that are outside the expectation limit.
Smoothing: Algorithms are used to ensure that the noisy field conditions are smooth and clear.

Data Analytics and Visualization
Data analytics tools and algorithms have been developed to analyze pre-processed data. These tools allow farmers to comprehend very complex datasets; hence, proper decisions are made regarding their strategies for crop management.

Predictive Analytics: Predictive models forecast the yield and disease of crops much in advance. Therefore, it can also optimize the irrigation schedule or fertilization by analyzing historical data and the current condition of the field.
Data Visualization: Heat maps, graphs, and dashboards in data visualization tools enable farmers to comprehend large data volumes and identify trends faster. For example, a heat map can highlight areas of a field that are water-stressed, and hence, irrigation can be localized to the stressed areas.

5.4.5 Challenges in Data Collection and Aggregation

There are several benefits of data collection and aggregation; however, it also possesses many challenges that must be addressed:

Data Overload: As the number of sensors and geospatial tools increases, farmers must face data overload issues. Large amounts of data must be handled and processed using the required powerful infrastructure and software platforms. This approach may increase computational complexity and operational costs.
Data Quality: The quality of data sensed from sensors and geospatial tools will determine the accuracy level of the insight. Faulty sensors, environmental interference, and connectivity problems may reduce the quality of the extracted data.

Data Accuracy and Reliability: Erratic or faulty data from sensors may lead to poor decision-making. Sensors must be repaired and calibrated to obtain high-quality data.
Interoperability: Proprietary data formats may be used by various IoT sensors and geospatial technologies. Hence, aggregation in a single system from disparate sources is difficult. Interoperability between devices and their platforms is crucial for effective data aggregation.
Data Privacy and Security: It is for this very reason that cyber threats pose a concern since farmers store ample amounts of data collected. Thus, data protection is a sensitive issue, as complete usage and realization of information privacy are maintained on cloud-based platforms. Real-time crop management involves the aggregation or collection of data. This helps monitor field conditions to optimize resource use and improve crop yields. IoT sensors and geospatial technologies create large volumes of valuable data, providing actionable insights into precision agriculture when aggregated and analyzed. Notwithstanding data overload and quality concerns, improved data processing and analytics tools make it easier for farmers to use real-time data for smarter farming more sustainably.

5.5 Crop Monitoring and Analytics

Using advanced analytics tools, farmers can convert raw data into actionable insights for timely interventions to enhance productivity. This section focuses on the methodologies and technologies for crop monitoring, analytics that are applied to optimize crop management, and the challenges in adopting these systems. In this way, the farmer learns about the environmental and soil conditions that may affect the crops. The following are certain aspects of IoT-based crop monitoring.

5.5.1 Monitoring of Environmental Conditions

IoT sensors mounted in the field monitor all environmental factors that affect growth. Some of these are:

Temperature: With the help of air and soil temperature monitoring and tracking, crop development stages can be predicted in advance, along with the risk of frost.
Humidity: Atmospheric humidity sensors can help identify conditions that may favor pest infestations and disease outbreaks, thus taking the necessary precautions by the farmers.

Soil Moisture: Real-time soil moisture sensors provide an excellent source of information regarding water availability in the root zone, thus allowing the optimization of irrigation.
Light Intensity: Light sensors monitor solar radiation, which allows decisions such as crop shading, supplementation of light in greenhouses, and optimal photosynthesis.

5.5.2 Crop Health Monitoring

IoT sensors can also be used to measure physiological crop health parameters. These include:

Plant Stress: Sensors measure plant transpiration and leaf temperature to understand the onset of water stress.
Nutrient Levels: Soil sensors monitor nitrogen, phosphorus, and potassium availability, providing a fertilizer application schedule.
Pest and disease detection: Sensors combined with predictive models use environmental data to predict the chance of an outbreak or infestation of disease or pests.

5.5.3 Remote Sensing for Crop Monitoring

Drones, satellites, and aerial imagery use remote sensing technologies that have gained significant momentum over the last few years for large-scale crop monitoring. These technologies capture the entire field from a bird's-eye view perspective; hence, comprehensive data on the spatial variability of crops and the performance of crops at different locations can be easily obtained. The main remote sensing applications of crop monitoring are as follows:

Multispectral Imaging: Images from multispectral cameras flown on drones or satellites at a particular wavelength (red, green, or near infrared) are captured. These images were then processed to assess the health of the plants, determine nutrient deficiencies, and assess water stress.
Hyperspectral Imaging: Hyperspectral sensors can capture hundreds of spectral bands that contain a high level of detail about the health and composition of crops. Such information can enable the detection of diseases or nutritional imbalances that may in the early stages.

Normalized Difference Vegetation Index (NDVI): Probably the most used index, this index computes the difference in reflectance between near-infrared and red light. Higher values indicate strong growth. Lower values could indicate stress, disease, or poor growth.
Enhanced Vegetation Index (EVI): EVI is an improved version of NDVI that reduces atmospheric conditions and soil background noise. It provides better results in approximating vegetation cover and health, especially in complex terrain.

Cameras on thermal drones or satellites measure the temperatures of crop canopies. This information goes a long way in pinpointing portions of the fields suffering from heat stress or where irrigation may be lacking. They can detect heat stress early enough to rectify it and avoid large losses.

5.5.4 Crop Monitoring through Data Analytics

Advanced algorithms, machine learning, and artificial intelligence applied to them further enable farmers to make more appropriate and timely decisions regarding crops.

Descriptive Analytics
Descriptive analytics describe the history and present data in a study of the current field conditions. This covers the following aspects:
Data Visualization: Graphs, Heat Maps, Dashboards Crop health, soil moisture, nutrient availability, and all other critical parameters will be graphically represented; the farmer can easily spot the emergence of trends and anomalies.
Field Mapping: The GIS tools create highly detailed maps of the field, showing changes in soil properties, crop health, and environmental conditions. This will allow targeted interventions.

Predictive Analytics
Predictive analytics can predict future crop performance, possible risks, and resource requirements based on historical data and machine learning models. The applications of predictive analytics are shown in Figure 5.4.
Yield Prediction: Using these machine learning models, weather conditions, soil moisture, nutrient levels, and health status of crops are analyzed to estimate probable yields to help the farmer decide when to harvest and how much resources to put in the field.

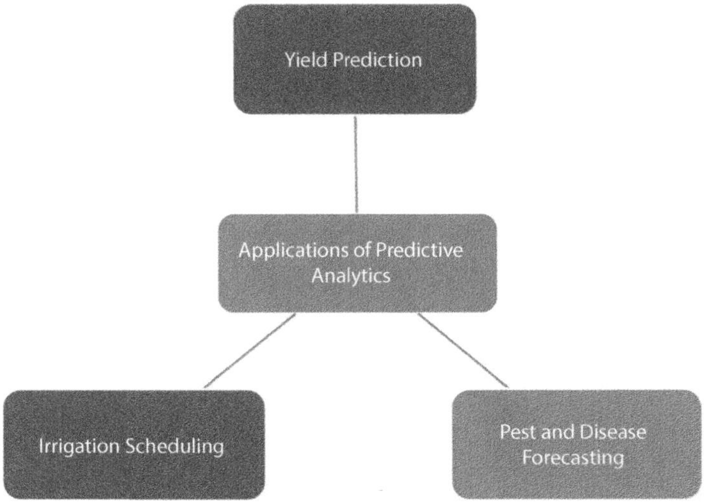

Figure 5.4 Applications of predictive analytics.

Pest and Disease Forecasting: Applying artificial intelligence models to environmental and sensor data, the system anticipates the breakout and appearance of diseases, allowing farmers to take precautions in time.
Irrigation Scheduling: Predictive models analyze soil moisture and weather forecasts to optimize irrigation scheduling. This saves a lot of water while the crops healthy.

Prescriptive Analytics

Prescriptive analytics takes it a notch higher by providing action recommendations for the best management of crops. Analytics can advise farmers on the best strategies for irrigation, fertilization, and pest control by considering several data points and eventualities. For instance:
Fertilization Recommendations: With live nutrient data, prescriptive analytics tools provide quantity and type recommendations for fertilizers across each field zone.
Optimization of Irrigation: This involves watering crops precisely, achieved through a prescriptive analytics tool that utilizes actual data from soil moisture sensors and weather forecasts, ensuring that crops receive the exact amount of water at the optimal time.

Table 5.1 Comparative case studies of precision agriculture across diverse global contexts: technologies, outcomes, and key challenges.

Case study	Context	Technology application	Results	Challenges
Precision agriculture in U.S. corn farms	A corn farm in Illinois adopted IoT sensors to address soil health and optimize water usage.	-Soil moisture and temperature sensors -Drones and satellite imagery -Data analytics for irrigation and fertilization	-12% increase in yield -20% reduction in water use -15% reduction in fertilizer costs	-Significant initial setup costs -Need for technical expertise
Vineyard management in France	A vineyard in Bordeaux used technologies to improve grape production and water usage.	-Soil and climate sensors -Drones with multispectral cameras -AI-based analytics for irrigation	-15% improvement in grape quality -25% reduction in water use -20% increase in profitability	-Maintenance costs for sensors -Challenges in data management
Rice farming in India	Smallholder rice farmers in Tamil Nadu implemented IoT sensors for water optimization.	-Low-cost soil moisture sensors -Mobile data platform for real-time alerts	-30% reduction in water consumption -Stable yields despite variable rainfall -Expansion to neighboring farms benefiting over 300 farmers	-Connectivity issues in rural areas -Initial investment costs

(*Continued*)

Table 5.1 Comparative case studies of precision agriculture across diverse global contexts: technologies, outcomes, and key challenges. (*Continued*)

Case study	Context	Technology application	Results	Challenges
High-tech greenhouses in the Netherlands	A greenhouse operation adopted IoT and AI technologies to automate climate control.	-Environmental sensors -AI-driven automation for ventilation and irrigation	-40% reduction in water use -30% reduction in energy use -25% increase in year-round production -35% rise in profitability	-High initial costs -Need for specialized technical expertise
Wheat farming in Australia	Wheat farmers used IoT sensors and satellite imaging to manage water scarcity.	-Soil moisture sensors -Satellite data for stress identification -Automated irrigation system	-35% reduction in water use -Stable yields despite drought conditions	-Durability issues of sensors -Challenges in data integration

5.5.5 Crop Monitoring and Analytics Innovations

More innovations are in the pipeline to make crop monitoring and analytics more efficient and accessible to farmers. Crop monitoring and analytics innovations include:

AI-Based Crop Monitoring: AI-based systems have been gaining momentum with the automatic analysis of enormous amounts of sensor and remote-sensing data. These systems can detect the slightest change in crop health well before visual presence.

Drone-Based Analytics: Drones can collect ultra-high-resolution data on crop health, water stress, and nutrient deficiencies. AI algorithms are combined with advanced sensors to enable routine crop monitoring without the need for human intervention.

Crop Monitoring and Analytics: A highly powerful toolkit for improving farmer decision-making and maximizing the usage of resources while increasing yields. Farmers can now rely on real-time insights into field conditions that predict potential risks and inform the precise interventions required. Some of the key challenges, including data integration, cost, and expertise, will take more time to overcome; however, innovation is continuous, and these technologies will become accessible and effective for farmers worldwide. Table 5.1 presents a comparative overview of precision agriculture applications across five diverse agricultural contexts, ranging from corn farms in the United States to rice farming in India and greenhouse operations in the Netherlands.

5.6 The Future of Smart Farming

This section provides an overview of the major trends, innovations, and potential challenges that will reshape the future landscape of smart farming (Figure 5.5).

Introduction of IoT and Sensor Technologies: With increasing adoption of IoT devices and sensors, data on soil health, weather conditions, crop status, and resource usage data can be transmitted in real time to farmers, depending on the availability of communication infrastructure. Using these data, one can manage inputs appropriately with higher productivity and greater sustainability.

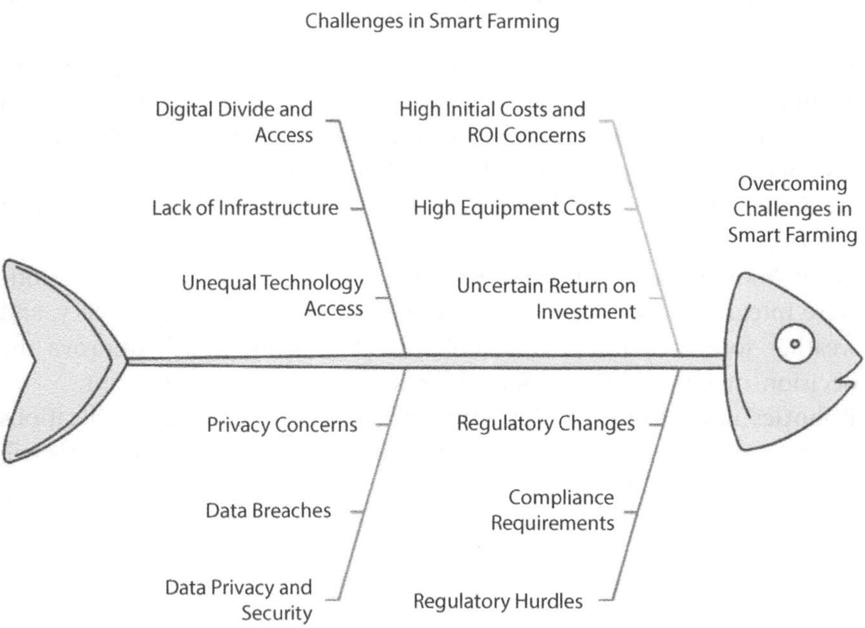

Figure 5.5 Challenges in smart farming.

AI/ML Integration: AI and ML algorithms will be used on a large scale for processing the volumetric data generated from IoT devices. The above would help in predictive analytics, thereby enabling farmers to make informed decisions regarding planting, irrigation, pest control, and harvest timing using data-driven insights.

Autonomous Machinery: Autonomous tractors, drones, and harvesters herald a new revolution in field operations. Labor costs will be reduced, while efficiency will increase, and detailed management of any field will enable superior use of resources and avoid waste.

Geospatial Technologies and Precision Agriculture: Geospatial technologies will be further used in enhancing satellite imagery and remote sensing to inform farmers of much more detailed information on crop conditions, soil, and environmental conditions within the farm, thus encouraging precision agriculture practices, targeted interventions, and resource usage management [18].

5.6.1 Smart Farming Innovations

Blockchain for Transparency and Traceability: Blockchain technology will increase transparency and traceability in supply chain management, allowing end-users to trace the origin of their food. This can establish trust between farmers, retailers, and consumers, in addition to increasing food safety.

Vertical Farming and Controlled Environment Agriculture: Technological advancements in vertical farming and CEA will make food production feasible in urban spaces with an optimization of the decreasing distance that food travels from farm to table. These technologies will also facilitate year-round production, moving away from weather-dependent practices.

Farm Management Software (FMS): Advanced farm management software integrates data from all sources (IoT sensors, satellite imagery, and weather forecasts) under one umbrella [14]. Thus, it will improve the decision-making ability, resource planning, and farm management.

Robotics and Drones in Agriculture: Robotics and drone applications will take over, making planting, monitoring, and harvesting easy. This will remove the dependency on manpower and boost operational efficiency [15].

5.6.2 Challenges to Overcome

Digital Divide and Access: The deployment of smart farming technologies may widen the inequalities between technologically advanced farms and those with limited technological resources. Training, support, and inclusive access to technology are critical to ensuring that no inequality is further amplified using such systems in the agricultural sector.
Data Privacy and Security: As more institutions embrace data-driven decision-making on their farms, concerns over data privacy and security will arise. Frameworks for data governance must be established to ensure the protection of sensitive information.
High Initial Costs and ROI Concerns: A huge upfront investment in smart farm-related technologies can scare most farmers off. Establishing return on investment and offering financial support or incentives will be essential to promote their uptake.
Regulatory Hurdles: Any modification in the related smart farming technologies will need to involve the concerned regulatory bodies, which should draft policies that ensure adequate food safety, environmental friendliness, and data privacy, among others. Stakeholders, including farmers, technology providers, and regulators, must engage to ensure the development of effective policy. The prospects of smart farming are promising; as such, agricultural practices can increase food security and make it more sustainable, an improvement that should enhance agricultural productivity [19]. By leveraging advanced technologies such as IoT, AI, robotics, and blockchain, farmers will be able to make decisions in smarter ways that will result in better crop yields and improvements on the environment and society at large [4]. However, the realization of smart farming would require improvements in accessibility, data privacy, and regulatory frameworks. As the face of agriculture changes, stakeholder collaboration will determine a future for farmers that is resilient and innovative.

5.7 Conclusion

Finally, the integration of IoT sensors and geospatial technologies in agriculture represents a major milestone in the real-time management of crops. The applications dealt with in this book chapter suggest examples wherein diversity in agricultural practices enhances productivity, optimizes resource usage, and promotes sustainability. Smart farming will expand further as the future unfolds with continuous innovations in technology and data analytics. The increased adoption of autonomous machinery and AI-driven decision-making, together with the applications of precision agriculture, will ensure that farmers make choices that remain consistent with both economic and environmental objectives. Sustainable farming practices will be a source of food security while protecting natural resources for future use. However, challenges remain in the broader acceptance of smart farming technologies. Urgent attention to critical issues such as the digital divide and data privacy issues, and support by favorable regulatory frameworks, must be given to realize greater benefits from such advancements. By embracing innovative technology and fostering a culture of collaboration and knowledge sharing, we can construct a more sustainable and efficient agricultural system to meet the needs of the increasing global population while safeguarding resources on our planet. Transitioning toward smart farming is not an option but an imperative for future agriculture.

AI disclosure statement: The author acknowledges the use of AI-based tools for minor language editing or grammatical correction. However, all intellectual contributions, analyses, and conclusions presented in this chapter are entirely the author's own.

References

1. Freeda, R.A. and Sharmila, R.N., A review of bulk data dissemination protocols for reprogramming in WSN, in: *2016 International Conference on Information Communication and Embedded Systems (ICICES)*, pp. 1–4, 2016, DOI:10.1109/ICICES.2016.7519846.
2. Dhaya, R. and Kanthavel, R., Energy efficient resource allocation algorithm for agriculture IoT. *Wireless Pers. Commun.*, 125, 2, 1361–1383, 2022.
3. Dhaya, R. and Kanthavel, R., IoE based private multi-data center cloud architecture framework. *Comput. Electr. Eng.*, 100, 107933, 2022.
4. Bhati, N. and Vyas, N., Elevating IoT Sensor Data Management and Security Through Blockchain Solutions. *Advances in Systems Analysis, Software Engineering, and High Performance Computing*, pp. 58–77, IGI Global, Hershey, PA, USA, Mar. 18, 2024.
5. Rejeb, A., Abdollahi, A., Rejeb, K., Treiblmaier, H., Drones in agriculture: A review and bibliometric analysis. *Comput. Electron. Agric.*, 198, 107017, Jul. 2022, doi:10.1016/j.compag.2022.107017.

6. Dhaya, R., Ujwal, U.J., Sharma, T., Singh, P., Kanthavel, R., Selvan, S., Krah, D., Energy-efficient resource allocation and migration in private cloud data centre. *Wirel. Commun. Mob. Comput.*, 2022, 2022.
7. Kanthavel, R., Indra Priyadharshini, S., Sudha, D., Sundara Velrani, K., Dhaya, R., Multi-hoped cooperative communication-based wireless underground sensor network design. *Int. J. Commun. Syst.*, 35, 10, e5174, 2022.
8. Rajak, P., Ganguly, A., Adhikary, S., Bhattacharya, S., Internet of Things and smart sensors in agriculture: Scopes and challenges. *J. Agric. Food Res.*, 14, 100776, Dec. 2023, doi:10.1016/j.jafr.2023.100776.
9. Elbeltagi, M.F.M.E. and Elhoseny, A.S.M., An IoT-Based Model for Smart Agriculture. *IEEE Access*, 9, 64334–64343, 2021, DOI:10.1109/ACCESS.2021.3070597.
10. Saleh, M.M., El-Shafie, H.A., Abdelhafez, K.M.A.S.N.N., Innovations in Smart Agriculture: A Comprehensive Review. *IEEE Access*, 9, 27653–27673, 2021, DOI:10.1109/ACCESS.2021.3051190.
11. Valenzuela, A.J.M. and Cuadros, A.M.M.E., Integration of IoT Technologies in Smart Agriculture: A Review. *Comput. Electron. Agric.*, 174, 105557, 2020, DOI:10.1016/j.compag.2020.105557.
12. Santhosh, K.R.B.K.R. and Manikandan, S.S.S., Smart Farming Using IoT and Geospatial Technology, in: *IEEE 2nd International Conference on Innovations in Information, Embedded and Communication Systems (ICIIECS)*, pp. 1–6, 2015, DOI:10.1109/ICIIECS.2015.7193140.
13. de Souza, F.R., Silva, T.G.A.R., de Almeida, R.C., Geospatial Technology and IoT in Agriculture: A Review. *J. Agric. Sci. Technol.*, 20, 6, 1385–1399, 2020, DOI:10.17265/2161-6256/2020.06.011.
14. de Lima, G.G.E.A., de Sousa, M.A.M., Ferreira, A.C.C., IoT Applications in Agriculture: A Systematic Review. *IEEE Access*, 8, 194455–194478, 2020, DOI:10.1109/ACCESS.2020.3031051.
15. Petrovic, B., Kononets, Y., Csambalik, L., Adoption of drone, sensor, and robotic technologies in organic farming systems of Visegrad countries. *Heliyon*, 11, 1, e41408, Jan. 2025, doi:10.1016/j.heliyon.2024.e41408.
16. Gonçalves, L.B.S.L.U.G. and Oliveira, D.H.T.F.D., The Impact of IoT on Agriculture: An Overview. *Int. J. Inf. Technol.*, 13, 1, 55–64, 2021, DOI:10.1007/s41870-020-00509-6.
17. Abdullah, M.A.H., Ali, S.H.M., Rahman, S.A.B.A., Smart Agriculture: IoT-Based Framework for Monitoring and Management. *IEEE Access*, 8, 71257–71267, 2020, DOI:10.1109/ACCESS.2020.2989317.
18. Ali, M.S., Rahman, M.F., Noor, M.S.M., IoT-Based Smart Farming: A Review. *IEEE Access*, 8, 113697–113708, 2020, DOI:10.1109/ACCESS.2020.3006268.
19. Mandal, S., Yadav, A., Panme, F.A., Devi, K.M., Kumar, S.S.M., Adaption of smart applications in agriculture to enhance production. *Smart Agric. Technol.*, 7, 100431, Mar. 2024, doi:10.1016/j.atech.2024.100431.

20. Taştan, M., Machine Learning–Based Calibration and Performance Evaluation of Low-Cost Internet of Things Air Quality Sensors. *Sensors*, 25, 10, 3183, May 2025, doi:10.3390/s25103183.
21. Patil, P.D.D.A. and Dhoble, M.S.N.A., Smart Agriculture using IoT: An Overview, in: *2020 IEEE International Conference on Technology, Software and Applications (ICTSA)*, pp. 116–120, 2020, DOI:10.1109/ICTSA49814.2020.9131531.

6

Enhancing Irrigation Efficiency with GIS and Precision Mapping Techniques

Himanshu and Deepika Saxena*

Department of Horticulture, Lovely Professional University, Phagwara, Punjab, India

Abstract:
Agricultural water management is a serious challenge in the era of drought and environmental issues. Precision irrigation systems remain one of the most effective methods for ensuring sustainable agricultural practices and efficient water consumption. Therefore, this chapter focuses on the use of remote sensing and Geographical Information System (GIS) to enhance farmers' decision-making based on timely and accurate information. Remote sensing techniques provide valuable insights into soil moisture, crop health, and water stress. The GIS platform helps integrate and analyze data, enabling farmers to better understand the need for irrigation while minimizing irrigation costs. Precision mapping, combined with GIS, allows farmers to design irrigation systems tailored to the different zones within their fields. This helps optimize the amount of water used for irrigation and simultaneously increases crop production. This technology supports informed decision-making by identifying areas that require more water while preventing over-irrigation and flooding. In addition, mapping is enhanced through the integration of remote sensing data, including satellite images and drone surveillance, which incorporate real-time information on plant growth, soil moisture content, and climatic conditions. GIS and precision mapping help maximize the use of available water resources by reducing operating costs and minimizing environmental impact. The ability to provide food security and promote water conservation through compatible and sustainable agricultural techniques supports global initiatives. In this chapter, we present detailed information on the application of GIS and precision mapping in irrigation water management for researchers, irrigators, and policymakers, aimed at enhancing water-use efficiency.

**Corresponding author*: deepika.20662@lpu.co.in

Gurwinder Singh, Vishakha Sood, Narayan Vyas, Surendra Yadav and Dankan Gowda V. (eds.) Geospatial Intelligence for Precision Agriculture, (109–140) © 2026 Scrivener Publishing LLC

***Keywords*:** Water resource management, smart agriculture, soil moisture monitoring, variable rate irrigation

6.1 Introduction

Climate change has lengthened drought periods, resulting in low agricultural productivity, particularly in arid and semi-arid regions. Through irrigation, 40% of the world's food production is supplied, accounting for 25% of the total land cover on Earth. Nevertheless, 70% of all freshwater withdrawals are used for irrigation, of which 60% is lost to evaporation, seepage, and deep percolation. Irrigation is problematic in many countries because of the lack of water. Most regions worldwide are experiencing increased drought indices coupled with rapid population growth, which poses a serious threat to future food and water supplies. Irrigated agriculture suffers from several issues, including governance, infrastructure, decision-making, siltation of dams, and scheduling of irrigation, among others, which undermine its effectiveness [37]. Installing and maintaining sprinkler and drip irrigation systems, which are more water-efficient than surface irrigation practices, incurs relatively high costs owing to labor, repairs to broken components, leaks from broken pipes, or even differences in soils that cause geographical variation in irrigation. In this way, further irrigation systems can be achieved through the sustainable management of various technologies and software to enhance production and water-use efficiency. Water is an essential resource for crop growth, and there is a high demand for it in agriculture. Nevertheless, the aim of enhancing irrigation efficiency is to use less water in agricultural activities while minimizing crop yield losses [22]. There are also several advantages in social, economic, and environmental terms associated with irrigation practices that consume less energy and water.

High irrigation efficiency is becoming increasingly important with the rise in agricultural activities, which are driven by population growth and the ongoing depletion of available water resources. Such enhancing technology can be illustrated by improved irrigation systems, which are equipped with adjustable water dispensation facilities that deliver only the required moisture to crops (drip or pressurized irrigation systems). Some modern irrigation systems are self-propelled and integrate wireless sensors and GPS technology, which can increase site-specific and volumetric efficiency in water applications to match the needs of both soil and crops. Changing agricultural practices, such as crop rotation, planting crops according to seasonal and soil conditions, and conservation tillage, where crop residue from the previous year's crop is left in the field to reduce soil erosion and runoff, can also increase irrigation efficiency [25]. The advancement of Geographical Information System (GIS) technology has been necessitated by the need to manage various spatial and temporal scenarios related to irrigation and water resources effectively. GIS can be defined as the ability to capture, store, manipulate, manage, analyze, and output geographically referenced data. Therefore, it is a crucial technique in processes such as resource allocation, modeling, optimization, and decision-making. Remote sensing (RS) and GIS are significant and effective in assessing land use and land cover changes. GIS has also been employed to study plant health, the presence or lack of pests, expected yield, soil, crop types, crop area, water management processes, and other methods [51].

6.2 Fundamentals of GIS and Precision Agriculture

6.2.1 GIS in Agriculture

This section aims to introduce the concept of GIS in agriculture, its elements, and technologies, including information collection methods such as field sensors and remote sensing, as well as the integration concept of mapping and spatial data [54]. For instance, GIS is a technology comprising various components, including hardware, software, data, and personnel, with an input–output function to collect, store, process, and present spatial or geographical data. This includes the use of field sensors, aerial drones, global positioning system (GPS) receivers, and satellite systems for data acquisition. These devices are used to capture spatial data, such as position, height, and other environmental conditions associated with locations.

GIS software provides functions for data manipulation, spatial analysis, and graphics. Some commonly used GIS applications include the Google Earth Engine (GEE), QGIS, and Esri's ArcGIS, which offer mapping, modeling, and agricultural decision support systems. Other forms of data used in GIS include spatial geographical elements, such as points, lines, and polygons, and attribute data, which are descriptive information about spatial characteristics. Imagery includes satellite, aerial, and drone images. In agriculture, GIS employs a range of data collection techniques, providing spatial data that are crucial for decision-making processes. Remote sensing is a method of data collection in GIS for agriculture, whereby data are collected remotely, primarily from space through satellites or sensor-equipped aeroplanes [59].

Satellite imaging can provide high-resolution details on land use, soil moisture, and vegetation health. This information is significant, especially for the remote sensing of crop progression history, insect outbreak ranges, and geographical shifts over a large area. Farmers can also assess the health of crops, nutrient levels, and field conditions without being physically present, thanks to multi- and thermal infrared satellite images. Advanced satellite platforms such as Sentinel-2 enable open-access data and frequent revisits for real-time monitoring and analysis. In-field sensors collect data on temperature, pH, nutrients, and soil moisture content. The information provided by sensors is essential for managing soil health and the appropriate application of fertilizers and irrigation. By utilizing sensor data with GIS, farmers can create spatially explicit maps that prompt specific actions based on the prevailing conditions in that area [2]. Precision irrigation (PI) involves controlling the input of water and the necessary nutrients in a specific crop growth cycle in various regions of an open field or field, thereby reducing production costs. Therefore, PI can generate both economic benefits and water savings. Smart irrigation technology utilizes soil moisture sensors to measure the state of the soil, compute its moisture level, and relay this data to the irrigation controller. At the scheduled time, the sensor checks the moisture content or moisture level in a particular zone, and only water in that zone is allowed when the content is low. Two key advantages of automating irrigation systems are the reduction in water usage and increase in crop yield, as shown in Figure 6.1 [61].

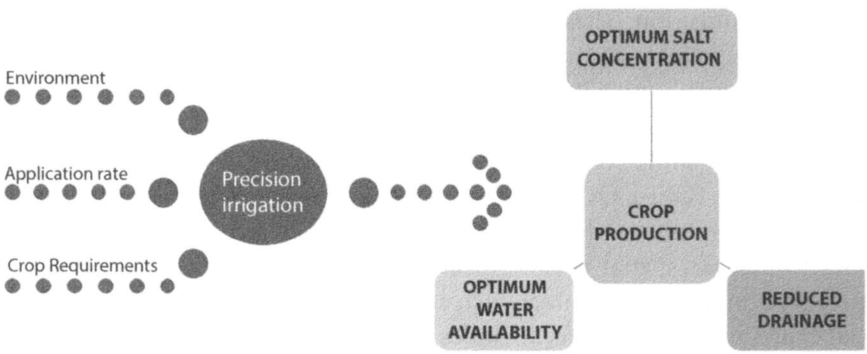

Figure 6.1 Framework of precision irrigation for enhanced crop production.

6.3 Precision Mapping Techniques for Irrigation

Precision irrigation enables the tailored management of agricultural zones by mapping the spatial variability in soil and crop conditions, as shown in Figure 6.2.

6.3.1 Soil Moisture Mapping

Most new commercial enhancements in irrigation management using soil moisture sensors have been developed by independent research institutions and high-tech companies worldwide. A significant portion of these innovations stems from the rapid decline in the cost of data storage and monitoring devices used in the field. Although the availability of sensors has increased recently, the underlying principles of sensor operation remain the same. These methods integrate physical, chemical, and mechanical approaches with diverse sensing modalities including electrical, electromagnetic, optical, radiometric, acoustic, pneumatic, and electrochemical techniques to acquire the desired information [65]. The most common indirect (non-destructive) methods for measuring the quantity of water in soil are electrical resistance (Boyoucus), tensiometry, neutron

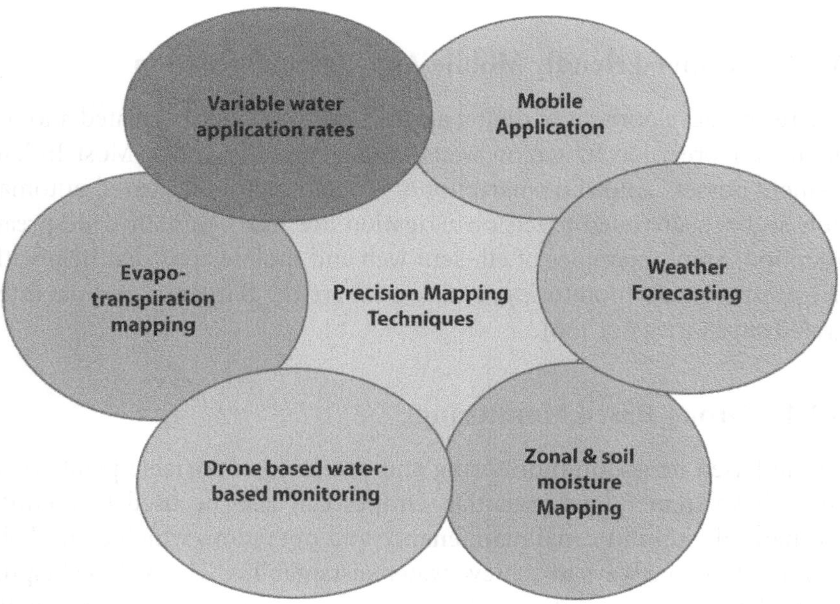

Figure 6.2 Innovative approaches to irrigation using precision mapping.

moderation, and time-domain reflectometry (TDR). The most widely used and simplest indirect (non-destructive) method for measuring soil moisture involves resistive sensors. These sensors measure the difference in electrical resistance between two electrodes buried in the soil. Their basic advantages include lower purchasing costs, easy sensor operation, and market availability; however, a disadvantage is their low accuracy in terms of soil moisture. The models consisted of SEN0114 and FC-28. The remaining sensors were capacitive. Some of them comprise the SHT10 and EC-5. They measured the dielectric constant of the soil by timing the emission of an electromagnetic pulse produced by metal rods buried in the soil. This sensor has a very high reading accuracy; however, because it contains complex and expensive electronic components in its construction, it can carry a high purchase price [36].

6.3.2 Weather Forecasting

A weather forecast for the next 14 days was provided for each field under EOSDA monitoring. Accurate weather forecasts determine when irrigation is necessary to maintain soil moisture at optimal levels for crops. Farmers anticipating upcoming rain will conserve resources by postponing their watering efforts. Notifications help avoid missing important times when conducting field duties [29].

6.3.3 Farmer-Friendly Mobile App-Based Operation

The recent development of digital agricultural systems has created a foundation for precision irrigation water management in India. Most Indian farmers possess Android smartphones. Therefore, mobile-based automation and sensor-based precision irrigation are likely to gain widespread adoption. Some examples of efficient web and mobile apps that help with irrigation, remote monitoring, and control are the Blink app and the integrated smart irrigator [66].

6.3.4 Drone-Based Monitoring

Drone-based irrigation monitoring and detection of variable plant stress zones have tremendous potential. Drones can also be used to identify mechanical issues in canal maintenance and operation, which could help identify potential ways to achieve water sustainability. This will be helpful in canal maintenance and rehabilitation, as it can inspect a larger area while significantly reducing the inspection time. However, although drones are rarely used for irrigation with very high precision, there is still scope for improvement. This technology may be adopted in India to apply accurate irrigation water control within canal commands [7].

6.3.5 Evapotranspiration Mapping

A computer model developed by the FAO is called CROPWAT, which supports crop water and irrigation demands that are dependent on crop and climate data. It is also possible to calculate the water requirements of the system across the entire area and design irrigation schedules for various management scenarios. It is based on water balance, where inputs such as rainfall and computed evapotranspiration are used to estimate the desired soil moisture level. The software utilizes climate data, including the highest and lowest temperatures, relative humidity, wind speed, sunshine hours, and rainfall information, to function properly. The Penman–Monteith method was used to compute the rate of evapotranspiration [18].

6.3.6 Zonal Mapping

Management zones (MZ) are units created to manage the physical, chemical, and/or biological variations within a growing region. MZs refer to smaller portions of a cultivated area that are each distinctly different from one another, yet similar enough to require the implementation of a specific agronomic management practice. To delineate management zones, information on the area under investigation must include, but is not limited to, agricultural land use, soil properties, landscape relief, biomass indices, and satellite images. These separate or joined pieces of data inform the creation of thematic maps, which are then used to analyze geographical data. However, owing to several reasons, such as planting, field topography, soil type hierarchy, and precipitation sequences, the field soil moisture content is not homogeneous. Enhanced productivity resulting from effective irrigation management within each management zone leads to a more efficient system. Less water is used because of minimized wastage and reduced soil nutrient leaching. Management zones, which divide non-uniform regions into uniform regions under the same irrigation management practice, have been applied to improve zoned irrigation systems [48].

6.3.7 Variable Water Application Rates

Among the various agricultural practices that rely on the use of freshwater, irrigated agriculture is the most critical. Therefore, precision irrigation is an essential management practice for promoting efficient and sustainable agricultural practices. The application of SS-VRI as a management practice enables farmers to enhance productivity while simultaneously conserving water, reducing the leaching of nutrients and agrochemicals from the soil, and optimizing inputs. This can be achieved by applying water according to spatial variability, considering elements such as soil and culture conditions. Unlike other standard sprinkler systems, SS-VRI requires variable-rate sprinklers, which are primarily determined by the angle of departure and arrival points, sprinkler head, and pulse modulation [20]. The control of variable-rate irrigation systems has utilized spectral images captured by either satellites or unmanned aerial vehicles (UAVs). The basis for this use is the knowledge of the leaf's spectral response, which has been applied at the canopy level and employed to map areas of variable-rate irrigation depths using a central pivot system. Management zones were established through a mapping process using satellite-provided images. To achieve the same objective, multispectral vegetation images were captured using UAVs [39].

6.4 Challenges in Traditional Irrigation Systems

6.4.1 Water Wastage

A major issue with traditional irrigation is the loss of water. In flood irrigation, large quantities of water are applied to the entire field surface. Water is primarily lost through runoff, evaporation, and deep percolation outside the root zone. It can be deduced from the literature, depending on the soil type, climatic conditions, and irrigation management methods. Water waste also arises from furrow and sprinkler irrigation. Poor water supply, runoff, and evaporation, are some disadvantages arising from these conditions. Water losses add pressure to the water supply, in addition to losses through nutrient leaching, waterlogging, and soil erosion, each of which detrimentally affects crop development and yield potential [6].

6.4.2 Energy Consumption

Traditional irrigation methods often require large amounts of energy to distribute, pump, and transport water. Water is often diverted from rivers, reservoirs, or groundwater through fossil fuel-based or electricity-powered pumps in most flood and furrow irrigation systems. High-altitude or long-distance pumping consumes a significant amount of energy and contributes to greenhouse gas emissions and environmental degradation. However, this also makes sprinkler irrigation systems dependent on electricity, as an energy source is necessary for transferring water through pipes and sprinkler heads using pumps and pressure regulators. Excessive energy use often exacerbates the environmental impact of agriculture and increases irrigation operating costs, particularly when fossil fuels are used for production [10].

6.4.3 Soil Degradation

Traditional irrigation methods often lead to degradation *via* soil erosion, salinization, and waterlogging. Flood and furrow irrigation can cause soil erosion problems, particularly on slopes, where running water can remove essential organic matter, nutrients, and the topsoil. Excessive irrigation with inadequate drainage leads to waterlogging and soil compaction, inhibiting root development and reducing soil fertility. Soil salinization, or increased soil salt concentration, is a prominent problem in dry and semi-arid areas, often associated with flood irrigation and high groundwater tables. The impact of erosion on crops is highly significant. However, the threat to the long-term productive viability of agricultural land poses a serious problem for rural livelihoods and food security [16].

6.5 Optimizing Irrigation Efficiency with GIS

As a consequence of the increasing need for resource optimization and water conservation in agriculture, new irrigation strategies have been implemented. These include GIS, which has revolutionized water usage in farming activities. Through GIS, mapping, and real-time observations, farmers can optimize their irrigation practices, save water, and increase their yields. Enhancing irrigation efficiency through GIS represents a significant technological advancement, marking a shift in farming practices in an era where environmental concerns are a primary concern [50].

6.5.1 Soil Moisture Mapping and Crop Water Need

The greatest strength of using GIS in irrigation is its ability to map soil moisture over a field. GIS enables the receipt of real-time soil moisture data from sensors placed in the ground or remotely sensed data acquired by satellites and drones. Therefore, farmers can apply irrigation precisely where it is needed by using these data to identify sections of the field that require more or less water. This information enables farmers to implement site-specific irrigation, ensuring that the field area receives the exact amount of water it needs. This means that the field is not properly watered, thereby avoiding the risk of over-irrigation or under-irrigation of the crops. This enhances crop performance while ensuring water availability [3].

6.5.2 Site-Specific Irrigation Scheduling

The schedules of the traditional irrigation system is based on fixed and usually inflexible timetables, irrespective of the actual time water is required for crops. GIS enhances irrigation scheduling through the real-time integration of weather, soil moisture, and crop growth stage data. This can be achieved by importing station data and satellite images into a GIS to calculate the rate of evapotranspiration, which measures the loss of water through evaporation and plant transpiration. Furthermore, the irrigation schedule is adjusted according to the actual crop water requirements and environmental conditions so that water is provided only at those times and places where it is most required. Moreover, GIS-based irrigation scheduling can be performed autonomously, where real-time data are fed into decision support systems, allowing the irrigation timing and amount to be adjusted accordingly. This, in turn, ensures optimal water use, lowers labor costs, and increases operational efficiency [52].

6.5.3 Zoning of Water Management

The heterogeneity of soil types, terrain, and crops within a field can alter the moisture requirements of different parts of the field. For these reasons, agricultural land can be separated using GIS into several zones. Each zone is managed separately, and the irrigation supply is adjusted according to the needs of each zone. Using GIS, this zonal management plan facilitates the more efficient utilization of water resources by minimizing wastage in areas with low water requirements and preventing excessive irrigation in regions with limited drainage. As a result, GIS-based zonal irrigation is more efficient in water use across different zones, resulting in better yields and reduced environmental degradation. In addition to saving water, farmers can reduce production costs and maintain the quality of crops and soils for many years [68].

6.5.4 Real-Time Monitoring and Decision Support System

Another advantage of GIS is its ability to integrate all data sources and provide real-time monitoring of field conditions. IoT sensors, drones, and satellite photography, for instance, are used to provide instant updates on soil moisture, weather conditions, crop health, and water usage through GIS platforms. Farm operators can remotely monitor their fields, gaining detailed information on the performance of their irrigation systems and crops [1]. By leveraging real-time data, DSS can feed predictive models that offer actionable insights to farmers. For instance, based on immediate moisture levels and weather predictions, a DSS can predict when a field will require irrigation. The software can also recommend adjustments to irrigation schedules based on the growth stages of various crops [12].

6.5.5 Water Flow Mapping and Drainage Planning

GIS can be applied in drainage system management and mapping of water flow patterns. Highly detailed descriptions and topographic maps are essential for the application of GIS to identify areas susceptible to waterlogging and to subsequently improve drainage system design. The regulation of water flow prevents soil deterioration, nutrient loss, and erosion, all of which can be exacerbated by inadequate irrigation methods. For instance, GIS can show how water flows through a field after irrigation or precipitation, thereby enabling farmers to plan irrigation systems that maximize water infiltration into the soil, reduce runoff, and minimize water loss. GIS helps farmers manage their soil health while reducing irrigation-related environmental impacts, as improved drainage and control over water flow help to address these issues [26].

6.6 Remote Sensing in Precision Irrigation

6.6.1 Role of Remote Sensing in Detecting Water Stress

A limited water supply causes water stress, which reduces the health of flora owing to changes in physiological and biochemical processes. Remote sensing can identify zones of water stress and monitor plant health by detecting these changes. The identification of water stress in plants can be carried out by measuring variations in the reflectance of visible and near-infrared light. Other methods for detecting water-stressed plants include measuring changes in the thermal properties of plants using thermal infrared sensors. As water stress reduces the efficiency of leaf transpiration for cooling and leads to heat generation, water-stressed plants may also exhibit higher leaf temperatures. Different spectral bands and vegetation spectral reflectance indices can be employed to monitor the chlorophyll and water content of plants [72].

6.6.1.1 *Vegetative Indices*

Several VIs have been formulated based on the reflection properties of plant spectra that were determined using several imaging techniques that integrated reflectance at particular spectral wavelengths (Table 6.1). These VIs are used to monitor changes in vegetation and related physiological processes. Several indicators have been used to detect water stress in plants. Typical VIs include water VIs, pigment VIs, and temperature VIs. While Water VIs rely on reflectance in the SWIR bands, typical Vis are based on reflectance in the red (600 nm–700 nm) and part of the NIR (700 nm–900 nm) spectral range. In addition, Pigment VIs are based on red-edge and green-edge reflectance, which reflect the amounts of leaf pigments, mainly chlorophyll [57].

6.6.1.2 *Leaf Chlorophyll*

Chlorophyll is a pigment involved in photosynthesis. It is one of the key pigments that facilitates the conversion of light energy into chemical energy through photosynthesis. Chloroplasts, which contain a component of the chlorophyll molecule, comprise a part of the plant cell. These molecules absorb light energy and serve as an input to power the production of organic products, such as glucose, which plant require for growth and metabolism. Chlorophyll is the pigment responsible for photosynthesis, absorbing light best in the two extremes of the visible light spectrum, red and blue, and reflecting green light; therefore, it also accounts for the green color of plants. A reduced concentration of chlorophyll in the leaves of stressed plants alters their ability to absorb solar radiation, thereby altering the normal spectrum reflectance patterns that favor red and blue reflections over green reflections. Thus, establishing the normal spectral reflectance patterns of healthy plants as a reference is crucial for the remote-sensing diagnosis of water stress in plants. Some red and near-infrared reflectance wavelengths are sensitive to changes in chlorophyll pigment concentrations. The 550 nm and 700 nm wavelengths have been found to be the maximum reflectance sensitivity to chlorophyll concentrations. Many spectral

Table 6.1 Key vegetative indices and their uses in monitoring vegetation health.

Vegetative index	Uses	Source
Normalized Different Vegetative Index (NDVI)	Mostly used to track the overall health of the vegetation	[40]
Normalized Difference Water Index (NDWI)	To measure the water content in the crop	[34]
Soil Adjusted Vegetation Index (SAVI)	To detect water stress in semi-arid and arid regions where the soil may affect NDVI readings	[42]
Water Band Index (WBI)	To assess early signs of drought stress in crops	[9]
Photochemical Reflectance Index (PRI)	Useful in evaluating short-term physiological responses to stress	[44]
Thermal indices	Useful in precision agriculture for real-time monitoring of plant needs and irrigation scheduling	[24]
Enhanced Vegetation Index (EVI)	Useful in areas with dense vegetation where NDVI may become saturated	[75]
Greenness index (GI)	Used in the early detection of plant stress	[15]
Normalized Difference Infrared Index (NDII)	Similar to NDWI but is often applied to monitor drought stress in forests and large-scale agricultural fields	[4]
Canopy water content index	Useful in assessing water stress in agricultural fields, especially during prolonged dry periods	[13]
Green Normalized Difference Vegetation Index (GNDVI)	Particularly sensitive to chlorophyll content and can be used to detect water stress in the early stages	[21]

(*Continued*)

Table 6.1 Key vegetative indices and their uses in monitoring vegetation health. (*Continued*)

Vegetative index	Uses	Source
Shortwave Infrared Water Stress Index (SIWSI)	Useful for detecting water stress in both agricultural and natural vegetation during drought conditions	[5]
Modified Perpendicular Drought Index (MPDI)	To monitor vegetation drought stress and soil moisture using satellite data	[23]
Vegetation Water Supply Index (VWSI)	To assess plant water stress by measuring the water supply conditions affecting vegetation health	[56]
Normalized Moisture Index (NMI)	To assess the moisture content in vegetation and soil, helping to monitor water stress, drought, and irrigation needs	[19]

indices have been developed based on the combination of reflectance at these wavelengths to describe the relationship between the reflectance value and chlorophyll content in leaves. These include popular indices, such as the normalized difference vegetation index (NDVI) and different types of chlorophyll indices (CIs) [71].

6.6.1.3 Canopy Temperature

The thermal radiation produced by plants over different parts of the electromagnetic spectrum as a function of their temperature is known as the spectral features of canopy temperature. The temperature of a plant canopy depends on various factors, such as air temperature, humidity, sun radiation, and watering of plants. Plants emit thermal radiation in the thermal infrared (TIR) region (8–14 μm), which is used to estimate plant temperature. Due to water stress, plants close their stomata to prevent further loss of water. Consequently, the amount of cooling through evaporation decreases, and it remains balanced in terms of temperature with the surrounding air. Well-watered plants maintain better canopy temperatures and, as such, provide continued transpiration and evaporative cooling.

The TVDI is a spectral index used to estimate vegetation water stress *via* canopy temperature and vegetation cover. The calculation of the TVDI can be performed based on the following: the difference between the surface temperature, which is sensed through thermal sensors, and the ambient temperature; the division of the difference by the difference in surface temperature over the reference temperature, which represents the maximum transpiration under the same conditions of atmospheric setup. A more significant value of TVDI represents greater criticality of water stress, and vice versa: less critical means an adequate supply. More critical values of TVDI are related to more severe water stress, whereas lower values reflect adequate water supply [38].

6.6.1.4 *Plant Photosynthetic Efficiency*

Many features of the spectrum can be used to estimate the photosynthetic efficiency of a plant. An important feature of chlorophyll concentration, measured in a standard way, is the photochemical reflectance index (PRI), which considers changes in the pigment pool of the xanthophyll cycle and protects the plant from excessive light. As plants usually regulate their photosynthetic apparatus according to variations in light intensity, the spectral response in the red and far-red wavelengths, including chlorophyll fluorescence emission, might also reflect variations in photosynthetic activity. Leveraging the distinct reflectance at 531 nm and 570 nm wavelengths, the PRI vegetation index computes variations in the pigment pool involved in the xanthophyll cycle. Chlorophyll fluorescence has a set of characteristic spectral features that can be measured using remote sensing. During photosynthesis, some of the absorbed light energy is converted to heat, and the remainder is used as fuel for the synthesis of organic molecules from carbon dioxide and water. However, if excess light energy is absorbed beyond that required for photosynthesis, it is released as a fluorescence signal. Chlorophyll fluorescence originates within the red and far-red parts of the spectrum, primarily between 650 nm and 750 nm, peaking within this region [45].

6.7 UAVs for High-Resolution Water-Use Mapping

The use of UAV platforms might open up new avenues in crop scouting, soil moisture management, land classification, and detecting changes in land use and other areas of agriculture. Drones are used for crop water status surveillance, water demand for crops, and characterization of water bodies.

6.7.1 Soil Moisture, Crop Water Demand, and Irrigation

Recent advancements in UAVs and remote sensing technologies have allowed for the rapid acquisition of vast amounts of data usable for different hydrologic research objectives. Soil moisture within a watershed, continent, and even globally, is one of the most important climatic factors impacting precipitation and heat received, carbon processes, and climate change. Soil moisture data are critical for addressing issues of food and water shortage today and in the future because of decision support tools and solutions such as flood runoff projections, water-efficient irrigation, and aquifer storage and recovery. For example, UAVs have been incorporated into systems in which precision irrigation is performed through smart water management platforms (SWAMP). Aerial photographs provided by drones were used to map the area in question and help orient the selection of sampling points by the different management zones. Aircraft are equipped with augmented reality systems [33].

6.7.2 Crop Water Status Monitoring

Various parameters, such as plant cover, leaf water content, leaf area index, and leaf chlorophyll, are monitored by precision farmers using UAVs with multispectral and other cameras. Such vegetative indices were arrived at because it has been established that the levels of light reflected by surfaces of plants vary with the type and composition, and morphological and chemical characteristics of the plants (e.g., water, chlorophyll, sugar, and carbohydrate). Indices such as the Soil-Adjusted Vegetation Index (SAVI), Enhanced Vegetation Index (EVI), Leaf Water Content Index (LWCI), Normalized Difference Vegetation Index (NDVI), Normalized Difference Red-Edge Index (NDRE), and Plant Water Status are derived from the wavebands used to acquire images using MS sensors [14]. Hyperspectral sensors are also employed in mapping, differentiation, and forecasting of leaf area index in greenery as they can collect information at different wavelengths within the range of 10 nm–20 nm primarily (Table 6.2). This, in turn, makes vegetative indices an effective means of monitoring the state of plants and moisture in the body. In particular, the state of the health of crops is studied, and plans for irrigation and combating droughts are designed and executed using the spectral indices of the crop canopy.

Table 6.2 Integration of sensors, models, and applications in precision agriculture.

Sensors	Application model used	Application	Source
Multispectral	Ensemble learning (EL)	To achieve accurate inversion mapping of Soil moisture content in Kiwifruit	[74]
Thermal infrared, multispectral, and LiDAR	surface energy balance system (SEBS)	Optimized for surface heat roughness length at different land cover types	[67]
Hyperspectral	Linear Equation Model	For mapping of hailstorm damage on winter wheat	[17]
Hyperspectral	Enhanced random forest model (ERFM)	For estimation of soil salinity in cotton field	[58]
Multispectral	Digital Surface Model (DSM)	For detecting the spatial-temporal fluctuations of soil moisture content	[28]
Multispectral	Random Forest Model	Understanding of 2D/3D urban structure on PM2.5 concentration	[35]
Hyperspectral	Digital terrain models (DTMs) and digital surface models (DSMs)	To monitor crop response to irrigation and rainfall in cotton field	[49]
Multispectral	Linear Regression Model	To determine the water status and delineate the management zone of citrus	[31]

(Continued)

Table 6.2 Integration of sensors, models, and applications in precision agriculture. (*Continued*)

Sensors	Application model used	Application	Source
Multispectral	Two-source energy source balance (TSEB)	For estimation of evapotranspiration (ET) in sorghum crop	[60]
Hyperspectral	Latent Variable model	To visualize the distribution of small-scale spatial variability of water composition	[64]
Multispectral	Advanced machine and deep learning regression models	To develop drought-tolerant wheat varieties	[55]
Multispectral	Digital Surface Model (DSM)	For measuring the primary growth parameters of hazelnut cultivars	[63]
Multispectral	Partial least squares (PLS), support vector machines (SVM), and random forest (RF)	To estimate the crop water stress proxy in maize crop	[27]
Hyperspectral	Digital elevation model (DEM)	For the classification of grapevine varieties	[32]
Hyperspectral	Convolutional neural network (CNN) model	Differentiation between invasive noxious weed species (INWS) and native alpine grass species	[69]

To enhance the process of identifying and choosing plant types for reproductive development and agricultural advancement, UAVs can also be used to assess water efficiency in plants. However, there are some challenges associated with assessing water status and water stress that need to be addressed. These studies have focused on the effects of solar zenith angle effects on canopy NDVI and the bidirectional reflectance distribution function, as well as its characteristics and effects [46].

6.8 Challenges and Barriers to Adoption

6.8.1 High Initial Cost and Financial Barriers

The major barrier to the adoption of GIS and precision mapping technologies in irrigation projects is the relatively high initial capital costs associated with them. Remote-controlled aerial vehicles, aerial photographs, soil moisture sensors, and computerized data analysis are among the technologies used in precision irrigation systems. These technologies often require considerable capital in small- and medium-sized farms, which often have no substantial funding. Moreover, installing equipment necessary to support the implementation of this directive, such as improved irrigation management systems integrated with GIS data, is very costly. This may be a bottleneck for many farmers, especially those in the developing world, as they may not be in a position to justify the cost because the returns may not be conspicuous in the near future. General precision irrigation systems are suitable for large farms. For large farm fields, this is attributed to their commercially viable size in terms of realizing the cost of investment. However, India's incredibly small average land parcel sizes are the first impediment to the application of effective precision-irrigation systems. Precision irrigation systems are unsuitable for small land holdings because their main parts, including sprinkler and drip systems, require farmers to spend most of their money on the installation of irrigation equipment [30].

6.8.2 Limited Access to Technology and Infrastructure

Another major hindrance is access to the required infrastructure and technology. While precision mapping and GIS technology are easily available in developed countries, access is more restricted in poorer countries.

Farmers are often deprived of satellite services, Internet coverage, and up-to-date technological advancements. Adequate digital infrastructure, such as cloud computing services and high-speed Internet, is necessary for specific mapping and collecting real-time data, which is usually not very common in agricultural areas, especially in rural or remote areas. Additionally, Legislative, logistical, or financial constraints may limit the availability of high-quality satellite data or drone services in some areas. The problem is that the lack of the above-mentioned infrastructure makes it difficult for the farmers to reap the full benefits of these technologies, as the adoption of precise mapping is hindered. Infrastructure, such as computers and controllers, is one of the elements of precision irrigation systems, and it has not yet advanced as much as rural areas require. Furthermore, they explain why the demand for precision irrigation systems in rural areas is not prevalent. However, this quick installment of technology and investment regimens require vast up-front capital that most rural people are unable to bear [53].

6.8.3 Technology Complexity and Lack of Skilled Labor

There is also the issue of GIS and the complexity of precision mapping. These technologies are designed to precisely forecast the need for irrigation in systems using combinations of geographical analysis, data interpretation, and machine learning (ML) or artificial intelligence (AI) algorithms. However, some farmers or farm managers may not have the technical know-how to operate such devices. Adoption might be challenging even when farmers have the necessary equipment in place because it is often difficult for them to get enough trained workers to handle and analyze data. Geographical data require specific knowledge for the operation of GIS software to analyze and draw conclusions from the data, which is lacking in farmers. This knowledge gap requires more than familiarity with the subtleties of data analytics, remote sensing, and the integration of many data streams toward a harmonious irrigation strategy. However, even where precision agriculture is feasible, its promise often remains unrealized owing to a dearth of qualified personnel, especially in rural areas of developing countries. More funding needs to be allocated to training programs, agricultural education, and capacity building to ensure that farmers and agricultural workers can use GIS and precision mapping tools efficiently [47].

6.9 Future Trends and Innovation in Irrigation Efficiency

6.9.1 Robotic Irrigation System

These approaches aim to improve the intelligence, effectiveness, and accuracy of water management in agricultural systems. Moreover, AI, or the automation of processes, has advanced to the extent that intelligent robots can be designed to control and fine-tune all irrigation operations on farms. Additionally, such robotic systems are expected to control elements such as pipes, nozzles, and other apparatus fitted with irrigation systems, regulate the water flow directed to the crops, and even adjust the watering robotic schedules. This significantly increases plant yield while minimizing water wastage. It also offers the opportunity to learn and interpret the implications of farmland use over time. For instance, drones and UAVs are machines that perform aerial scans to collect relevant information about agriculture, such as crop health, soil content, and diseases present in crops. This information is used to determine the requisite irrigation. For example, in arid areas, a drone that measures soil moisture can ensure that water is provided at the right time and in the right amounts. These machines perform irrigation processes efficiently with low power consumption owing to smart energy management. In addition, water savings reduce energy consumption. In the future, we will build a smarter and more advanced irrigation system using robotic irrigation technology. These approaches enhance the efficiency of the available water resources while at the same time ensuring that agriculture is ready and able to face the impending challenges. Robotic irrigation systems are one of the innovations that will change the manner in which farming is practiced in the future [41].

6.9.2 Advanced AI and IoT-Supported Irrigation

The agricultural sector is looking for suitable and feasible solutions to the looming water shortage problem. The fundamental components of these solutions are smart irrigation systems that use Internet of Things (IoT) and AI-based irrigation technologies. Another promising area in the agricultural irrigation process is the IoT; the agricultural sector itself has undergone a gigantic change owing to this new era phenomenon. The IoT presents the idea of a far-enhanced water management process on farms. Using AI, normal data processing, the management of irrigation and farms, and other methods of farm monitoring on a regular basis can be achieved. Future irrigation systems will have the ability to work in conjunction with improved climate and weather prediction systems. Through weather forecasts, irrigation plans can change automatically, especially during dry periods caused by climate change, which are becoming increasingly frequent. This benefits both water conservation and plant productivity [43].

6.9.3 Holographic Irrigation Systems

There is a need for fresh approaches to more sustainable water management and better use of water resources in agriculture in the future. This advanced technology has tremendous potential for ensuring that water usage is more accurate with the inclusion of holography and improved water management. Water applied to plants is provided in a manner befitting the specific needs of the plants with three-dimensional holographic projections due to holographic irrigation. These agricultural areas had their growth monitored, and water consumption was measured using real-time projections above the agricultural areas. This information automatically adjusts the irrigation system accordingly. This minimizes the chances of encountering salt problems or negatively affecting plant roots. The understanding gleaned from holographic irrigation techniques may lead to significant effects as future trends in irrigation technologies and agricultural water management are likely to unfold. Thus, this technology can alleviate the impact of the water crisis and increase industry robustness through a more efficient and sustainable approach to water utilization. With reference to food safety and environmental preservation, these innovations present a glimmer of hope for agriculture and farming [8].

6.9.4 Irrigation with Sound Waves

Sound wave irrigation applies acoustic vibrations to the root zone to potentially enhance water infiltration and soil–water interactions. The microcracks caused by these waves help water to spread and penetrate down to the plant roots. This technique prevents water from evaporating on the surface and blowing away, thereby decreasing water waste. When used, it enables plants to respond to water appropriately and use it more optimally. It can also allow for monitoring the salinity of the water and prevent the accumulation of salt. This can increase yield regardless of how unfavorable the ground in question might be, for instance, salty soils. Environmental conservation is one of the advantages that is likely to be realized using this technique. It is more efficient than traditional irrigation methods in saving water and reducing the use of chemical fertilizers. This is important for soil and water conservation. The future of water management for agricultural purposes involves the use of sound waves in irrigation as an epoch-making phenomenon. By facilitating the more sustainable and effective use of irreplaceable water resources , this technology may alleviate the current water shortage crisis and help the agricultural sector cope with any existing challenges in the future. The future of agricultural practices will be characterized by the use of innovative and efficient irrigation technologies, and sound wave irrigation could lead to innovation in this context [70].

6.9.5 3D GIS

Compared to traditional GIS, 3D GIS adds an extra dimension (z) to provide more exposure and definition to items. One of the components that cannot be seen on the 2D maps is the elevation component that is incorporated into objects using the 3D GIS. When designing GIS maps with 3D graphics, real-world object scales can be matched with the explanatory images. The three-dimensional models yield a good look at conducting surveys in different spheres of life. Integrated 3D technologies help irrigation engineers envision the changes that can occur in irrigation schemes, the feasibility of such changes, and other relevant impacts, as well as the sustainability of these changes. A 3D model of a normal scenario would contain statistical data, satellite imagery, and data regarding existing irrigation systems to support the designers involved with irrigation systems in deciding the path of the lowest cost for irrigation conveyance systems. Hence, to enhance the 2D and 3D representations and functionalities of the irrigation system and the corresponding GIS-based simulations, ArcGIS can be linked to the model. With this integration, the impacts of floods on irrigated agriculture in three dimensions can be assessed using GIS 3D spatial analysis tools using remote sensing (RS) and a 3-D geographic information system (3-DGIS). Together with a hydrological facility map, ASTER GDEM, and high-resolution remote sensing images from Google Earth (GE), 3D terrain and ground feature models were created interactively. Subsequently, the texture data were employed to develop both models and were incorporated into ArcGIS [11].

6.9.6 Mobile GIS

The application of mobile GIS eases the feasibility study of irrigation development implementation because it facilitates the storage, addition, alteration, analysis, and display of geographical data and information by the field staff. This is possible by the integration of GPS, GIS software, handheld devices, and wireless connections, which allow Internet GIS access. Data can be collected in real time, and changes can be made in real time using mobile GIS. In addition, the processing time was reduced, and the results were more accurate. Mobile GIS applications have been developed for various purposes. Owing to the stunning changes in mobile technology, these applications do much more than mapping. Some of these apps were developed for commercial purposes by private sector organizations [73].

6.9.7 Big Data and ML

Large amounts of data from an irrigated field can be used to determine irrigation and other factors of farming. The literature has shown that through the use of data, the productivity of agriculture is enhanced, costs are reduced, and sustainability is achieved. By incorporating GIS, RS, big data, and ML, the automation of agricultural data collection, analysis, and decision-making has been promoted. The development of a GIS-based data management system assisted farmers in executing a variety of irrigation tasks, including weather forecasting, mapping, scheduling, automated field management, Big Data, and ML tools. Additionally, the high increase in computing power has enabled computing algorithms to use data gathered from sensors, UAVs, and satellites, and helped render them to facilitate decision-making. These data-driven ML models for irrigation have been relatively more discussed for their potential in soil moisture content estimation, rather than the less discussed physical/mechanistic models that tend to consume more calibration data than data-driven models. Meanwhile, ML models also show better performance regarding prediction aspects than the physical model and can process both spatial and temporal data rapidly and easily [62].

6.10 Conclusion

Precision mapping and GIS in sustainable irrigation will enhance data gathering, analysis, and decision-making processes in the future. Trends in the technological applications of irrigation control focus on ideas synonymous with concepts connected to emerging technologies. Land suitability analyses, crop water consumption estimations, scheduling applications of irrigation, tracking yields, and checking the effectiveness of irrigation have all found a launch pad in geographic information systems. With a huge quantity of geographical and temporal data, the integration of GIS with current irrigation management systems would be much stronger and more effective in handling such data volumes. Geographic data for irrigation at the field or basin scale were organized, retrieved, and visualized using the system. Furthermore, more techniques are required to assess the spatial water demand for irrigation. In addition, few systems and tools have been developed for simulating irrigation models. These tools can be confidently developed from existing GIS software applications. Accordingly, several applications of GIS, as demonstrated in this study, the use of these geospatial technologies in irrigation systems in the future would be influenced by advancements in GIS technology and related applications. This is because aerial vehicles that employ remote sensing and GIS technology will be a major part of Agriculture 5.0.

Automated irrigation for precision watering includes site-specific drip and sprinkler irrigation. Governments and policymakers should promote the adoption of GIS and precision mapping tools by providing subsidies and financial incentives to farmers, particularly smallholder farmers. Another goal of research and development expenditures should be aimed at trying to reduce the cost of such technologies so that they can be used. Training programs that help farmers understand and properly use precision irrigation systems should be among the leading priorities in educational work. The private sector, particularly software firms, can be particularly useful in the advancement of economical and sound solutions for a variety of farming predicaments. To share knowledge and improve best practices, agricultural organizations should facilitate the information flow between farmers, the research fraternity, and technology providers. When implemented sequentially, the proposed plan will create a more sustainable irrigation management system that is more efficient and reliable.

AI disclosure statement: The author acknowledges the use of AI-based tools for minor language editing or grammatical correction. However, all intellectual contributions, analyses, and conclusions presented in this chapter are entirely the author's own.

References

1. Ali, A., Hadeed, M.Z., Safavi, S., Ahmad, M., Leveraging GIS for Environmental Planning and Management, in: *Global Challenges for the Environment and Climate Change*, pp. 308–331, IGI Global, 2024.
2. Vyas, N., Sundar, K., Sharma, A., Adhikari, P.S., Leveraging Sentinel-2 Multispectral Data and Machine Learning Algorithms for Land Use Land Cover Mapping in Semi-Arid Regions, in: *2025 3rd International Conference on Advancement in Computation & Computer Technologies (InCACCT)*, Gharuan, India, pp. 439–443, 2025, doi:10.1109/InCACCT65424.2025.11011472.
3. Barrile, V. and Genovese, E., GIS, Remote Sensing, and Forecasting Systems for Precision Agriculture Development, in: *International Conference on Computational Science and Its Applications*, Springer Nature Switzerland, Cham, pp. 302–318, 2024.
4. Bartold, M., Wróblewski, K., Kluczek, M., Dąbrowska-Zielińska, K., Goliński, P., Examining the Sensitivity of Satellite-Derived Vegetation Indices to Plant Drought Stress in Grasslands in Poland. *Plants*, 13, 16, 2319, 2024.
5. Boninsenha, Í., Mantovani, E.C., Rudnick, D.R., Ribeiro, H.D.Q., Revealing irrigation uniformity with remote sensing: A comparative analysis of satellite-derived uniformity coefficient. *Agric. Water Manage.*, 301, 108944, 2024.
6. Boru, G.W., Birhan, T.S., Beyene, T.L., Integrated effects of furrow irrigation systems and irrigation regimes on water productivity of onion (Allium cepa L.) crop in a semi-arid area. *Heliyon*, 10, 21, 2024.
7. Carracelas, G., Ballester, C., Marchesi, C., Roel, A., Hornbuckle, J., Assessing Drone-Based Remote Sensing Indices for Monitoring Rice Nitrogen Plant Status Under Different Irrigation Techniques. *Agronomy*, 14, 12, 2976, 2024.
8. Chakravarthy, S.P., 6G wireless communication. *6G Communication Network*, pp. 306–314, CRC Press, Sep. 30, 2024. doi: 10.1201/9781003522003-22.
9. Chandrasekar, K., Srikanth, P., Chakraborty, A., Choudhary, K., Ramana, K.V., Response of crop water indices to soil wetness and vegetation water content. *Adv. Space Res.*, 73, 2, 1316–1330, 2024.
10. Chen, C., Hussain, Z., Liu, J., Zaman, M., Akhlaq, M., Innovative technologies in sprinkler irrigation: An overview. *Agrociencia*, 1–18, 2024.
11. Cho, J., Kim, C., Lim, K.J., Kim, J., Ji, B., Yeon, J., Web-based agricultural infrastructure digital twin system integrated with GIS and BIM concepts. *Comput. Electron.Agric.*, 215, 108441, 2023.
12. Conde, G., Guzmán, S.M., Athelly, A., Adaptive and predictive decision support system for irrigation scheduling: An approach integrating humans in the control loop. *Comput. Electron. Agric.*, 217, 108640, 2024.
13. Das, S., Kaur, S., Sharma, V., Determination of threshold crop water stress index for sub-surface drip irrigated maize-wheat cropping sequence in the semi-arid region of Punjab. *Agric. Water Manage.*, 301, 108957, 2024.

14. Dong, H., Dong, J., Sun, S., Bai, T., Zhao, D., Yin, Y., Wang, Y., Crop water stress detection based on UAV remote sensing systems. *Agric. Water Manage.*, 303, 109059, 2024.
15. Fan, X., Zhang, H., Zhou, L., Bian, L., Jin, X., Tang, L., Ge, Y., Evaluating drought stress response of poplar seedlings using a proximal sensing platform *via* multi-parameter phenotyping and two-stage machine learning. *Comput. Electron. Agric.*, 225, 109261, 2024.
16. Farooqi, Z.U.R., *et al.*, Management of Soil Degradation. *Ecosystem Management*, pp. 55–78, Wiley, Nov. 08, 2024. doi: 10.1002/9781394231249.ch3.
17. Furlanetto, J., Dal Ferro, N., Caceffo, D., Morari, F., Mapping hailstorm damage on winter wheat (*Triticum aestivum* L.) using a microscale UAV hyperspectral approach. *Precis. Agric.*, 25, 2, 681–703, 2024.
18. Gaddikeri, V., Rajput, J., Jatav, M.S., Kumari, A., Rana, L., Rai, A., Gangwar, A., Estimating crop water requirement in Madhya Pradesh's agro-climatic regions: A CROPWAT and CLIMWAT software case study. *Environ. Conserv. J.*, 25, 1, 308–326, 2024.
19. Getahun, Y.S. and Li, M.H., Flash drought evaluation using evaporative stress and evaporative demand drought indices: a case study from Awash River Basin (ARB), Ethiopia. *Theor. Appl. Climatol.*, 155, 1, 85–104, 2024.
20. Gonzalez-Dugo, V., Mateos, L., Quemada, M., Torres-Sánchez, J., Delgado, A., Villalobos, F.J., Site-Specific Agriculture, in: *Principles of Agronomy for Sustainable Agriculture*, pp. 583–597, Springer International Publishing, Cham, 2024.
21. Gu, H., Mills, C., Ritchie, G.L., Guo, W., Water Stress Assessment of Cotton Cultivars Using Unmanned Aerial System Images. *Remote Sens.*, 16, 14, 2609, 2024.
22. Hamze, M., Cheviron, B., Baghdadi, N., Lo, M., Courault, D., Zribi, M., Detection of irrigation dates and amounts on maize plots from the integration of Sentinel-2 derived Leaf Area Index values in the Optirrig crop model. *Agric. Water Manage.*, 283, 108315, 2023.
23. Hua, Z.H.A.N.G., Haiting, Y.A., Cungang, X.U., Remote sensing retrieval of soil moisture and estimation of vegetation water requirements in the north and south mountains of Lanzhou City. *Arid Zone Res.*, 41, 4, 566–580, 2024.
24. Javed, Z., Optimizing Water Resources: A Sensor-Based Framework for Precision Irrigation Scheduling. *Magna Carta: Contemporary Social Science*, 2, 1, 1–12, 2024.
25. Jensen, T.A., Antille, D.L., Tullberg, J.N., Improving On-farm Energy Use Efficiency by Optimizing Machinery Operations and Management: A Review. *Agric. Res.*, 1–19, 2024.
26. Kaliraj, S., Shunmugapriya, S., Lakshumanan, C., Suresh, D., Arun Prasad, K., Srinivas, R., Flood risk zone mapping and future projections for the Thamirabarani river basin, Southern India: insights from decadal rainfall

trends and GIS-based analytical hierarchy process technique. *Nat. Hazard.*, 1–35, 2024.
27. Kapari, M., Sibanda, M., Magidi, J., Mabhaudhi, T., Nhamo, L., Mpandeli, S., Comparing Machine Learning Algorithms for Estimating the Maize Crop Water Stress Index (CWSI) Using UAV-Acquired Remotely Sensed Data in Smallholder Croplands. *Drones*, 8, 2, 61, 2024.
28. Khose, S.B. and Mailapalli, D.R., Spatial mapping of soil moisture content using very-high-resolution UAV-based multispectral image analytics. *Smart Agric. Technol.*, 8, 100467, 2024.
29. Kim, J.Y., Open-Source Software for Satellite-Based Crop Health Monitoring. *J. Biosyst. Eng.*, 1–15, 2024.
30. Lakhiar, I.A., Yan, H., Zhang, C., Wang, G., He, B., Hao, B., Rakibuzzaman, M., A review of precision irrigation water-saving technology under changing climate for enhancing water use efficiency, crop yield, and environmental footprints. *Agriculture*, 14, 7, 1141, 2024.
31. Longo-Minnolo, G., Consoli, S., Vanella, D., Pappalardo, S., Guarrera, S., Manetto, G., Cerruto, E., Delineating citrus management zones using spatial interpolation and UAV-based multispectral approaches. *Comput. Electron. Agric.*, 222, 109098, 2024.
32. López, A., Ogayar, C.J., Feito, F.R., Sousa, J.J., Classification of grapevine varieties using UAV hyperspectral imaging. *Remote Sens.*, 16, 12, 2103, 2024.
33. Lubczynski, M.W., Leblanc, M., Batelaan, O., Remote sensing and hydrogeophysics give a new impetus to integrated hydrological models: A review. *J. Hydrol.*, 130901, 2024.
34. Lykhovyd, P.V. and Sharii, V.O., Normalised difference moisture index in water stress assessment of maize crops. *Agrology*, 7, 1, 21–26, 2024.
35. Lyu, R., Zhang, J., Pang, J., Zhang, J., Modeling the impacts of 2D/3D urban structure on PM2. 5 at high resolution by combining UAV multispectral/LiDAR measurements and multi-source remote sensing images. *J. Clean. Prod.*, 437, 140613, 2024.
36. Mane, S., Das, N., Singh, G., Cosh, M., Dong, Y., Advancements in dielectric soil moisture sensor Calibration: A comprehensive review of methods and techniques. *Comput. Electron. Agric.*, 218, 108686, 2024.
37. McDermid, S., Nocco, M., Lawston-Parker, P., Keune, J., Pokhrel, Y., Jain, M., Yokohata, T., Irrigation in the Earth system. *Nat. Rev. Earth Environ.*, 4, 7, 435–453, 2023.
38. Medauar, C.C., Silva, S.D.A., Carvalho, L.C.C., Moraes, W.B., Masson, M.V., Martins, R.P., Hott, J.M.C., Zoning the occurrence of Thyrinteina arnobia in eucalypts and its relationship with the spectral behavior of canopies and soil attributes. *South. For.*, 86, 1, 53–64, 2024.
39. Mehedi, I.M., Hanif, M.S., Bilal, M., Vellingiri, M.T., Palaniswamy, T., Remote Sensing and Decision Support System Applications in Precision Agriculture: Challenges and Possibilities. *IEEE Access*, 2024.

40. Mehmood, K., Anees, S.A., Muhammad, S., Hussain, K., Shahzad, F., Liu, Q., Khan, W.R., Analyzing vegetation health dynamics across seasons and regions through NDVI and climatic variables. *Sci. Rep.*, 14, 1, 11775, 2024.
41. Muhammad, M.B., Alwan, S.I., Raddad, M.Q., Hamidi, M.Q., Owaid, Z.S., Innovation in Agriculture Advantages of a Remote Irrigation Monitoring and Control System. *South Asian Res. J. Eng. Tech.*, 6, 5, 110–123, 2024.
42. Navyasree, P., Optimized Soil Adjusted Vegetation Index Mapping of Pune District using Google Earth Engine, 2024.
43. Nsoh, B., Katimbo, A., Guo, H., Heeren, D.M., Nakabuye, H.N., Qiao, X., Kiraga, S., Internet of Things-Based Automated Solutions Utilizing Machine Learning for Smart and Real-Time Irrigation Management: A Review. *Sensors (basel, Switzerland)*, 24, 23, 7480, 2024.
44. Ogawa, T., Tamaki, M., Usui, T., Hikosaka, K., Hyperspectral image extraction to evaluate the photosynthetic and stress status of plants, using a photochemical reflectance index (PRI). *Sci. Hortic.*, 336, 113349, 2024.
45. Parida, P.K., Eagan, S., Ramanujam, K., Sengodan, R., Uthandi, S., Ettiyagounder, P., Rajagounder, R., Machine learning approaches for estimation of the fraction of absorbed photosynthetically active radiation and net photosynthesis rate of maize using multi-spectral sensor. *Heliyon*, 10, 13, 2024.
46. Pirbasti, M.A., McArdle, G., Akbari, V., Hedgerows Monitoring in Remote Sensing: A Comprehensive Review. *IEEE Access*, 2024.
47. Raihan, A., A comprehensive review of the recent advancement in integrating deep learning with geographic information systems. *Research Briefs on Information and Communication Technology Evolution*, 9, 98–115, 2023.
48. Rezaee, L., Davatgar, N., Moosavi, A.A., Sepaskhah, A.R., Implications of spatial variability of soil physical attributes in delineating site-specific irrigation management zones for rice crop. *J. Soil Sci. Plant Nutr.*, 23, 4, 6596–6611, 2023.
49. Risal, A., Niu, H., Landivar-Scott, J.L., Maeda, M.M., Bednarz, C.W., Landivar-Bowles, J., Bhandari, M., Improving Irrigation Management of Cotton with Small Unmanned Aerial Vehicle (UAV) in Texas High Plains. *Water*, 16, 9, 1300, 2024.
50. Sangeetha, C., Moond, V., Raje7sh, G.M., Damor, J.S., Pandey, S.K., Kumar, P., Singh, B., Remote Sensing and Geographic Information Systems for Precision Agriculture: A Review. *Int. J. Environ. Clim. Change*, 14, 2, 287–309, 2024.
51. Sapitang, M., Dullah, H., Latif, S.D., Ng, J.L., Huang, Y.F., Malek, M.B.A., Ahmed, A.N., Application of Integrated Artificial Intelligence Geographical Information System in Managing Water Resources: A Review. *Remote Sens. Appl.: Soc. Environ.*, 101236, 2024.
52. Shaikh, M. and Birajdar, F., Advancements in remote sensing and GIS for sustainable groundwater monitoring: applications, challenges, and future directions. *Int. J. Res. Eng. Sci. Manag.*, 7, 3, 16–24, 2024.

53. Sharma, A., Prakash, A., Bhambota, S., Kumar, S., Investigations of precision agriculture technologies with application to developing countries. *Environ. Dev. Sustainability*, 1–37, 2024.
54. Sharma, S., Beslity, J.O., Rustad, L., Shelby, L.J., Manos, P.T., Khanal, P., Khanal, C., *Remote Sensing* and GIS in Natural Resource Management: Comparing tools and emphasizing the importance of *in-situ* data. *Remote Sens.*, 16, 22, 4161, 2024.
55. Sharma, V., Honkavaara, E., Hayden, M., Kant, S., UAV remote sensing phenotyping of wheat collection for response to water stress and yield prediction using machine learning. *Plant Stress*, 12, 100464, 2024.
56. Shekede, M.D., Gwitira, I., Chemura, A., Climate variability and agriculture, in: *Remote Sensing of Climate*, pp. 51–83, Elsevier, 2024, doi: 10.1016/b978-0-443-21731-9.00010-7.
57. Song, C., *et al.*, Ecological Characterization of Vegetation Using Multi-Sensor Remote Sensing in the Solar Reflective Spectrum. *Remote Sensing Handbook*, vol. IV, pp. 249–308, CRC Press, Oct. 30, 2024. doi: 10.1201/9781003541172-10.
58. Tan, J., Ding, J., Wang, Z., Han, L., Wang, X., Li, Y., Hong, Y., Estimating soil salinity in mulched cotton fields using UAV-based hyperspectral remote sensing and a Seagull Optimization Algorithm-Enhanced Random Forest Model. *Comput. Electron. Agric.*, 221, 109017, 2024.
59. Toromade, A.S. and Chiekezie, N.R., GIS-driven agriculture: Pioneering precision farming and promoting sustainable agricultural practices. *World J. Adv. Sci. Technol.*, 6, 1, 2024.
60. Tunca, E., Evaluating the performance of the TSEB model for sorghum evapotranspiration estimation using time series UAV imagery. *Irrig. Sci.*, 42, 5, 977–994, 2024.
61. Van de Zande, G.D., *Bringing the Water-Efficiency Benefits of Precision Irrigation to Resource-Constrained Farms Through an Automatic Scheduling-Manual Operation Irrigation Tool,* Doctoral dissertation, Massachusetts Institute of Technology, 2023.
62. Veeramanju, K.T., Predictive Models for Optimal Irrigation Scheduling and water management: A Review of AI and ML Approaches. *Int. J. Manag. Technol. Soc. Sci.*, 9, 2, 94–110, 2024.
63. Vinci, A., Baiocchi, V., Brigante, R., Traini, C., Farinelli, D., Use of Drones with Multispectral and Thermal Cameras to Assess the Biometric Characteristics and Water Status of Different Hazelnut Cultivars, 2024.
64. Waczak, J., Aker, A., Wijeratne, L.O., Talebi, S., Fernando, A., Dewage, P.M., Lary, D.J., Unsupervised Characterization of Water Composition with UAV-Based Hyperspectral Imaging and Generative Topographic Mapping. *Remote Sens.*, 16, 13, 2430, 2024.
65. Wanyama, J., Bwambale, E., Kiraga, S., Katimbo, A., Nakawuka, P., Kabenge, I., Oluk, I., A systematic review of fourth industrial revolution technologies

in smart irrigation: constraints, opportunities, and future prospects for sub-Saharan Africa. *Smart Agric. Technol.*, 100412, 2024.

66. Wei, H., Xu, W., Kang, B., Eisner, R., Muleke, A., Rodriguez, D., Harrison, M.T., Irrigation with Artificial Intelligence: Problems, Premises, Promises. *Hum. Centric Intell. Syst.*, 1–19, 2024.
67. Wei, J., Dong, W., Liu, S., Song, L., Zhou, J., Xu, Z., Sun, J., Mapping super high resolution evapotranspiration in oasis-desert areas using UAV multi-sensor data. *Agric. Water Manage.*, 287, 108466, 2023.
68. Xiao, J., Qiao, J., Han, D., Ma, Y., Zhu, Q., Wang, W., Spatial distribution and transformation mechanism of specialized villages in typical agricultural areas: Case study of Henan province, China. *Habitat Int.*, 146, 103034, 2024.
69. Xing, F., An, R., Guo, X., Shen, X., Mapping invasive noxious weed species in the alpine grassland ecosystems using very high spatial resolution UAV hyperspectral imagery and a novel deep learning model. *Gisci. Remote Sens.*, 61, 1, 2327146, 2024.
70. Xu, Y., Smirnov, M., Kohler, M.C., Dong, Z., Amineh, R.K., Li, F., Rojas-Cessa, R., A Survey of Wireless Soil Sensing Technologies. *IEEE Access*, 2024.
71. Yasir, Q.M., Zhang, Z., Ren, J., Wang, G., Naveed, M., Jahangir, Z., Rahman, A.U., Spectral index for estimating leaf water content across diverse plant species using multiple viewing angles. *J. Appl. Remote Sens.*, 18, 4, 042603–042603, 2024.
72. Zahir, S.A.D.M., Jamlos, M.F., Omar, A.F., Jamlos, M.A., Mamat, R., Muncan, J., Tsenkova, R., Review–Plant nutritional status analysis employing the visible and near-infrared spectroscopy spectral sensor. *Spectrochim. Acta A Mol. Biomol. Spectrosc.*, 304, 123273, 2024.
73. Zhao, W., Wang, M., Pham, V.T., Unmanned aerial vehicle and geospatial analysis in smart irrigation and crop monitoring on IoT platform. *Mob. Inf. Syst.*, 2023, 1, 4213645, 2023.
74. Zhu, S., Cui, N., Jin, H., Jin, X., Guo, L., Jiang, S., Wang, M., Optimization of multi-dimensional indices for kiwifruit orchard soil moisture content estimation using UAV and ground multi-sensors. *Agric. Water Manage.*, 294, 108705, 2024.
75. Zhu, X., Li, Q., Guo, C., Evaluation of the monitoring capability of various vegetation indices and mainstream satellite band settings for grassland drought. *Ecol. Inf.*, 82, 102717, 2024.

7

Integrating SAR and Multispectral Data with HSV-Based Fusion for Accurate Land Cover Classification

Narayan Vyas[1,2]

[1]*Department of Computer Science and Engineering, Vivekananda Global University, Jaipur, Rajasthan, India*
[2]*Centre of Excellence in Geospatial Intelligence for Climate Studies, Vivekananda Global University, Jaipur, Rajasthan, India*

Abstract

Resource monitoring, agricultural planning, and sustainable environmental management depend on accurate land use land cover (LULC) classification. The integration of multi-sensor satellite data has significantly enhanced the precision and reliability of these classifications. This study proposes a Hue, Saturation, Value (HSV)-based fusion approach for integrating Sentinel-1 Synthetic Aperture Radar (SAR) data with Sentinel-2 multispectral imagery to generate a pan-sharpened image for LULC mapping in the Doaba region of Punjab, India. Data spanning from 1 to 28 February 2025, were accessed and processed using Google Earth Engine (GEE). A median composite of Sentinel-2 bands B11 (SWIR), B8 (NIR), and B2 (Blue) was created and converted to the HSV color space. The fused dataset was classified using a support vector machine (SVM) trained on manually labeled samples. The approach achieved high accuracy, with an overall classification accuracy of 96.77% and a kappa coefficient of 95.69%, indicating strong model performance. In addition, visual validation against high-resolution base maps confirmed the spatial coherence and thematic integrity of the classified outputs. This study demonstrates the potential of HSV-based fusion for improving land cover classification and offers a scalable methodology for multisensor analysis in heterogeneous agricultural landscapes.

Keywords: Environmental management, Sentinel-1, classification, Sentinel-2, fusion, agriculture

Email: narayanvyas87@gmail.com; ORCID: 0000-0003-4225-0764

Gurwinder Singh, Vishakha Sood, Narayan Vyas, Surendra Yadav and Dankan Gowda V. (eds.)
Geospatial Intelligence for Precision Agriculture, (141–158) © 2026 Scrivener Publishing LLC

7.1 Introduction

Earth Observation (EO) provides critical spatial and temporal data for monitoring the Earth's surface and atmosphere, supporting data-driven decisions [1]. Remote sensing refers to the collection and analysis of geospatial data from various platforms, including satellites, aerial sensors, and unmanned aerial vehicles (UAVs) [2]. The greatest strength of EO is its capacity to track natural and anthropogenic changes over large spatial extents and then repeat that observation over a range of time scales [3]. EO data are critical for monitoring climate change, deforestation, urban development, agriculture, water resources, and disaster risk reduction [4]. There are many categories of EO applications in several disciplines. For example, in agriculture, EO can contribute to monitoring crop health, scheduling irrigation, forecasting yields, and identifying pests or disease outbreaks [5].

Recent studies have shown that combining multi-sensor imagery can improve the classification of land cover and the identification of surface changes over time [6–8]. The synergetic combination of Sentinel-1 and Sentinel-2 data provides greater spatial detail and spectral resolution, resulting in improved thematic maps [9]. For classification tasks, the complementary nature of radar and optical imagery is enhanced by the use of machine learning (ML)-based algorithms, such as support vector machines (SVM) [10], random forests (RF) [11], and Naïve Bayes [12], particularly when distinguishing spectrally similar classes (e.g., urban–rural fringes or mixed vegetation). Finally, fused datasets are particularly useful for detecting seasonal changes, facilitating effective monitoring of seasonal variations in agricultural fields, deforestation, and water bodies [13]. Radar is not affected by weather and enables near-continuous monitoring of a given area, whereas optical data contain valuable spectral information [14]. Deep learning (DL)-based techniques are also gaining attention from researchers worldwide for EO-based applications [15–18].

Fusing satellite sensor data improves land use land cover (LULC) classification accuracy, attracting significant research interest [19]. The combination of Sentinel-1 (SAR) and Sentinel-2 (optical) imagery is particularly useful in the context of EO [16]. Advanced pan-sharpening approaches, among which the Hue, Saturation, Value (HSV) color space-based method is the most effective, are applied to maximize the spatial characteristics of SAR and spectral richness of optical data [20]. In this approach, the optical image with B11, B8, and B2 bands is transformed from the RGB color space to the HSV color space [21]. The original "Hue" and "Saturation" channels in HSV maintain the color information; therefore, to obtain

Table 7.1 Comparative study of fusion-based techniques for geospatial analysis.

Dataset	Methodology	Results	Limitations	Application	Source
GF-5 (HSI) + S2 (MSI)	SFIM fusion + ViT-DGCN	OA: 97%; robust with only 1% labeled data	HSI has low spatial resolution; ViT is compute-heavy	Lithological classification (Cuonadong Dome, Tibet)	[22]
S2 (MSI) + Vision-1	Image fusion + MLP-ANN, SVM, RF classifiers	OA: 98.5%, κ = 0.97; stable class separation	Vision-1 data not free; expensive fusion processing	Forest damage mapping	[23]
S2 (MSI) + PlanetScope (MSI)	NDVI-based fusion + Feedforward ANN	Yield error reduced to 0.2 t/ha–0.5 t/ha	Cross-sensor inconsistency, high computational cost	Soybean yield prediction (Hungary fields)	[24]

(*Continued*)

Table 7.1 Comparative study of fusion-based techniques for geospatial analysis. (*Continued*)

Dataset	Methodology	Results	Limitations	Application	Source
S1 (SAR) + S2 (MSI) + Landsat 8/9	Temporal thresholding fusion	κ: 0.80–0.95; revisit rate 3.5–7 days	Lower accuracy in dense vegetation	Surface water mapping (Global scale)	[25]
S1 (VV, SAR) + S2 (13 bands, MSI)	CVTNet (CNN + ViT) with attention layers	Outperformed ViT, RF, HybridSN on test samples	Bog–fen confusion; swamp area misclassification	Wetland classification (Newfoundland, Canada)	[26]
S1 (VV, VH) + S2 (L2A MSI)	GS, PCS fusion + SVM, MLC, MDC (FPCD framework)	OA = 96.5%, κ = 0.957; stable change maps	Spectral loss from GS and PCS fusion methods	Multitemporal agriculture monitoring (Punjab, India)	[27]

GF-5, Gaofen-5 (Hyperspectral satellite); HSI, Hyperspectral Image; S2, Sentinel-2; MSI, Multispectral Image; SFIM, Smoothing Filter-based Intensity Modulation; ViT, Vision Transformer; DGCN, Deep Graph Convolutional Network; OA, Overall Accuracy; S1, Sentinel-1; SAR, Synthetic Aperture Radar; Vision-1, High-resolution commercial optical satellite; MLP-ANN, Multi-layer Perceptron Artificial Neural Network; SVM, Support Vector Machine; RF, Random Forest; κ, Kappa Coefficient; NDVI, Normalized Difference Vegetation Index; ANN, Artificial Neural Network; t/ha, Tons per hectare; Landsat 8/9; USGS optical satellite missions; CVTNet, Convolutional Vision Transformer Network; CNN, Convolutional Neural Network; HybridSN, Hybrid Spectral Net; VV/VH, Vertical transmit/Vertical receive and Vertical transmit/Horizontal receive polarizations in SAR; L2A, Level-2A (Atmospherically corrected surface reflectance product); GS, Gram–Schmidt fusion; PCS, Principal Component Spectral fusion; MLC, Maximum Likelihood Classifier; MDC, Minimum Distance Classifier; FPCD, Fusion-based Post-classification Change Detection.

high-resolution simulated color images, the "Value" (brightness) channel can be replaced with a high-resolution intensity image, which in this case was the normalized Sentinel-1 VV backscatter. Replacing the "Value" channel with the SAR image (after brightness adjustment) and converting it back into the RGB space yields an image that preserves the original color semantics and adds spatial details. The fused image is clearer and enables better discrimination of land features, particularly in complex or heterogeneous environments, such as agricultural landscapes found within interspersed urban and forest patches. Table 7.1 provides a comprehensive literature review of recent advancements in geospatial analysis using multisensor data fusion.

This study utilized the Google Earth Engine (GEE) to demonstrate a scalable methodology for integrating Sentinel-1 and Sentinel-2 data for LULC classification. The specific objectives of this study are as follows: (a) To implement a fusion methodology using HSV-based pan sharpening to integrate Sentinel-1 SAR data with Sentinel-2 multispectral imagery, thereby generating a high-resolution, spectrally rich image for land surface analysis. (b) The fused image dataset was applied for supervised LULC classification using an SVM classifier, categorizing the landscape into several classes. (c) Evaluate the performance of the classification process using both quantitative accuracy metrics and qualitative visual interpretation through overlay analysis on high-resolution base maps. By achieving these objectives, this study seeks to highlight the operational benefits of sensor fusion in remote sensing applications and provide a workflow that can be replicated for similar agro-ecological landscapes.

7.2 Study Area and Satellite Dataset

7.2.1 Study Area

This study focuses on the Doaba region, a fertile tract of land situated between the Beas and Sutlej rivers in the Indian state of Punjab [28], as shown in Figure 7.1. The word "Doaba" is of Persian origin, meaning "two waters," denoting the interfluvial zone of the region. Geographically, Doaba is situated between latitudes 30°57′ to 32°7′ N and longitudes 75°4′ to 76°30′ E, covering an area of approximately 8,915 km^2. It encompasses all the districts of Jalandhar, Hoshiarpur, Kapurthala, and Shaheed Bhagat Singh Nagar (formerly, Nawanshahr). Jalandhar is an important city known for sports equipment, and Hoshiarpur has high literacy rates.

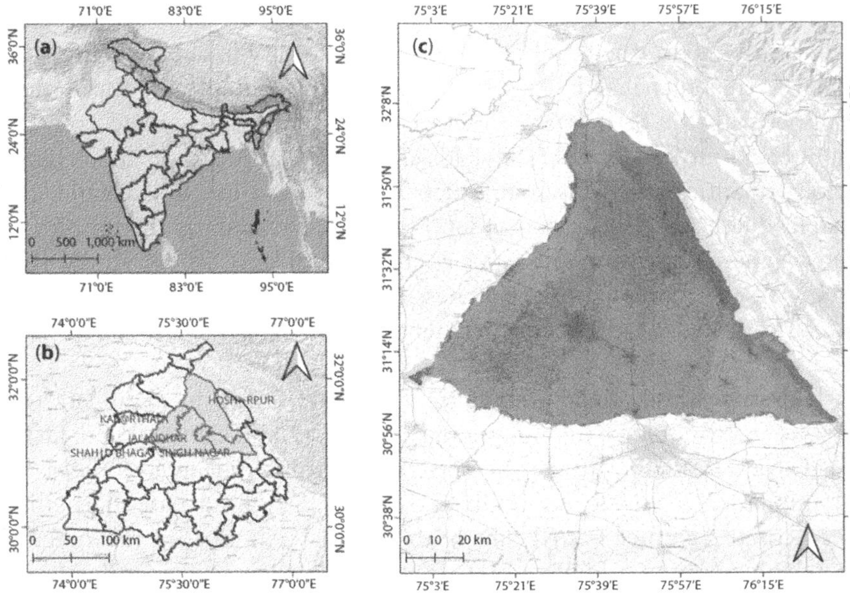

Figure 7.1 Study area where (a) shows a map of India highlighting Punjab state, (b) shows a map of Punjab highlighting the study area, and (c) shows Sentinel-2 satellite imagery of the study area.

7.2.2 Satellite Dataset

This study utilized multispectral and radar satellite data from the Sentinel mission to create a high-resolution pan-sharpened land classification map for the study area. Sentinel-1-based free datasets were obtained using GEE. GEE is an efficient cloud-based geospatial processing platform that provides free access to a wide variety of satellite datasets from which the data were obtained [29]. Imagery of the Sentinel-2 Surface Reflectance (S2_SR) product was used for optical data. The analysis date range was the Rabi cropping season in Punjab, from 1 February to 28 February 2025. To reduce the effects of ephemeral cloud cover and atmospheric noise, a median composite was created from the filtered images by applying the *median()* reducer function in GEE. It provides a representative image by averaging multiple observations throughout the period and selecting the median value for every pixel. For band combinations, a false-color composite of the shortwave infrared (SWIR) band B11, near-infrared (NIR) band B8, and band B2 was utilized [30]. The bands were rescaled, normalized, and

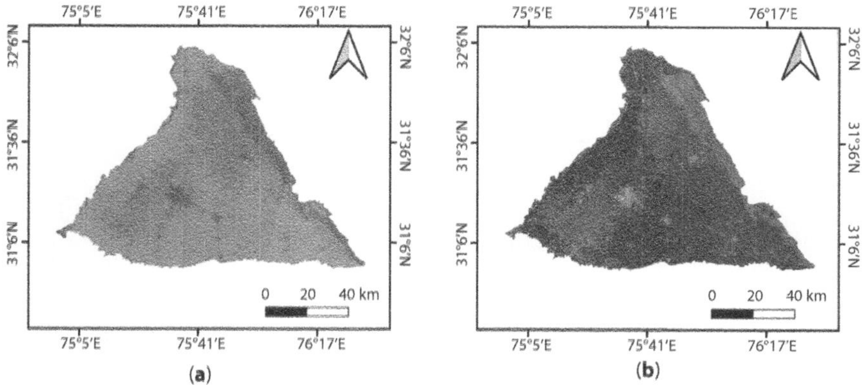

Figure 7.2 Input images where (a) shows Sentinel-2 imagery, and (b) shows Sentinel-1 imagery.

transformed into the HSV color space prior to fusion. The fusion improved the spatial and spectral quality of the optical image, facilitating accurate SVM-based classification. The input satellite images are shown in Figure 7.2.

7.3 Methodology

This study presents a methodological framework that utilizes HSV-based fusion to integrate multisensor satellite data to enhance LULC classification. The workflow is based on the cloud-based processing power offered by GEE and includes four main steps: preprocessing, data fusion, classification, and accuracy assessment. The Sentinel-1 and Sentinel-2 datasets were selected for their complementary capabilities: the SAR backscatter from Sentinel-1 provides precise structural information, whereas the multispectral image data from Sentinel-2 offers rich spectral information [31]. HSV pan-sharpening fused spatially precise SAR data with spectrally rich optical data. This was followed by supervised classification using an SVM classifier, and the classification accuracy was verified both, statistically and visually. The workflow shown in Figure 7.3 offers a scalable and reproducible method for detailed land cover mapping in complex agricultural regions.

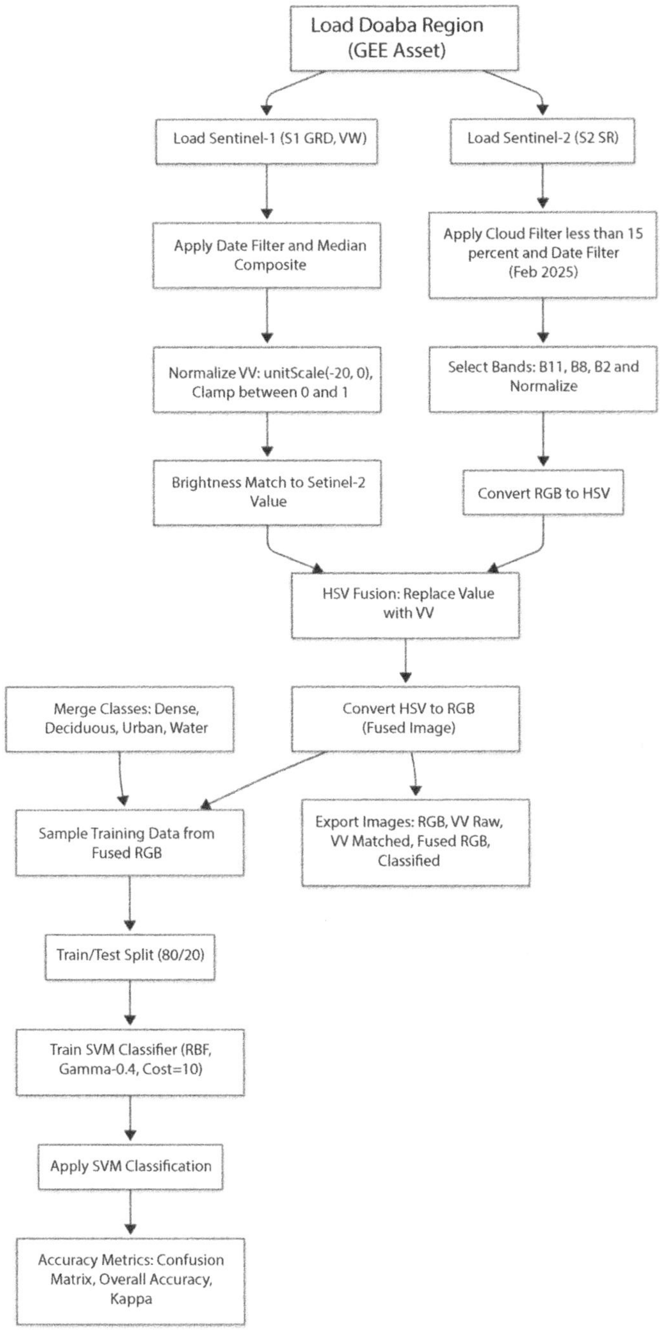

Figure 7.3 Methodology of the proposed framework.

7.3.1 Preprocessing

The preprocessing phase consists of acquiring, filtering, and preparing optical and radar data. A user-uploaded asset defined the region of interest (ROI) as the Doaba region in Punjab, India. The Sentinel-2 Surface Reflectance (S2_SR) image collection was filtered to obtain optical imagery of the area of interest. To ensure that the images were not significantly affected by clouds, filtered images were created by setting a threshold for the percentage of cloudy pixels to be less than 15% [32]. This was then used to generate a median composite from the filtered data. Three spectral bands (B11 (SWIR), B8 (NIR), and B2 (Blue)) were selected from this composite. This band combination is useful for discriminating urban structures, vegetation stress, and soil moisture. The bands were originally rescaled to reflectance values by multiplying them by 10,000. The Sentinel-1 Ground Range Detected (GRD) collection was used for the radar imagery. This was filtered for the same date range and spatial extent as Sentinel-2 and additionally filtered for images with VV polarization in the IW mode. The VV band was calculated as a median composite and was rescaled. The *unitScale(-20, 0)* standardizes the backscatter to the 0–1 range.

7.3.2 Satellite Data Fusion

The fusion process aimed to enhance the spatial details of the multispectral image by integrating the radar-derived texture information. This was achieved using a pan-sharpening technique based on the HSV transformation. The Sentinel-2 RGB composite (B11, B8, and B2) was transformed from RGB to HSV color space using GEE's *.rgbToHsv()* method. The "value" channel, which corresponds to brightness, was extracted from the HSV image. To match the radar brightness to the optical image, the mean of the Sentinel-2 "value" channel was computed over the entire region using *.reduceRegion()* with a mean reducer. This scalar mean value was then multiplied by the normalized Sentinel-1 VV backscatter image to produce a VV-matched image, ensuring that the radar brightness approximately matched the median optical brightness of the image. The fused HSV image was constructed by combining the "hue" and "saturation" channels from the Sentinel-2 HSV image, and the VV-matched backscatter was the new "value" channel. The resulting image was converted back to RGB using the *.hsvToRgb()*, generating a pan-sharpened, fused RGB image that combines the spectral richness from Sentinel-2 and spatial texture from Sentinel-1.

7.3.3 LULC Classification

LULC classification was performed using a supervised approach based on an SVM classifier with a radial basis function (RBF) kernel [33]. The fused RGB image was sampled using a merged ground-truth feature collection consisting of four land cover types: dense vegetation, deciduous vegetation, urban, and water. Each class polygon has an associated class property representing integer-coded labels. The samples were extracted using *.sampleRegions()* at a 10-meter scale. The sampled data were randomly split into training (80%) and testing (20%) subsets using a *.randomColumn()* followed by filtering the data. The classifier was trained using the SVM algorithm with an RBF kernel, where the gamma parameter was set to 0.4 and the cost parameter to 10, reflecting the balance between the margin width and misclassification tolerance. The input properties for the model were the RGB bands of the fused image, and the model was trained on the class property. After training, the model was applied to classify the entire fused image. The classified image was then visualized using custom color palettes corresponding to the four LULC classes and exported for further analysis.

7.3.4 Accuracy Assessment

Accuracy of LULC classification is assessed using both quantitative metrics (confusion matrix) and qualitative visual analysis. For quantitative assessment, a classic confusion matrix was utilized with a test image set taken from the fused RGB image. The dataset was randomly extracted from 20% of all sample points that had been previously digitized and manually labeled among the four LULC classes. We then used the *errorMatrix()* function, which produces a confusion matrix that shows the number of samples correctly and incorrectly classified in the respective classes. Two important evaluation metrics were derived from this matrix: Overall Accuracy (i.e., the percentage of all correctly classified samples) and Kappa (i.e., a measure of classification performance adjusted for any agreement that occurs by chance) [34]. In addition to the numeric metrics, the classified thematic map was visually inspected. This involved superimposing the classified output on Google Earth, retrieving high-resolution imagery, and visually comparing the spatial distribution, class partitioning, and known land features with actual field observations. This step was essential to confirm the dependability of the classification, mainly in complex areas such as urban-agricultural edges and water-surface cover changes. A visual inspection of the thematic map, fused image, and intermediate layers (VV-matched, raw VV, and original RGB) was conducted to confirm the semantic congruency. Thus, by combining statistical and visual validation, dual validation confirmed that the classification had both technical accuracy and contextual value.

7.4 Results

7.4.1 Quantitative Analysis of Classified Maps

The LULC classification performance was assessed using a confusion matrix derived from a test dataset representing 20% of the ground truth samples. The classification employed an SVM with an RBF kernel on a fused Sentinel-1 and Sentinel-2 image using HSV-based integration. The confusion matrix shown in Figure 7.4 reveals a strong performance across all classes. Dense and deciduous vegetation were perfectly classified, with seven correct predictions each. The urban class had eight correct classifications, with one misclassified as deciduous vegetation, whereas the water class was classified with complete accuracy, yielding eight correct

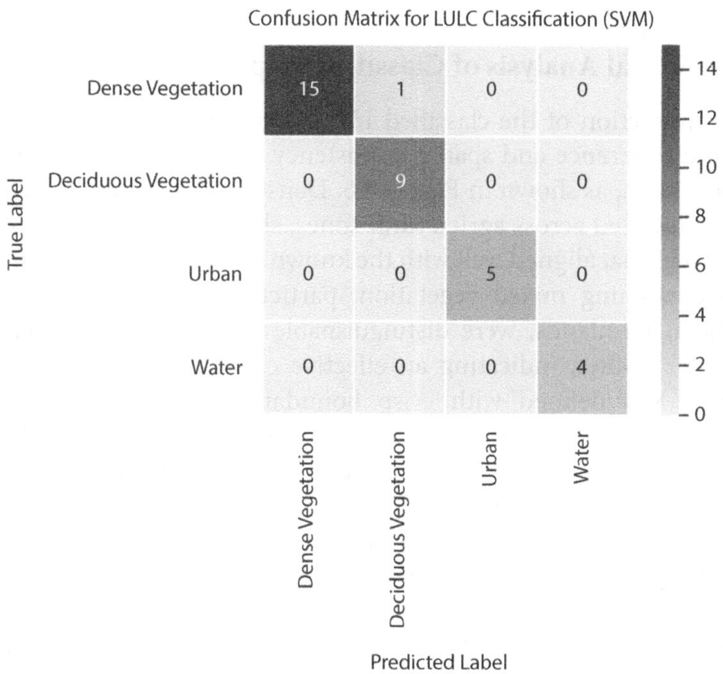

Figure 7.4 Confusion matrix of the classified map.

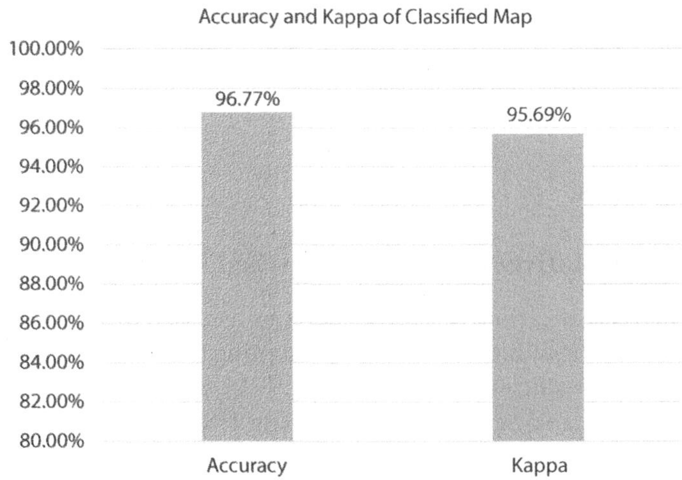

Figure 7.5 Accuracy and Kappa values of the classified map.

predictions. The overall classification accuracy was 96.77%, and the Kappa coefficient was 95.69%, indicating an almost perfect agreement beyond chance (Figure 7.5).

7.4.2 Visual Analysis of Classified Maps

Visual inspection of the classified LULC map revealed a high degree of thematic coherence and spatial consistency with the actual landscape of the study area, as shown in Figure 7.6. Dense vegetation areas were accurately delineated across agricultural zones, showing smooth and continuous patches that aligned well with the known cultivation belts. Transitional zones containing mixed vegetation, particularly along riverbanks and peripheral farmlands, were distinguishable from both dense vegetation and water bodies, indicating an effective class separation. Urban areas appeared well-defined with crisp boundaries and minimal classification noise, benefiting from the enhanced spatial texture introduced by Sentinel-1 VV backscatter. Water bodies, such as canals and river channels, were distinctly mapped with no significant spectral overlap with the surrounding vegetation or built-up areas. The map exhibited minimal pixel-level misclassifications, with clear class boundaries validated against high-resolution Google Earth imagery. These observations support the reliability of the fused image and classification methodology in capturing land-cover heterogeneity across the region.

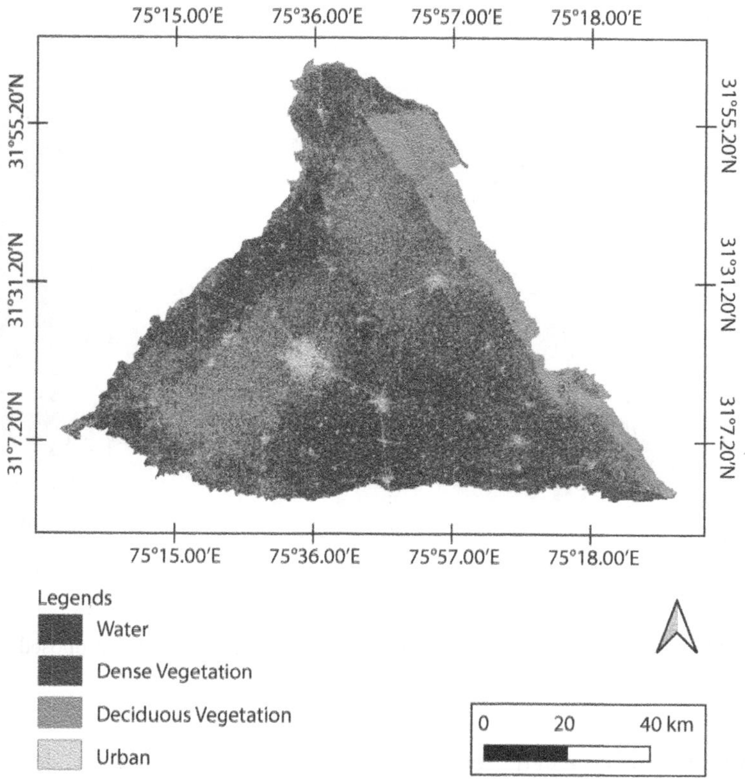

Figure 7.6 Classified image using the SVM algorithm.

7.5 Discussion

The fusion-based LULC classification presented in this study demonstrates the utility of integrating Sentinel-1 SAR and Sentinel-2 multispectral data through HSV-based pan-sharpening for improved thematic mapping in LULC. Figure 7.7 provides a comparative visual assessment across three sub-areas within the Doaba region, highlighting the outputs from Sentinel-2, Sentinel-1, fused imagery, classified maps, and the corresponding high-resolution Google Earth imagery. Sentinel-2 offers detailed spectral data, whereas Sentinel-1 enhances the spatial resolution in urban and diverse agricultural areas. For instance, in Small Area-1, the central urban cluster was better delineated in the fused image and classified map than in the raw Sentinel-2 imagery alone. Similarly, Small Area-2 accurately identified vegetation patches and water bodies, benefiting from the spatial clarity provided by the radar backscatter. Small Area-3, a more agriculturally

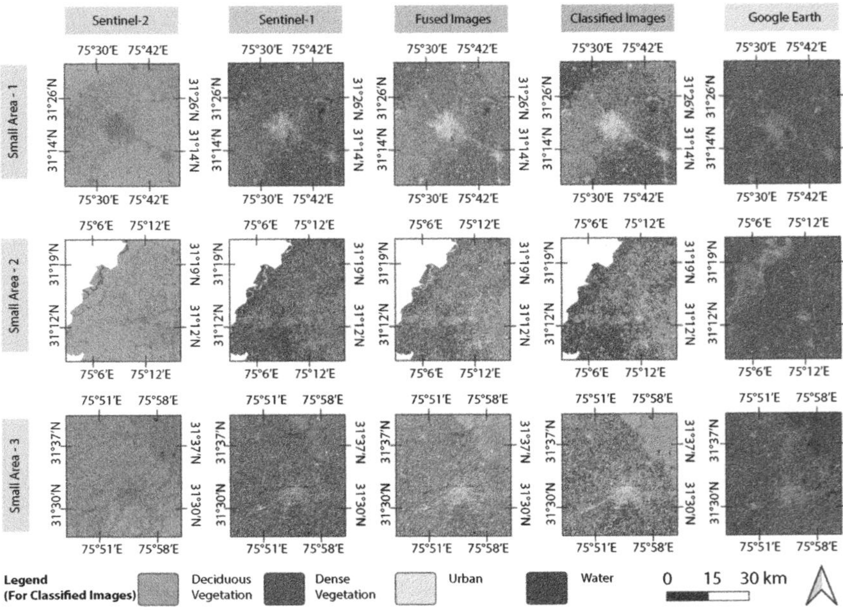

Figure 7.7 Small area of the study is to showcase the effectiveness of the proposed framework.

dominant zone, revealed precise boundaries between dense and deciduous vegetation, showcasing the advantage of HSV-based integration in resolving subtle intraclass variations. The classified maps exhibited high spatial fidelity, with well-defined class boundaries and minimal spectral confusion, validated through visual comparison with Google Earth imagery.

7.6 Conclusion

This study demonstrated the effectiveness of HSV-based data fusion for enhancing LULC classification by integrating optical and radar imagery from Sentinel-2 and Sentinel-1, respectively. By employing a median composite of cloud-free Sentinel-2 images and combining it with Sentinel-1 VV polarization backscatter through the HSV pan-sharpening technique, a spectrally rich and spatially detailed fused image was created. This approach addressed the limitations associated with individual sensors, specifically, the lack of spatial detail in optical imagery and the absence of spectral diversity in radar data, by leveraging their complementary strengths. The fused imagery was classified using an SVM classifier, which effectively categorized the Doaba region into four major land cover types: dense and deciduous vegetation, urban areas, and water bodies. The classification achieved a high overall accuracy and kappa score, validated both quantitatively through a confusion matrix and qualitatively *via* visual inspection against high-resolution base images. The classification output displayed strong spatial coherence and thematic accuracy, particularly in complex zones, such as urban–rural interfaces and fragmented vegetation patterns. The study showed that HSV-based fusion boosts classification accuracy by integrating multisensor data.

AI disclosure statement: The author acknowledges the use of AI-based tools for minor language editing or grammatical correction. However, all intellectual contributions, analyses, and conclusions presented in this chapter are entirely the author's own.

References

1. Zhao, Q., *et al.*, An Overview of the Applications of Earth Observation Satellite Data: Impacts and Future Trends. *Remote Sens.*, 14, 8, Jan. 2022, doi:10.3390/rs14081863.
2. Jafarbiglu, H. and Pourreza, A., A comprehensive review of remote sensing platforms, sensors, and applications in nut crops. *Comput. Electron. Agric.*, 197, 106844, Jun. 2022, doi:10.1016/j.compag.2022.106844.
3. Ozigis, M., *et al.*, Utility of Earth Observation data in mapping post-disaster impact: A case of Hurricane Dorian in the Bahamas. *Remote Sens. Appl. Soc. Environ.*, 37, 101466, Jan. 2025, doi:10.1016/j.rsase.2025.101466.
4. Anderson, K., Ryan, B., Sonntag, W., Kavvada, A., L., Friedl, Earth observation in service of the 2030 Agenda for Sustainable Development. *Geo-spatial Inf. Sci.*, 20, 2, 77–96, Apr. 2017, doi:10.1080/10095020.2017.1333230.
5. Aziz, D., *et al.*, Remote sensing and artificial intelligence: revolutionizing pest management in agriculture. *Front. Sustainable Food Syst.*, 9, Feb. 2025, doi:10.3389/fsufs.2025.1551460.
6. Prudente, V.H.R., *et al.*, Multisensor approach to land use and land cover mapping in Brazilian Amazon. *ISPRS J. Photogramm. Remote Sens.*, 189, 95–109, Jul. 2022, doi:10.1016/j.isprsjprs.2022.04.025.
7. Wang, B., Choi, J., Choi, S., Lee, S., Wu, P., Gao, Y., Image Fusion-Based Land Cover Change Detection Using Multi-Temporal High-Resolution Satellite Images. *Remote Sens.*, 9, 8, Aug. 2017, doi:10.3390/rs9080804.
8. Aljenaid, S.S., Kadhem, G.R., AlKhuzaei, M.F., Alam, J.B., Detecting and Assessing the Spatio-Temporal Land Use Land Cover Changes of Bahrain Island During 1986–2020 Using Remote Sensing and GIS. *Earth Syst. Environ.*, 6, 4, 787–802, Dec. 2022, doi:10.1007/s41748–022-00315-z.

9. Fang, G., Xu, H., Yang, S.-I., Lou, X., Fang, L., Synergistic use of Sentinel-1, Sentinel-2, and Landsat 8 in predicting forest variables. *Ecol. Indic.*, 151, 110296, Jul. 2023, doi:10.1016/j.ecolind.2023.110296.
10. Awad, M. and Khanna, R., Support Vector Machines for Classification, in: *Efficient Learning Machines: Theories, Concepts, and Applications for Engineers and System Designers*, M., Awad and R., Khanna, (Eds.), pp. 39–66, Apress, Berkeley, CA, 2015, doi:10.1007/978–1-4302–5990-9_3.
11. Belgiu, M. and Drăguţ, L., Random forest in remote sensing: A review of applications and future directions. *ISPRS J. Photogramm. Remote Sens.*, 114, 24–31, Apr. 2016, doi:10.1016/j.isprsjprs.2016.01.011.
12. Sharma, V., Mir, A.A., Sarwr, D.A., Detection of Rice Disease Using Bayes' Classifier and Minimum Distance Classifier. *J. Multimed. Inf. Syst.*, 7, 1, 17–24, 2020, doi:10.33851/JMIS.2020.7.1.17.
13. Zhu, Z., Qiu, S., Ye, S., Remote sensing of land change: A multifaceted perspective. *Remote Sens. Environ.*, 282, 113266, Dec. 2022, doi:10.1016/j.rse.2022.113266.
14. Joshi, N., *et al.*, A Review of the Application of Optical and Radar *Remote Sensing* Data Fusion to Land Use Mapping and Monitoring. *Remote Sens.*, 8, 1, Jan. 2016, doi:10.3390/rs8010070.
15. Adegun, A.A., Viriri, S., Tapamo, J.-R., Review of deep learning methods for remote sensing satellite images classification: experimental survey and comparative analysis. *J. Big Data*, 10, 1, 93, Jun. 2023, doi:10.1186/s40537–023-00772-x.
16. Adrian, J., Sagan, V., Maimaitijiang, M., Sentinel SAR-optical fusion for crop type mapping using deep learning and Google Earth Engine. *ISPRS J. Photogramm. Remote Sens.*, 175, 215–235, May 2021, doi:10.1016/j.isprsjprs.2021.02.018.
17. Bai, C., Bai, X., Wu, K., A Review: Remote Sensing Image Object Detection Algorithm Based on Deep Learning. *Electronics*, 12, 24, Jan. 2023, doi:10.3390/electronics12244902.
18. Ahmed, S.F., *et al.*, Deep learning modelling techniques: current progress, applications, advantages, and challenges. *Artif. Intell. Rev.*, 56, 11, 13521–13617, Nov. 2023, doi:10.1007/s10462–023-10466–8.
19. Manandhar, R., Odeh, I.O.A., Ancev, T., Improving the Accuracy of Land Use and Land Cover Classification of Landsat Data Using Post-Classification Enhancement. *Remote Sens.*, 1, 3, Sep. 2009, doi:10.3390/rs1030330.
20. Singh, S. and Tiwari, K.C., Exploring the optimal combination of image fusion and classification techniques. *Remote Sens. Appl. Soc. Environ.*, 24, 100642, Nov. 2021, doi:10.1016/j.rsase.2021.100642.
21. Kau, L.-J. and Lee, T.-L., An HSV Model-Based Approach for the Sharpening of Color Images, in: *2013 IEEE International Conference on Systems, Man, and Cybernetics*, pp. 150–155, Oct. 2013, doi:10.1109/SMC.2013.33.

22. Dong, Y., Yang, Z., Liu, Q., Zuo, R., Wang, Z., Fusion of GaoFen-5 and Sentinel-2B data for lithological mapping using vision transformer dynamic graph convolutional network. *Int. J. Appl. Earth Obs. Geoinf.*, 129, 103780, May 2024, doi:10.1016/j.jag.2024.103780.
23. Mateen, S., Nuthammachot, N., Techato, K., Random forest and artificial neural network-based tsunami forests classification using data fusion of Sentinel-2 and Airbus Vision-1 satellites: A case study of Garhi Chandan, Pakistan. *Open Geosci.*, 16, 1, Jan. 2024, doi:10.1515/geo-2022–0595.
24. Amankulova, K., *et al.*, A Novel Fusion Method for Soybean Yield Prediction Using Sentinel-2 and PlanetScope Imagery. *IEEE J. Sel. Top. Appl. Earth Obs. Remote Sens.*, 17, 13694–13707, 2024, doi:10.1109/JSTARS.2024.3402114.
25. Declaro, A. and Kanae, S., Enhancing Surface Water Monitoring through Multi-Satellite Data-Fusion of Landsat-8/9, Sentinel-2, and Sentinel-1 SAR. *Remote Sens.*, 16, 17, Jan. 2024, doi:10.3390/rs16173329.
26. Marjani, M., Mahdianpari, M., Mohammadimanesh, F., Gill, E.W., CVTNet: A Fusion of Convolutional Neural Networks and Vision Transformer for Wetland Mapping Using Sentinel-1 and Sentinel-2 Satellite Data. *Remote Sens.*, 16, 13, Jan. 2024, doi:10.3390/rs16132427.
27. Vyas, N., Singh, S., Sethi, G.K., A novel image fusion-based post classification framework for agricultural variations detection using Sentinel-1 and Sentinel-2 data. *Earth Sci. Inf.*, 18, 1, 40, Dec. 2024, doi:10.1007/s12145–024-01506–0.
28. Kaur, K., Sidhu, M.C., Ahluwalia, A.S., Angiosperm diversity in Doaba region of Punjab, India. *J. Threat. Taxa*, 9, 8, Aug. 2017, doi:10.11609/jott.2748.9.8.10551–10564.
29. Tamiminia, H., Salehi, B., Mahdianpari, M., Quackenbush, L., Adeli, S., Brisco, B., Google Earth Engine for geo-big data applications: A meta-analysis and systematic review. *ISPRS J. Photogramm. Remote Sens.*, 164, 152–170, Jun. 2020, doi:10.1016/j.isprsjprs.2020.04.001.
30. Efthimiou, N., Psomiadis, E., Papanikolaou, I., Soulis, K.X., Borrelli, P., Panagos, P., Developing a high-resolution land use/land cover map by upgrading CORINE's agricultural components using detailed national and pan-European datasets. *Geocarto Int.*, 37, 25, 10871–10906, Dec. 2022, doi:10.1080/10106049.2022.2041107.
31. Frantz, D., *et al.*, National-scale mapping of building height using Sentinel-1 and Sentinel-2 time series. *Remote Sens. Environ.*, 252, 112128, Jan. 2021, doi:10.1016/j.rse.2020.112128.
32. Prudente, V.H.R., Martins, V.S., Vieira, D.C., de F. e Silva, N.R., Adami, M., Sanches, I.D., Limitations of cloud cover for optical remote sensing of agricultural areas across South America. *Remote Sens. Appl. Soc. Environ.*, 20, 100414, Nov. 2020, doi:10.1016/j.rsase.2020.100414.

33. Khurana, Y. and Gupta, S., LCLU Classification using SVM, MLC and ANN of Multispectral Imagery from Sentinel-2. *Int. J. Eng. Res. Technol.*, 11, 10, Oct. 2022, doi:10.17577/IJERTV11IS100056.
34. Dahiya, N., Singh, S., Gupta, S., Comparative Analysis and Implication of Hyperion Hyperspectral and Landsat-8 Multispectral Dataset in Land Classification. *J. Indian Soc. Remote Sens.*, 51, 11, 2201–2213, Nov. 2023, doi:10.1007/s12524–023-01760–7.

8

Precision Agriculture through IoT and Geospatial Technologies for Sustainable Crop Management

Rupinder Singh*, Jaswinder Singh and Amanpreet Singh

Chitkara University Institute of Engineering and Technology, Chitkara University, Rajpura, Punjab, India

Abstract

A new technological direction has emerged in agriculture that combines Internet of Things (IoT) sensors and geospatial technologies, particularly in real-time crop monitoring. This study further examines how these technologies are utilized in crop management, offering detailed insights into the management practices employed by farmers. IoT devices, such as soil moisture sensors, temperature sensors, and weather stations, collect precise real-time data essential for making informed decisions regarding irrigation, fertigation, and pest control. Moreover, the application of geographic information systems (GIS) and remote sensing capabilities in the overall program strengthens the visual and analytical evaluation of field heterogeneity and crop status. This review illustrates how the integration of IoT and geospatial technologies enables data-driven decision-making by providing real-time insights into the crop and environmental conditions. Unmanned aerial vehicles (UAVs) equipped with multispectral or thermal imaging sensors can provide detailed aerial assessments of crop conditions, enabling timely interventions to improve yield. Furthermore, machine learning analysis of data gathered from the use of these technologies can predict how well crops will perform and identify problem areas before they become significant. This study aims to address these existing gaps by providing an overview of current approaches, case studies, and prospects in smart agriculture, with a focus on IoT and geospatial technologies as key enablers of sustainable and efficient crop production.

Keywords: IoT sensors, geographical information systems (GIS), crop monitoring, remote sensing, sustainable farming

**Corresponding author*: dca.rupinder@gmail.com

Gurwinder Singh, Vishakha Sood, Narayan Vyas, Surendra Yadav and Dankan Gowda V. (eds.) Geospatial Intelligence for Precision Agriculture, (159–176) © 2026 Scrivener Publishing LLC

8.1 Introduction

The agricultural sector is perhaps undergoing one of the most significant transformations today, driven mainly by technology, especially the Internet of Things (IoT) and geospatial technologies. These innovations are transformative to conventional farming paradigms, allowing farmers to fully integrate the continuous monitoring, analysis, and optimization of resources. The IoT enables the gathering of extensive data from multiple sources, including on-the-field soil moisture sensors, local weather stations, and crop health checkers, to ensure the optimal control of agricultural resources. On the other hand, geographical information systems (GIS) and remote sensing provide farmers with the necessary information about spatial data, which forms the basis for them to view the variability of fields and crop conditions. This study analyzes the primary applications of IoT sensors and geospatial technologies in real-time crop management, illustrates their various uses, discusses their advantages, and highlights the limitations of using IoT sensors and geospatial technologies in crop management. This study aims to analyze case studies and present the current practices and future opportunities of these technologies, emphasizing their positive impacts on sustainable farming systems that can meet the growing global food requirements without incurring high costs in terms of environmental degradation.

8.1.1 Smart Agriculture: Transforming Traditional Practices

Agriculture is a vital subsector of the global economy, delivering food, strategic materials, and fiber to different industries. However, the sector is often challenged by numerous social issues, including climate change, resource scarcity, and sustainability imperatives. Traditional farming practices, often reliant on limited or outdated information, lead to inefficient resource use, waste, and suboptimal results. This means that the pressure to produce more food worldwide is ever-increasing owing to the growing human population; hence, the need to adopt new technologies. The adoption of IoT and geospatial technologies may be a practical approach to addressing these challenges, offering the perspective of precision agriculture. Precision agriculture is the management and application of inputs and outputs that are in sync with the actual field situation to increase the production efficiency. In this process, IoT devices, such as soil moisture sensors, weather stations, and drones, are used to capture a massive volume of data related to various environmental and climate parameters.

This information can be used to provide options for when to water crops and apply fertilizers, as well as when to control pests, thereby boosting crop productivity and rationality [1].

Furthermore, advanced technologies such as GIS and remote sensing make precision agriculture more precise by embracing spatial analysis and presentation. These technologies enable farmers to understand field complexity, inspect crop status, and assess environmental conditions. For instance, remote sensing technologies can capture high-resolution images of crops, allowing farmers to gain insights into areas that require attention, such as the application of fertilizers, water, and weed control. Smart agriculture not only solves existing problems in the agricultural sector but also creates conditions for sustainable development in the future. Therefore, by adopting a data-backed approach, farmers can reduce their environmental impact, restore water resources, and reduce chemical use. Such a transition is crucial when considering climate change at the global level because it seeks to minimize the agricultural sector's harm to the environment [2].

In addition, unstable farming operations can be significantly enhanced by applying the IoT and geospatial technologies. When efficiency and productivity are enhanced, there are improved margins and returns on investment, which will favor farmers. These technologies enable future planning and investment opportunities to enhance the profitability and sustainability of farming. Therefore, this study demonstrates that the integration of IoT and geospatial technologies in crop production is not a recent phenomenon driven by advanced technologies but rather a process of adapting to new realities in food production and the management of natural resources. As the agricultural sector advances, it is crucial that such innovations are adopted to meet the challenges of the new century, including food security and sustainable agriculture, for the actual benefit of farmers and the environment [3].

8.1.2 The Role of IoT in Agriculture

The IoT is a new generation of computing technology that forms a network of connected objects capable of communicating with each other and exchanging data *via* the internet. In the agricultural domain, IoT devices are gaining popularity for measuring various parameters, including moisture content, temperature, humidity, and crop health, in real time [10]. By constantly monitoring the environment, farmers can adjust as needed, which in turn leads to significantly less waste. With the help of IoT technologies, farmers can make informed decisions that impact crop production and resource management [7]. In addition, IoT devices can be integrated with other technologies, such as machine learning and data analytics, to provide predictive information. For instance, based on data obtained from past inputs along with present-day inputs, farmers can predict the yields of their produce, check for incoming pests, and determine the best time to plant. This predictive capability is crucial in today's rapidly evolving agricultural environment, which is increasingly influenced by climate and market shifts that affect yields. Technology ownership is also a challenge, as smallholder farmers may not afford the initial capital outlay required to acquire it. Nevertheless, several challenges remain, including data management, system integration, and a lack of farmer training on these technologies. Overcoming these challenges will require various stakeholders, including governments, technology suppliers, and other players in the agricultural value chain, to design and implement cost-effective solutions with adequate backup for farmers [8].

Additionally, the authors note that the digital divide in rural contexts can hinder access to suitable infrastructure, including reliable connectivity, which is essential for IoT applications in agriculture. Some rural areas lack broadband facilities to support the transfer of large amounts of data used by IoT devices, thereby hindering IoT development. Thus, investments in rural broadband and programs that would familiarize farmers with the workings and the value of IoT systems are required. Dependence on IoT in agriculture is crucial for advancing the sector and improving efficiency, sustainability, and productivity. With the help of IoT technologies, farmers can manage their work in real time and use data to address issues that affect farming. The future of agriculture lies in the ability to integrate IoT technologies into practice, which will help transform conventional farming into a bright and environmentally friendly system for both farmers and the environment.

8.1.3 Geospatial Technologies in Crop Management

Remote sensing and GIS technologies, which are essential modern tools in agriculture, offer considerable support in terms of the wealth of geographical information. Through GIS, farmers can effectively evaluate data related to their fields to prepare management plans and practices. Aerial imagery is instrumental when utilizing remote-sensed imagery, which includes drone and satellite imagery, as it allows for easy visual inspection of crops and identification of those that are performing poorly [4]. Unmanned aerial vehicles (UAVs) equipped with sensors enable crop health monitoring and environmental analysis, relaying data to farmers for informed decision making. It can be useful in defining the management zones within a field that probably require different management strategies to optimize the use of available resources.

8.1.4 Advantages of Using IoT and Geospatial Technologies

As we have seen, several advantages would accrue from applying a combination of IoT and geospatial sciences to farmers and agriculture in general. First, they help in the acquisition of current information for decision-making and action by management with little delay [19]. This may result in higher crop production rates, minimized resource use, and efficiency in the management of farm activities. Sensors used in irrigation control, which secure audiences that can relay real-time information, increase the efficiency of water usage; hence, they improve water conservation. More importantly, the use of these technologies has the potential to enhance the economic feasibility of farming. Moreover, the obtained data show that different IoT device practices can be used in the future for planning and investing in long-term agricultural sustainability [5].

8.2 Literature Review

ICT for agriculture, specifically IoT and geospatial technologies, have proven to be vital tools in improving the agricultural sector, so that farmers can efficiently use resource produced by them. This study brings together some of the research articles in this field and presents different works that examine the prospects, uses, and drawbacks of these technologies. Table 8.1 presents a synthesis of various authors' research foci, methods, and significant results. This synthesis attempts to conduct a literature review and identify further research directions for IoT and geospatial technologies in agriculture today, as shown in Table 8.1.

8.3 The Role of IoT Sensors in Agriculture

8.3.1 Overview of IoT in Agriculture

The IoT has been touted as the key technology that has found its way into the agricultural industry and disrupted traditional models of farming. The IoT makes sustainable agriculture possible by enabling the real-time acquisition and processing of data from multiple sources. IoT technology

Table 8.1 Recent research contributions on IoT and emerging technologies in smart agriculture.

Technologies used	Key findings
Geospatial technologies, Satellite images	Deals primarily with crop categorization based on geospatial technologies and satellite images [11].
RFID technology	Provides an understanding of how and where RFID technology is used in its connection with IoT, in domains such as agriculture [12].
IoT systems, Literature review	Outlines the impediments to IoT uptake, costs, skills, and standardization, and explains the relatively sluggish rate of IoT technology and blockchain uptake in agriculture [13].
UAV, Machine Learning, IoT	Summarizes the UAVs and machine learning approaches to smart irrigation management with a focus on the role of data in water resources management [14].
UAV, IoT, Big Data Analytics	Introduces an IoT-supported Smart Farming Framework that employs UAVs in crop check and sprinkler control; shows substantial increases in crop production [15].
IoT, Machine Learning, artificial intelligence (AI)	Research presents the AgriFusion concept regarding the architecture of precision agriculture with diverse setups to improve farming practices.
IoT, AI, Sensors	Explores IoT and AI possibilities and challenges in integrated farming sectors using precision agriculture, focusing on the efficiency of farming practices and resource control [17].
Cloud Computing, IoT	Explores how cloud and IoT implement precision agriculture, paying particular attention to the efficiency of resources and preservation of the environment [18].

(*Continued*)

Table 8.1 Recent research contributions on IoT and emerging technologies in smart agriculture. (*Continued*)

Technologies used	Key findings
Geospatial technologies, Machine Learning	Evaluates the effects of drought on crops with the help of GIS and computational intelligence and expounds guidelines for improved farming [19].
IoT, Blockchain	Recapitulates the combined use of IoT and blockchain in supporting agriculture and discusses how they improve security and productivity in the use of agricultural practices [20].
IoT, Digital Image Processing	Discusses the application of the IoT system as well as the image processing approach to weed detection, proving the efficiency of the proposed Weed-ConvNet model for smart agriculture [21].
IoT, LoRaWAN	Introduces the low-cost IoT-based greenhouse monitoring to farmers of all sizes, to control the growth conditions within the greenhouses [22].
IoT, Deep Learning	Introduces an authentication scheme that may improve the security of data in IoT Agriculture systems and close data transmission gaps [23].
IoT, Wireless Sensor Networks	This work describes the process of applying an IoT-based system in agriculture based on the case of smart irrigation and weather systems [24].
IoT, Blockchain, Geospatial Technologies	Suggests a system solution of IoT, blockchain, and GIS for an improvement of efficient sustainable farming [25].

facilitates the application of fertilizers and pesticides to crops, enabling precise and accurate dosing to minimize environmental pollution and waste [6]. Furthermore, IoT systems allow remote monitoring and management of agricultural processes. This capability is particularly advantageous when there are multiple fields, as large-scale farming is becoming increasingly common. Further development is expected from the integration of IoT with other technologies, such as AI and Machine learning. These technologies can enable data analysis to determine crop yield potential, diagnose diseases at an early stage, and provide advice on appropriate farming practices. In conclusion, the application of IoT technology makes the adoption of smart farming more suitable for agricultural advancement. In this global environment characterized by limited resources, timely data and optimal analysis enable better utilization of resources, resulting in higher productivity and improved food security for farmers.

8.3.2 Applications of IoT Sensors

The introduction of IoT sensors in agriculture has opened up several highly beneficial possibilities. These applications not only increase productivity in agricultural systems but also support sustainable agriculture. The following are some key applications of IoT sensors in agriculture (Figure 8.1):

Soil Monitoring: IoT sensors are used in soil monitoring to measure parameters such as moisture, temperature, and nutrient composition. By frequently measuring the amount of moisture in the soil, farmers can

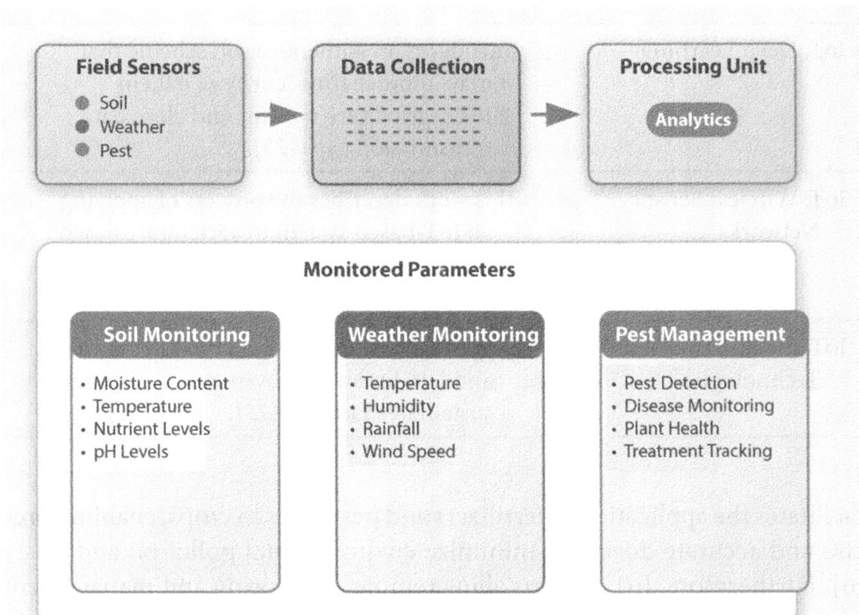

Figure 8.1 System architecture and parameter monitoring framework.

correctly program their irrigation and provide crops with the appropriate amount of moisture at the proper time. For example, when the moisture content of the soil drops below a specific value, automatic irrigation technology can be used to enhance healthy irrigation. Additionally, the assessment of soil nutrient status enables farmers to apply fertility inputs accurately, thereby reducing the application of excessive amounts of fertilizers that can pollute the environment [11].

Weather Monitoring: With IoT sensors, real-time agro-climatic events, such as frost, heavy rainfall, or drought, can be forecasted and relayed to farmers. Farmers, can set the proper dates for planting, watering, and harvesting crops by receiving relevant weather information. For instance, when a frost notice is issued, farmers can avoid planting crops that are likely to be damaged. In addition, weather prognostics can raise awareness of shifts in climate and enable farmers to adjust their crop-growing practices [9].

Pest and Disease Management: IoT sensors can also be used for pest and disease control in horticulture. Because these sensors help monitor the possible onset of pests and diseases, measures can be taken effectively before crops are affected, allowing for the use of chemical pesticides. For example, it can detect the status of foliage temperature and humidity, which may predict stress or infections within the plants. Once a problem is identified, farmers can be assigned notifications through their smartphones, which helps them respond appropriately by adding necessary treatments or adjusting environmental factors within their surroundings.

8.3.3 Case Studies

A case study involving the El-Salhia region of Egypt showed an impressive change when developing a real-time IoT system for site-specific potato-crop management. This experimental program, conducted over several growing seasons, demonstrated how technology can transform conventional resource-intensive farming in desert areas. The advanced IoT system utilized 85 wireless sensor nodes, which were strategically placed and optimized across various pieces of agricultural land. Analytical sensors create a network of data measurement points that record and constantly control the most important parameters, such as the moisture content in the soil horizon, changes in temperature, relative humidity, and concentration of certain nutrients. The structural design of the system incorporated underground instruments for soil monitoring and above ground devices to oversee atmospheric conditions, thereby establishing maximal environmental control [9]. Another important novelty of this study was the possibility of using a real-time data transmission system, which provided immediate access to key agricultural indicators. Staying connected was achieved using LPWAN technology, which provided a line of sight on the vast farmland. The data were then processed and sent to a command platform that could be accessed *via* a mobile application or desktop. This infrastructure enables farmers to monitor the growth of their crops from a suitable distance and receive real-time alerts when readings exceed acceptable ranges.

The algorithms designed by the research team were better able to analyze the collected data and generate meaningful insights. The seminal characteristic of these algorithms is their ability to process multiple parameters simultaneously, continually updating models of crop water demand or trace elements necessary for proper growth and development in response to current and past circumstances. The inbuilt predictive capability was highly effective during such key phases of growth, where active rather than passive measures could then be undertaken. The use of this IoT system yielded outstanding results in several key aspects. Water usage efficiency was enhanced by up to 30% through carefully scheduled regular irrigation practices, which integrated real-time control mechanisms with basic weather data to monitor the soil moisture content. In addition, the study noted increased yields and crop quality in various regions, accompanied by signs of increased investor confidence. Based on the findings of the economic analysis of the measures' implementation, the following benefits were established for the farmers. Although the initial capital investment costs are high for IoT devices, the savings from input expense reductions, which overlay yield enhancements, result in a brief payback period of less than two cropping cycles. Sustaining costs were reduced by approximately 28%, mainly due to improved resource efficiency and fewer person-hours required for supervision and control efforts.

This highlights the environmental impact associated with precision agriculture facilitated by IoT technology. Reduced water use and nitrogen application led to a lower environmental load, and efficiency in resource application demonstrated a sustainable practice in high-input production systems in water-limited environments. This possibility has boosted the adoption of similar systems in other neighboring agricultural practice areas, leading to a chain-like advancement of technology in the region's agricultural industry. Therefore, the results derived from this example, corroborates the need to integrate IoT technology into the advancement of a crucial world industry, notably agriculture, in areas experiencing harsh climate conditions. The use of sensors, data analysis, and decision support systems in farming demonstrates the potential for improved resource-use efficiency and productivity, which can inform future farming-related projects in similar climates.

8.4 Application of Geographical Information Systems (GIS) in Crop Production

8.4.1 Brief Introduction of Geospatial Technologies

Geospatial technologies are tools and techniques used for the acquisition, management, and display of geographical information. These technologies are crucial in many fields, especially in the agricultural industry, because of their ability to support decisions and aid in optimizing the working process. Two components are central to geospatial technologies: GIS and remote sensing, the synergy of which provides a stable foundation for appreciating and managing agricultural systems. It is an effective tool for recording, organizing, processing, assessing, maintaining, and representing geographical and spatial information. It enables the use of maps, satellite images, and field data, allowing farmers to easily comprehend multiple data inputs. This type of visualization aids in interpreting new data concerning the type of soil, crop yields, and any hindering factors such as weather [16]. Information about the Earth is obtained with minimal physical interaction between the acquiring instrument and the object being studied, a technique referred to as remote sensing. This can be achieved using satellite or aerial imagery, which captures data in the form of images at different wavelengths. The use of these technologies is synergistic, as they afford farmers an all-around view of the field, enabling precision agricultural practices to be optimized in terms of the available resources to produce more.

8.4.2 Remote Sensing Applications

Modern remote sensing methods have emerged in the agricultural sector, where information about crops and weather conditions is highly valuable. Satellite imagery and photography, combined with drones, provide detailed information about specific areas of fields to farmers, allowing them to access information without visiting the field. Crop vigor, a crucial component of crop yield, is one of the primary areas where the application of remote sensing techniques is most beneficial. Spectral data can provide farmers with information about differences in the state of plants, including the presence of diseases, pests, nutrient deficiencies, or water deficits.

For example, the Normalized Difference Vegetation Index (NDVI) is a remote sensing index used to assess vegetation health based on the reflection of near-infrared and visible light. Moreover, remote sensing results can provide information on the exact location in a field that requires attention. Thus, farmers use resources more efficiently, and fertilizers, pesticides, and water resources are applied only in areas where they are required. This granular application not only increases crop returns but also reduces the detrimental application of inputs. Additionally, remote sensing can be used in various applications to measure environmental phenomena, such as soil moisture, temperature differences, and weather information. This information is especially useful when scheduling tasks such as planting, watering, and harvesting, thereby improving sustainable agricultural production.

8.4.3 GIS

GIS technology is paramount in the location and analysis of numerous databases, offering a one-stop-shop approach to assessing agricultural terrains. With spatial data, farmers can identify patterns and cycles, which enable them to optimize the use of their assets and achieve greater results. This study focuses on one of the major uses of GIS in agriculture: creating management zone maps. These maps define the field sub-areas that need to be managed differently owing to factors such as the type of soil, health of the crops, or past yield performances. Through management zones, farming practices can be adjusted according to the requirements set by the fields generated from the sectors or regions, thus increasing yield levels. Additionally, the data can often be historical, allowing GIS to distinguish between temporal changes. For example, previous yield data analyzed in combination with climatic conditions ensure that farmers make proper rotational cropping programs, planting dates, and resource management. From this learnable historical angle, one is in a better position to grow and adapt to new climatic shocks and changes, as well as enhance sustainability.

Spatial data and analysis can thus be shared to improve efficiency in policy making and implementation, aligning with best practices from other producers, agronomists, and researchers. In conclusion, the successful implementation of geospatial technologies, such as remote sensing and GIS, has become the norm in agriculture. These technologies equip farmers with the correct tools to make informed decisions, manage resources effectively, and increase production sustainably. This is because the use of geospatial technologies is expected to expand in the future, leading to an enhancement of the capacity of agriculture to perform effectively.

8.5 Integration of IoT and Geospatial Technologies

8.5.1 Data Analytics and Decision Support Systems

IoT sensors and geospatial technologies facilitate the efficient collection and analysis of real-time crop data. Large amounts of data can be collected from the field using IoT devices, such as soil moisture sensors, weather stations, and crop health monitoring devices. These data are then used in combination with other related systems, such as GIS and remote sensing, to provide a comprehensive view of the agricultural status.

Real-Time Data Flow: The regular, on-time passing of data from the field to those who require that information is critical for developing timely responses to dynamic environments. For instance, IoT sensors can detect the humidity of the soil and notify when water is needed for the plants. When integrated with geospatial data, the database can be used to display the moisture levels in various regions of the fields, allowing for the application of efficient irrigation methods. This not only helps save water but also enables fertilizers to provide crops with the right amount of moisture for growth [21].

Advanced Decision Support System: Decision support systems (DSS) that utilize IoT and geospatial data enable farmers to make informed decisions based on real-time information. These systems can identify patterns in data, make predictions about what may occur, and suggest actions. Using the aforementioned analytic tools, farmers can position themselves to improve their operations, reduce production costs, and increase yields [22].

Predictive Analytics: Another important area of using IoT and geospatial technologies is predictive analytics. This analysis of past data, combined with the new feed, enables farmers to predict the occurrence of problems. such as pest or disease incidence. It also prevents crop failure that could otherwise occur, thereby increasing farm resilience.

8.5.2 Useful Examples and Case/Field Implementation

The findings of this study also reveal the following successes of IoT and geospatial technologies in the agricultural industry from case studies:

Green Water Drone Project Executive Summary, Greece: The Green Water Drone project in Greece is of particular interest, as drones equipped with thermal cameras were used to gauge the real-time water demand for

crops. They can determine the water requirements of crops from aerial views captured by drones. By observing the thermal image, farmers can differentiate between under-watered and over-watered areas of the field, leading to better irrigation. The project not only reduced water usage but also saved costs associated with irrigation. Irrigation involves applying water in the correct quantity and at the right time to improve crop yields while conserving water and other resources [26].
Wireless Sensor Networks (WSN) in Potato Crop Management in Egypt: Another strong example is the application of technology known as Wireless Sensor Networks (WSN) in potato crop management in Egypt. This was an actual project that involved the installation of several sensors in potato fields to measure the soil moisture, temperature, and humidity levels. The collected samples were sent to a leading network, where they were used to gather information on crop health [27]. Thus, the work of WSN enabled accurate observations and management of water irrigation strategies in the study area. Farmers can access streamed data to monitor real-time weather patterns and adjust irrigation schedules to ensure optimal water supply for potato crops. Thus, it was possible to demonstrate that the project enabled an increase in potato yield and a decrease in water consumption, which reflects the efficiency of Interactive Things and geospatial technologies in the agricultural sector [28].

Beyond these case studies, the integration of IoT and geospatial technologies has found applications in various agricultural practices, including:

- **Precision Fertilization**: By utilizing soil sensors and GIS mapping, farmers are better equipped to apply fertilizers with precision, thereby minimizing losses and negative environmental impacts.
- **Pest and Disease Management**: IoT devices can address pests and diseases through genuine and accurate detection, while reducing chemical applications.
- **Yield Monitoring**: Technology that integrates yield data with geographic information adds value for farmers to explain yield gaps in the fields for future use.

8.6 Challenges

Based on this knowledge, IoT sensors and geospatial technologies should be employed in agriculture; however, to harness their full potential, the following impediments must be considered. This is primarily due to challenges in data integration and compatibility, as farmers use multiple systems whose data cannot be easily integrated into a unified framework for comprehensive analysis. Therefore, incorporating these components is crucial for developing seamless data flows. This gap can only be closed, and adoption can be successful if accessible training programs and resources are available. The costs of these technologies also partly explain the extent of adoption, as smallholder producers can only afford to support them to a limited degree. However, for widespread adoption, they must be inexpensive, and their use must be financially supported. To address these issues, specific interfaces that connect the aforementioned technologies must be developed to enable farmers to analyze data with ease. In this sense, farming can benefit from formal training, such as training sessions or seminars, and hands-on training can enable farmers to adopt innovations effectively. The source of innovation and ideas about the use of technology must come from farmers using those technologies, technology suppliers, researchers, and other related actors. They will continue to promote policies that support IoT and geospatial technologies through government funding, incentives, and resources. There is also a need to increase spending on IoT and geospatial technology applications to support efficient agriculture and effectively address burgeoning issues.

8.7 Conclusion

The integration of IoT sensors and geospatial technologies represents a significant advancement in real-time crop monitoring and management. Because most technologies provide real-time input and output information, the IoT can help improve agricultural productivity and resource utilization sustainably. However, embracing these new production trends will be crucial as the farm sector navigates the 21st-century food production challenges. This study demonstrates that IoT and geospatial technologies enhance farming practices, promote sustainability, and improve the food security. In summary, the further development of agriculture depends on the refinement and implementation of reliable technologies and the effective use of data. The successful implementation of IoT and geospatial technologies is crucial in helping farmers address the growing concerns of modern-day farming as they move forward. Moving forward, there is a need to focus on collaboration, training, and the development of new solutions to achieve optimal value propositions from these disruptive technologies.

AI disclosure statement: The author acknowledges the use of AI-based tools for minor language editing or grammatical correction. However, all intellectual contributions, analyses, and conclusions presented in this chapter are entirely the author's own.

References

1. Mohammad EL-Basioni, B.M., Mohamed, E.S., Belal, A.A., Jalhoum, M.E.M., Abd EL-Kader, S.M., Zahran, M.B., A case study of a real-time internet of things system for site-specific potato crop management in El-Salhia Area-Egypt. *Sci. Rep.*, 12, 1, Nov. 2022, doi:10.1038/s41598-022-22690-3.
2. Belal, A.A., EL-Ramady, H., Jalhoum, M., Gad, A., Mohamed, E.S., Precision Farming Technologies to Increase Soil and Crop Productivity, in: *Springer Water*, pp. 117–154, Springer International Publishing, 2021, doi:10.1007/978-3-030-78574-1_6.
3. Sivakumar, V.G., Baskar, V.V., Vadivel, M., Vimal, S.P., Murugan, S., IoT and GIS Integration for Real-Time Monitoring of Soil Health and Nutrient Status, in: *2023 International Conference on Self Sustainable Artificial Intelligence Systems (ICSSAS)*, Erode, India, pp. 1265–1270, 2023, doi:10.1109/ICSSAS57918.2023.10331694.
4. Eze, V.H.U., Eze, E.C., Alaneme, G.U., BUBU, P.E., Nnadi, E.O.E., Okon, M.B., Integrating IoT sensors and machine learning for sustainable precision agroecology: enhancing crop resilience and resource efficiency through data-driven strategies, challenges, and future prospects. *Discov. Agric.*, 3, 1, May 2025, doi:10.1007/s44279-025-00247-y.
5. Shahab, H., Iqbal, M., Sohaib, A., Ullah Khan, F., Waqas, M., IoT-based agriculture management techniques for sustainable farming: A comprehensive review. *Comput. Electron. Agric.*, 220, 108851, May 2024, doi:10.1016/j.compag.2024.108851.
6. Latif, M.S., Kazmi, R., Khan, N., Majeed, R., Ikram, S., Ali-Shahid, M.M., Pest prediction in rice using IoT and feed forward neural network. *KSII Trans. Internet Inf. Syst.*, 16, 1, 133–52, 2022, https://doi.org/10.3837/tiis.2022.01.008.
7. Raza, S., Innovative technologies in agriculture: leveraging AI, ML, and IoT for sustainable food production and resource management. *Int. J. Agric. Sustainability Dev.*, 6, 3, 127–46, 2024.
8. Kumar, Y., Singh, R., Moudgil, M.R., Kamini, A Systematic Review of Different Categories of Plant Disease Detection Using Deep Learning-Based Approaches. *Arch. Comput. Methods Eng.*, 30, 8, 4757–4779, June 2023, doi:10.1007/11831-023-09958-1.
9. Anjaneyulu, M., *et al.*, Sustainable practices for plant nutrient control using IoT. *E3S Web Conf.*, 507, 01039, 2024, doi:10.1051/e3sconf/202450701039.

10. Ahmed, N. and Shakoor, N., Advancing agriculture through IoT, Big Data, and AI: A review of smart technologies enabling sustainability. *Smart Agric. Technol.*, 10, 100848, Mar. 2025, doi:10.1016/j.atech.2025.100848.
11. Kassim, M.R.M., IoT Applications in Smart Agriculture: Issues and Challenges, in: *2020 IEEE Conference on Open Systems (ICOS)*, IEEE, pp. 19–24, Nov. 17, 2020, doi:10.1109/icos50156.2020.9293672.
12. Soussi, A., Zero, E., Sacile, R., Trinchero, D., Fossa, M., Smart Sensors and Smart Data for Precision Agriculture: A Review. *Sensors (basel)*, 24, 8, 2647, 2024 Apr 21, doi:10.3390/s24082647.
13. Bhati, N. and Vyas, N., Elevating IoT Sensor Data Management and Security Through Blockchain Solutions, in: *Advances in Systems Analysis, Software Engineering, and High Performance Computing*, pp. 58–77, IGI Global, Mar. 18, 2024, doi:10.4018/979-8-3693-0968-1.ch003.
14. Sadowski, S. and Spachos, P., Wireless Technologies for Smart Agricultural Monitoring Using Internet of Things Devices with Energy Harvesting Capabilities. *Comput. Electron. Agric.*, 172, 105338, 2020, doi:10.1016/j.compag.2020.105338.
15. Rashid, A.B., Kausik, A.K., Khandoker, A., Siddque, S.N., Integration of Artificial Intelligence and IoT with UAVs for Precision Agriculture. *Hybrid Adv.*, 10, 100458, Sept. 2025, doi:10.1016/j.hybadv.2025.100458.
16. Hartono, R., Maulana Yoeseph, N., Aji Purnomo, F., Asri Safi'ie, M., Alim Tri Bawono, S., Portable internet of things-based soil nutrients monitoring for precision and efficient smart farming. *Bull. Electr. Eng. Inf.*, 13, 5, 3326–3333, Oct. 2024, doi:10.11591/eei.v13i5.7928.
17. A systematic review on monitoring and advanced control strategies in smart agriculture. *IEEE Access*, 9, 32517–32548, 2021. 10.1109/ACCESS.2021.3057865.
18. Li, C. and Niu, B., Design of smart agriculture based on big data and Internet of Things. *Int. J. Distrib. Sens. Netw.*, 16, 5, 155014772091706, May 2020, doi:10.1177/1550147720917065.
19. Prathibha, S.R., Hongal, A., Jyothi, M.P., IoT Based Monitoring System in Smart Agriculture, in: *2017 International Conference on Recent Advances in Electronics and Communication Technology (ICRAECT)*, Bangalore, India, pp. 81–84, 2017, doi:10.1109/ICRAECT.2017.52.
20. Anusha, A., Guptha, A., Sivanageswar Rao, G., Tenali Ravi Kumar, A Model for Smart Agriculture Using IoT. *Int. J. Innov. Technol. Explor. Eng. (IJITEE)*, 8, 2278-3075, April 2019.
21. Hoque, A. and Padhiary, M., Automation and AI in precision agriculture: innovations for enhanced crop management and sustainability. *Asian J. Res. Comput. Sci.*, 17, 95–109, 2024, doi:10.9734/ajrcos/2024/v17i10512.
22. Pachiappan, K., Anitha, K., Pitchai, R., Sangeetha, S., Satyanarayana, T.V.V., Boopathi, S., Intelligent Machines, IoT, and AI in Revolutionizing Agriculture for Water Processing, in: *Advances in Computational Intelligence*

and Robotics, B.B. Gupta, F. Colace (eds.), pp. 374–399, IGI Global, 2023, doi:10.4018/978-1-6684-9999-3.ch015.

23. Rajeswari, D., Venkatachalam Parthiban, A., Ponnusamy, S., Digital Twin-Based Crop Yield Prediction in Agriculture, in: *Advances in Business Information Systems and Analytics*, pp. 99–110, IGI Global, Feb. 02, 2024, doi:10.4018/979-8-3693-3234-4.ch008.
24. Hundal, G.S., Laux, C.M., Buckmaster, D., Sutton, M.J., Langemeier, M., Exploring Barriers to the Adoption of Internet of Things-Based Precision Agriculture Practices. *Agriculture*, 13, 1, 163, Jan. 2023, doi:10.3390/agriculture13010163.
25. Ibrahim, A.S., *et al.*, AI-IoT based smart agriculture pivot for plant diseases detection and treatment. *Sci. Rep.*, 15, 1, May 2025, doi:10.1038/s41598-025-98454-6.
26. Debauche, O., *et al.*, Edge AI-IoT pivot irrigation, plant diseases, and pests identification, in: *Int. Conf. Emerg. Ubiquitous Syst. Pervasive Netw. (EUSPN)*, vol. 177, pp. 40–48, 2020.
27. Qazi, S., Khawaja, B.A., Farooq, Q.U., IoT-Equipped and AI-Enabled Next Generation Smart Agriculture: A Critical Review, Current Challenges and Future Trends. *IEEE Access*, 10, 21219–21235, 2022, doi:10.1109/ACCESS.2022.3152544.
28. Navarro, E., Costa, N., Pereira, A., A Systematic Review of IoT Solutions for Smart Farming. *Sensors*, 20, 15, 4231, July 2020, doi:10.3390/s20154231.

9

Systematic Review on Agricultural Land Use Classification Based on Machine Learning and Deep Learning Approaches Using Satellite Dataset

Nisha Sharma[1]*, Kawaljit Kaur[1] and Sartajvir Singh[2]

[1]University Institute of Computing, Chandigarh University, Mohali, Punjab, India
[2]University Institute of Engineering, Chandigarh University, Mohali, Punjab, India

Abstract

Land cover (LC) refers to the physical elements on land, such as forests, water bodies, built-up areas, and vegetation. LC mapping involves the application of remote sensing information to identify and classify various land elements. Land use (LU) refers to the activities of people on land, including farming and industry, which are used to create products and services. Examples of LU classification are croplands for farming and built-up lands for industrial or commercial development. LC mapping and monitoring, which entails the classification of Land Use and Land Cover (LULC), majorly depends upon remote sensing data such as the Landsat series, Sentinel series, MODIS, and Radarsat 1-2, and other similar resources. These data are crucial for detecting and monitoring changes in land characteristics over time. LULC is employed in various administrative and research fields, including urban development, forest clearing, agricultural research, and disaster management, such as soil erosion, habitat loss, biodiversity loss, and responses to events such as floods, landslides, and forest fires. LULC analysis utilizes software such as ERDAS, Geographic Information System (GIS), QGIS, and ArcGIS, alongside tools such as Sen2Cor for Sentinel-2 data correction, to monitor land use and cover dynamics. This systematic review presents an overview of machine learning (ML) and deep learning (DL) methods for classifying agricultural land use using satellite imagery. It delves into the nature of the satellite data used, ML and DL algorithms implemented, and classification accuracies obtained.

**Corresponding author*: uic.nishasharma@gmail.com

Gurwinder Singh, Vishakha Sood, Narayan Vyas, Surendra Yadav and Dankan Gowda V. (eds.) Geospatial Intelligence for Precision Agriculture, (177–196) © 2026 Scrivener Publishing LLC

Keywords: Machine learning (ML), land use and land cover (LULC), deep learning (DL), satellite dataset, agricultural land, remote sensing data

9.1 Introduction

Agricultural research systematically studies the different aspects of farming practices, methods, and innovations to enhance the sustainability, productivity, and quality of food and fiber production [1]. As the world population continues to rise, the demand for a stable and efficient farming system is growing. Proper land-use classification for agriculture is essential for assisting data-driven decisions regarding resource allocation, yield forecasting, and ecosystem monitoring. Satellite remote sensing offers an economical and spatially extensive classification method. Over the past few years, machine learning (ML) and deep learning (DL) have evolved into highly capable tools for handling and interpreting satellite data [2]. This study provides an overview of contemporary approaches to land use and land cover (LULC) classification that focus on integrating geospatial analytics, ML, DL, and remote sensing technologies.

Remote sensing utilizes numerous image interpretation techniques, including pixel-level, sub-pixel, object-based [3], spectral analysis, vegetation indices, and temporal change detection. Pixel-level techniques examine every pixel in isolation, whereas sub-pixel techniques extract more sophisticated information from within individual pixels [4]. Object-based techniques improve pixel examination by grouping similar pixels to form coherent objects for a more meaningful examination. Fusion-based techniques, which combine multisensory datasets, are becoming increasingly popular [5]. Spectral methods focus on the unique spectral signatures of various land features. Vegetation indices, a category of spectral techniques, employ multiband combinations to estimate the health of plants. Simultaneously, change detection methods compare imagery at different times to determine the regions where changes are observable. Remote sensing is a process for gathering information concerning objects or surfaces from a distance, without physical contact. The use of satellite images is broad in agriculture for land use mapping and classification [6]. These data play critical roles across many fields, including weather forecasting, agriculture, urban planning, environmental monitoring, disaster relief, and scientific research [7]. Satellite-acquired data are sent to Earth, where they are analyzed to extract useful information to facilitate well-informed decision-making in various areas. Remote sensing systems acquire information in two fundamental modes: active and passive.

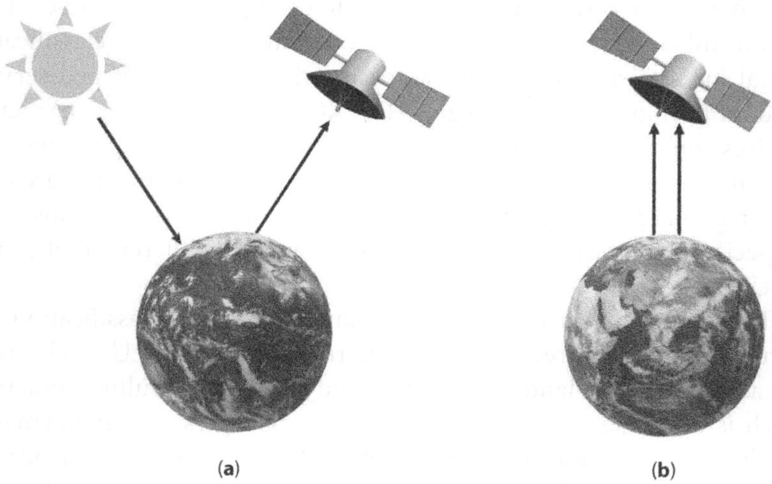

Figure 9.1 (a) Passive remote sensing (PRS) and (b) active remote sensing (ARS).

In passive remote sensing (PRS), sensors collect naturally available energy, such as sunlight reflected from the Earth's surface, without emitting energy. Examples include the Landsat-7 and Landsat-8 satellites. Another example is a satellite that captures images without its own source of light; the Sun is primarily used as a source of light. Sunlight reaches the Earth's surface and is reflected. The satellite sensor collects reflected satellite rays. Landsat-7 and Landsat-8 are examples of PRS satellites. Figure 9.1(a) shows the PRS. In the case of ARS, the system uses emitted energy to detect and capture objects. Radar satellites are an example of ARS. It sends radar and radio waves to the Earth's surface. The radar transmits an electromagnetic pulse, and the reflected signal from the object is received by the radar sensor itself. Figure 9.1 (b) shows the active remote sensing. RADARTSAT-1 [8] and RADARSAT-2 [9] are examples of active remote sensors [7].

9.1.1 Applications of LULC

Remote sensing has proven to be highly applicable in various sectors, including housing projects [10], and is instrumental in desertification studies [11]. For infrastructure development, such as road projects, detailed land-use classification surveys are crucial for identifying potential obstacles, such as wetlands, rivers, and deserts [12]. LULC data play a dual role in scientific research and administrative planning [13]. These LULC patterns are dynamic and evolve over time owing to economic, social, and environmental influences [14, 15]. In defense operations, land-use classification is vital for categorizing terrains across borders, allowing armed services to adopt terrain-specific strategies [16]. Road development, in particular, requires comprehensive land-use classification for effective planning [17]. Moreover, in agriculture, soil classification based on color or texture facilitates regional planning, enabling the selection of appropriate crops suited to specific soil types [18]. Figure 9.2 represents the different applications of LULC.

This review summarizes recent advances in LULC classification and discusses potential directions for future research. Proper LULC classification aids sustainable land management and effective agricultural practices, which is advantageous to researchers, land managers, and policymakers [19]. The identification and mapping of different types of LULC in agricultural areas are needed for several reasons related to agricultural planning, resource planning, and environmental monitoring [20]. This process is needed to document changes over time, monitor the spatial distribution of different land uses, and assist in making informed decisions for sustainable land management [21]. The use of remote sensing technology, particularly satellite data, is a standard method for performing LULC categorization

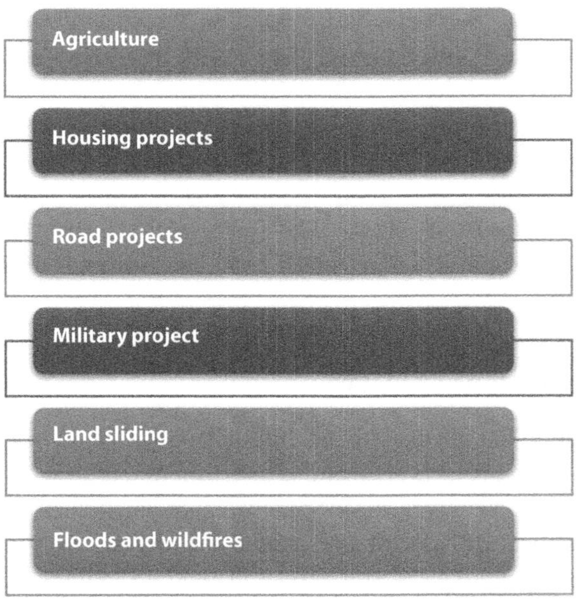

Figure 9.2 Application areas of LULC mapping.

in agricultural environments. Precise LULC classification necessary for interpreting environmental changes, observing temporal changes, and guiding sustainable land management policies. Here, we explain different research procedures agricultural fields using remote sensing datasets and details about how to procure satellite datasets. We then explain the recent advances in LULC mapping using DL and ML classifiers in agricultural land. The discussion and analysis section also encompasses methods of working on remote sensing datasets and performance levels of various classifiers.

9.2 Satellite Datasets for Agricultural LU Classification

This study targeted the use of different satellite imagery data sources for agricultural land use classification. Although there are various sources of satellite data, the Landsat and Sentinel series are the most widely used in agricultural applications. Each Sentinel mission generates particular types of data: Sentinel-1 offers radar-derived images, Sentinel-2 provides optical data, Sentinel-3 serves both ocean and land observations, Sentinel-4 and Sentinel-5 assist in atmospheric monitoring, and Sentinel-6 monitors sea surface parameters. However, Landsat missions utilize various sensors for each generation. Landsat-1 to Landsat-3 employed the Multispectral Scanner (MSS); Landsat-4 and Landsat-5 were equipped with MSS and Thematic Mapper (TM) sensors [22]; Landsat-7 was equipped with Enhanced Thematic Mapper Plus (ETM) [23]; Landsat-8 carried the Operational Land Imager (OLI) and Thermal Infrared Sensor (TIRS); and Landsat-9 employed OLI-2 and TIRS-2 sensors. These satellites provide critical data for agricultural remote sensing. Landsat data can be obtained from the USGS Earth Explorer portal (https://earthexplorer.usgs.gov) [24], and Sentinel-1, Sentinel-2, and other Sentinel data can be obtained from the European Space Agency's Copernicus portal (https://www.copernicus.eu/en) [25]. Both websites are the main sources for downloading satellite data employed in agricultural land use analysis.

9.3 Methodology

A search of the Scopus database identified 1581 research papers related to LULC classification, along with an additional 25 articles from other sources, including Google Scholar and institutional repositories. After eliminating duplicate entries, a total of 1,531 papers remained. Of the 1,531 papers identified, 1,347 were excluded as they were not directly related to agricultural LULC classification, based on strict inclusion criteria that focused on satellite-based ML and DL methods. One hundred papers could not be retrieved because of access restrictions or incomplete records, limiting their inclusion in the study. Only 84 papers were available, of which eight were related only to water bodies, 10 were related only to landslides, and 14 were related only to natural hazards; these were also excluded from the study. Only 52 papers were relevant to this study. These 52 papers were thoroughly reviewed and included in the current study. Figure 9.3 represents the Prisma flow of the referred papers.

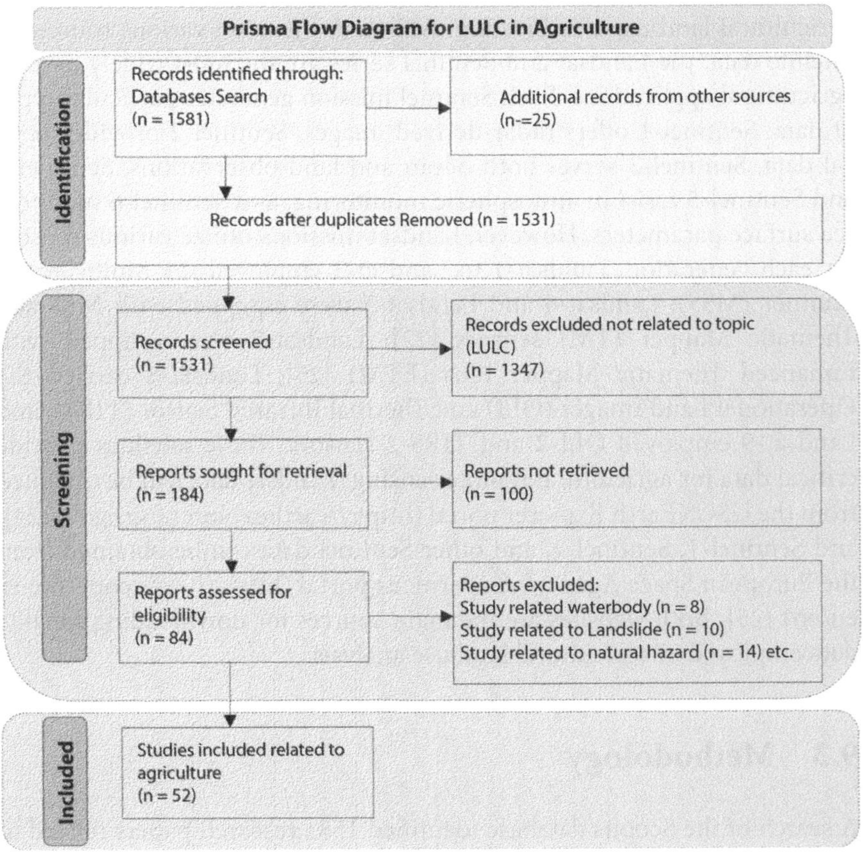

Figure 9.3 Prisma flow diagram for LULC in agriculture.

9.4 Related Work

Researchers have leveraged diverse remote sensing techniques, ML, DL algorithms, and geospatial tools to comprehend and map dynamic land use patterns in agricultural landscapes [26]. This study discusses different classifiers used in ML and DL models for LULC classification in agriculture. ML techniques using different classifiers, such as SVM, DT, and RF, and satellite dataset tools, such as Sentinel-1, Sentinel-2, Landsat-5, and Landsat-8, for conducting the study, as mentioned in Table 9.1 [27].

9.4.1 ML-Based LULC Classification

ML has revolutionized the LULC classification process using satellite datasets. This technique primarily relies on supervised learning using labeled data. In satellite image classification, every pixel in an image is classified as belonging to a particular LC type, such as forest or water. These classified images act as the training set, and based on the identified patterns, the model labels the subsequent images. Integral to improving the performance of such models is the selection and definition of informative features, such as spectral reflectance or spatial texture. The most frequently applied approaches to LULC classification are SVM and RF. These algorithms are appreciated for their high performance, level of automation, and insensitivity to the type of image data. In addition, their performance can be improved by training them with new data in conjunction with the current dataset. However, challenges persist. These models are highly sensitive to the quality and amount of labeled training data, and the choice of good features usually requires a lot of prior knowledge. However, the utility of ML for LULC classification is promising and versatile, making it possible for researchers to process and analyze large satellite data for various applications.

In 2022, a study was conducted in West Bengal using Sentinel-2 multispectral images with the agricultural class type with different summer crops and achieved an accuracy level of 95% using an RF classifier [27]. A study was conducted across Europe on different classes of woodland, grassland, shrubland, and 19 crop classes using pixel-based classifiers, SVM, and RF classifiers, along with Sentinel-2 and Landsat-8 dataset tools, achieving 77.9% overall accuracy for producing crop type maps [26]. A study conducted in 2022 using an RF classifier and Sentinel-1 and Sentinel-2 dataset tools for forest and non-forest land types in southern Portugal achieved 90% accuracy [28]. A study in Jharkhand and West Bengal used Landsat-8

OLI data to compare classifiers, finding that ANN and RF outperformed the non-parametric SVM in classifying agricultural, forest, industrial, and waterbody land types [24]. A study conducted in 2021 in Brittany, France, on croplands using a Sentinel-2 dataset tool, along with CNN, RF, and SVM classifiers, found a higher rate of urban development from 2007 to 2018 [29]. The multispectral satellite Sentinel-2 dataset tool was used to conduct this study in Greater Hungary. The neural network provides more accuracy in the present study for different land type classifications [30]. A study was conducted in Jharkhand and West Bengal using different classifiers, RF, ANN, SVM, SAM, MD, and ARTMAP, along with the Landsat-8 dataset tool, to determine the change detection in LULC. This study also determines the accuracy level of all algorithms and found that RF had the highest accuracy level of 89% [31].

Leveraging the cost-effectiveness, fine scale, and multi-temporal nature of remote sensing imagery, researchers have established LULC change mapping as a vital tool for historical and ongoing environmental monitoring [3]. It played a key role in studying phenomena such as floods and droughts associated with climate change in this study, and diverse ML approaches were utilized. It was discovered that Non-Nested Generalized Exemplars yielded the most favorable outcomes [32]. Table 9.1 presents the ML approach used in the agriculture domain. RF, Artificial Neural Network and SVM were used for change detection in Bangladesh from 2002 to 2022 using five land type classes and achieved 89% accuracy [33]. Researchers have worked on different ML techniques used to create crop maps in China with complex planting structures. The study used NDVI time series and the Sentinel-2 dataset tool to classify the crop structure [25]. Several studies have utilized satellite imagery and ML for land-use change detection. A recent study employed the Google Earth Engine (GEE), Landsat-5 and Landsat-9 datasets, and an RF classifier to analyze changes in the Chure region of Nepal. This study identified a significant increase in agricultural and forest land, along with a decrease in built-up land [34]. Additionally, supervised classification methods were applied in the Jabalpur district of Madhya Pradesh, India using a Sentinel-2 dataset [35].

9.4.2 DL-Based LULC Classification

Significant milestones have been achieved in the classification of LULC using satellite imagery with the aid of DL. Unlike the conventional approach, where the features must be defined, a DL model, such as a CNN, does this by itself. This enables them to identify numerous relations within images, which means that they are potentially more accurate in

Table 9.1 ML-based LULC classification.

Land type	Classifiers	Dataset tool	Study area	Best overall accuracy	Source
Crop land	ML algorithms-KNN and RF	Sentinel-2 multi-spectral image	West Bengal	95%	[27]
Woodland, grassland, shrubland	SVM, RF	Sentinel-2, Landset-8 compared with Sentinel-1	Across Europe	77.60%	[26]
Forestland	RF classifier	SAR Sentinel-1 SAR and optical Sentinel-2	Portugal.	90%	[28]
Cropland, water bodies, vegetation, and non-vegetation cropland with bare soil	SVM, RF, CNN	SPOT 5 and Sentinel 2	Crozon Peninsula (Brittany, France)	90%	[29]
Agriculture land, water bodies, Grassland, Forest land, Bare soil, Urban planning	RF, SVM, ANN	Sentinel 2	Great Hungarian Plain	90%	[30]
Water, soil, vegetation, Build-up	RF, DT, DTNB,	Landsat 8 OLI	north of Iran (Mazandaran Province)	>90%	[32]
Forest Land, Industrial Land, Residential area, Scrub area, Water body	RF, Fuzzy ARTMAP, SVM, SAM and MD	Landsat-8	West Bengal	89-92%	[24]
Agricultural land, built-up area, vegetation land, Rivers	RF, SVM, SAM and MD	Landsat 8	West Bengal	89%	[31]

(*Continued*)

Table 9.1 ML-based LULC classification. (*Continued*)

Land type	Classifiers	Dataset tool	Study area	Best overall accuracy	Source
Waterbody, Build-Up, Agriculture, Barren land Vegetation	ANN, RF SVM	Landsat-7	Bangladesh	89%	[33]
Cropland, Woodland, Grassland, Shrubland, wetland, water, Build-up area	RF, SVM	Sentinel-2	Handan city North China Plain	93%	[25]
forestland, Agriculture, Build-up, Barren land, Waterbody	RF	Google Earth Engine (GEE), Landsat-5 and Landsat-9	Nepal	90%	[34]
Agriculture land, Build-Up land, Water body	Supervised classification methods	Sentinel-2	Jabalpur District Madhya Pradesh	85.70%	[35]

KNN, K-Nearest Neighbors; RF, Random Forest; SVM, Support Vector Machine; CNN, Convolutional Neural Networks; ANN, Artificial Neural Network; Fuzzy ARTMAP, Fuzzy Adaptive Resonance Theory-Supervised Predictive Mapping; SAM, Spectral Angle Mapper; MD, Mahalanobis Distance; OA, Overall Accuracy.

the classification of land types, such as forests and water bodies. DL has a strong performance when it comes to big data, which also makes it convenient for dealing with the numerous amounts of satellite images. As tradeoffs, this approach offers advantages that include automation and flexibility in response to variations in land cover. However, challenges that need to be addressed include the high computational costs of training such models and the requirement for large labeled datasets. However, these challenges can be mitigated, and DL can be acknowledged as an effective approach to LULC classification, allowing researchers to unleash the potential of satellite imagery for various purposes.

A study conducted in Tehran utilized Sentinel-2 and Landsat-8 imagery within the Google Earth Engine platform. A Random Forest (RF) classifier was employed to generate LULC maps, achieving an overall accuracy of 89% [33]. Different classes of LULC, such as water bodies, building land, forest land, barren land, agricultural land, and wetlands, were used to study the Selangor area using Landsat-5 and Landsat-8 dataset tools, SVM, and CA-ANN models for change detection in the present study. The results showed that the number of water bodies, buildings, and barren land increased. Agricultural land, wetlands, and forest land decreased during the 1991–2021 period [36]. Another study in Portugal employed a combination of Sentinel-2 data, Recurrent Neural Networks (RNNs), and a decision support system to analyze agricultural land with peach and grape crops. The study suggests that low-pesticide-use agroecosystems can promote biodiversity in agricultural systems [37]. A different study in Ukraine also analyzed the categorization of major crops, such as wheat, sunflower, maize, soybeans, and others. This study used a mixture of different satellite data and a CNN for image classification [27].

A European study used Sentinel-2 data and a CNN to enhance geographical maps and produce LULC change detection with 98% accuracy [38]. In China, another study used the GF-1 satellite and a DCNN to suggest a rapid and precise process for large-scale land use classification. The process sought to automatically perform the work and reduce the resources needed, and by doing so, an accuracy of 89% was achieved [39]. Finally, research in India examined the mapping of four land cover classes (water bodies, barren land, greenery areas, and building areas) employing satellite datasets. This study used pre-pixel supervised methods to train the model with Indian satellite data [40].

A Canadian study compared Deep Convolutional Neural Networks (DCNNs) with conventional semi-automated approaches to LULC classification. The study utilized multispectral data from Landsat-5 and Landsat-7 satellites based on LC classes, such as forest land and vegetation land. The results showed that DCNNs can generate LULC maps much faster than the semi-automated techniques currently available [3]. In another study conducted in the Brazilian Amazon region in October 2020, researchers utilized a CNN for forest land classification using Landsat-8 data. A novel method or algorithm implemented with remote sensing datasets was proposed, achieving promising results [14]. Table 9.2 summarizes the satellite datasets, study areas, and DL approaches used in various agricultural applications.

A study was conducted in France using DL algorithms such as RNN, RF, and SVM on grassland agricultural land, forest land, and water bodies using a multispectral dataset, and it was found that RNN provides good results for classification compared to the state-of-the-art classifier [44]. Using building land, green land, and forest land studies were conducted in Germany using remote sensing datasets, such as Sentinel-2, and a 2D semantic dataset of high resolution for the encoder and decoder, which

Table 9.2 DL-based LULC classification.

Land type	Classifiers	dataset tool	Study area	Best overall accuracy	Source
Cropland, barren land, grassland, and water bodies	SVM, ANN classifier	Sentinel-2 and Landsat-8	Tehran and its suburbs	89%	[33]
Waterbodies, Developed land Forest land, Agriculture land	SVM, CA-ANN, ANN-MLP	Landset-5	Selangor	90%	[36]
Agriculture land	SVM, RNN	Sentinel-2	Portugal	–	[37]
Agriculture Land	NN, RF, CNN	Landsat-8, Sentinel-1	Ukraine	90%	[41]
Agriculture land	CNN	Sentinel-2	34 European countries	98%	[38]
Cropland, forest land, Grassland, Bare land, wetland	CNN	Labdsat-8, Sentinel-1, Sentinel-2	Ukraine	>92%	[42]
Garden Land, Forest land Bare land, Water land, Grass land	DCNN	GF-1 satellite	China	89%	[39]
Waterbody, build-up area, Greenery, Barren land	Per-pixel supervised techniques	Landsat-7,8 Sentinel-2	India	90%	[40]
Green land, Water bodies, Agriculture	CNN	Landsat-5/7, Multispectral satellite	Canada	88%	[3]
Forest land	CNN	Landsat-8	Brazilian Amazon region	99%	[43]
Grass land, Water body, Forest, Agriculture land	RNN, RF, SVM	–	France	–	[44]
Building lands, Green land	CNN	ISPRS and 2-d semantic	Germany	95%	[14]
Water bodies, Buildup Area	RF	Landsat-8 OLI	Brazil	94%	[45]
Forestland, vegetation land, Build-up area	CNN	Sentinel-2		95%	[46]

SVM, Support Vector Machine; CA-ANN, Cellular Automata-Artificial Neural Network; ANN-MLP, Artificial Neural Network-Model for Prediction; CNN, Convolutional Neural Networks; RNN, Recurrent Neural Networks; DCNN, Deep Convolutional Neural Networks.

reduces spatial dependencies [14, 46]. LULC classification performed in Brazil using the Landsat-8 OLI dataset tool and class types, including waterbodies and built-up areas, with an RF classifier for an environmental study, found that using the environmental variable with spectral variables improves monitoring in ecotone areas of neotropical regions [45].

9.5 Analysis and Discussion

The systematic review identified three primary learning paradigms used for LULC mapping in agriculture: supervised, unsupervised, and

Table 9.3 ML and DL-based LULC classification.

Categories	Properties	Classifiers	Source
Pixel-based	Pixel-based pixel is treated as the smallest unit; each pixel is classified into different categories based on its spectral characteristics.	SVM, RF, K-means clustering, MLC, NDVI	[3, 26]
Sub-pixel-based	where single pixel representing more than one object	LSU, Fuzzy Classification, MTMF	[51]
Object-based	pixels are grouped together to meaningful object	SVM, RF, NN	[44]
Spectral-based	Focus of unique properties of an object captured by sensor	NDVI, NDWI, PCA, SAM	[45]
Vegetation indices	combines the spectral bands to monitor vegetation health, density and cover	NDVI, NDWI, GNDVI	[33]
Change detection	Identify the difference in the phenomenon by observing it at different times	Image ratioing, image differencing, CVA	[40]

SVM, Support Vector Machine; RF, Random Forest; NDVI, Normalized Difference Vegetation Index; Fuzzy ARTMAP, Fuzzy Adaptive Resonance Theory-Supervised Predictive Mapping; SAM, Spectral Angle Mapper; PCA, Principal Component Analysis; LSU, Linear Spectral Unmixing; MTMF, Mixture Tuned Matched Filtering; NDWI, Normalized Difference Water Index; NN, Neural Network; GNDVI, Green Normalized Difference Vegetation Index; CVA, Change Vector Analysis.

semi-supervised learning paradigms. In the supervised method, a labeled dataset is used to train the model. In supervised methods, SVM, DT, LR, KNN, NN, and RF techniques or classifiers can be used for classification. The unsupervised method uses unlabeled data to train the model [47]. In unsupervised methods, different techniques can be used, such as K-mean clustering, PCA [48], independent component analysis, Gaussian mixture model [49], and self-organizing map [50]. Semi-supervised techniques use a combination of supervised and unsupervised techniques. It uses labeled and unlabeled data to train the model. This technique is used when it is difficult to obtain large amounts of expensive labeled data. Beyond learning paradigms, we explored various remote sensing techniques for image analysis, including sub-pixel-based, pixel-based, spectral-based, object-based, and vegetation indices. In addition, techniques such as change detection and LiDAR mapping were analyzed. Table 9.3 presents different remote sensing-based ML and DL approaches used for agricultural land.

9.6 Conclusion

This systematic review analyzed ML and DL approaches for agricultural LULC classification using satellite datasets. The study explored different satellite data sources and relevant remote sensing tools. This review examines various image analysis techniques in remote sensing, including pixel-based, subpixel-based, object-based, and spectral-based approaches. It highlights object-based techniques as the most effective method for LULC classification in agriculture. Remote sensing techniques enable applications in agriculture, including disease prediction, yield estimation, growth monitoring, soil moisture analysis, crop type classification, harvest planning, and environmental impact assessment, as supported by studies using Landsat and Sentinel data. Further studies are required in India in the field of agriculture.

AI disclosure statement: The author acknowledges the use of AI-based tools for minor language editing or grammatical correction. However, all intellectual contributions, analyses, and conclusions presented in this chapter are entirely the author's own.

References

1. Albahri, G., *et al.*, Enhancing essential grains yield for sustainable food security and bio-safe agriculture through latest innovative approaches. *Agronomy*, 13, 7, 1709, 2023.
2. Sharma, N. and Chawla, S., Digital Change Detection Analysis Criteria and Techniques used for Land Use and Land Cover Classification in Agriculture. Institute of Electrical and Electronics Engineers, IEEE, pp. 331–335, Jul. 2023, doi:10.1109/icacite57410.2023.10182604.
3. Alhassan, V., Henry, C., Ramanna, S., Storie, C., A deep learning framework for land-use/land-cover mapping and analysis using multispectral satellite imagery. *Neural Comput. Appl.*, 32, 12, 8529–8544, Jun. 2020, doi:10.1007/s00521-019-04349-9.
4. Hussain, M., Chen, D., Cheng, A., Wei, H., Stanley, D., Change detection from remotely sensed images: From pixel-based to object-based approaches. *ISPRS J. Photogramm. Remote Sens.*, 80, 91–106, 2013.
5. Vyas, N., Singh, S., Sethi, G.K., A novel image fusion-based post classification framework for agricultural variations detection using Sentinel-1 and Sentinel-2 data. *Earth Sci. Inf.*, 18, 1, Springer Science and Business Media LLC, Dec. 13, 2024, doi:10.1007/s12145-024-01506-0.
6. Rajmohan, G., Chinnappan, C.V., William, A.D.J., Balakrishnan, S.C., Muthu, B.A., Manogaran, G., Revamping land coverage analysis using aerial satellite image mapping. *Trans. Emerging Telecommun. Technol.*, 32, 7, e3927, 2021.
7. Zhao, Q., *et al.*, An overview of the applications of earth observation satellite data: impacts and future trends. *Remote Sens. (basel)*, 14, 8, 1863, 2022.
8. Álvarez-Mozos, J., Casal\'\i, J., González-Aud\'\icana, M., Verhoest, N.E.C., Correlation between ground measured soil moisture and RADARSAT-1 derived backscattering coefficient over an agricultural catchment of Navarre (North of Spain). *Biosyst. Eng.*, 92, 1, 119–133, 2005.
9. Wiseman, G., McNairn, H., Homayouni, S., Shang, J., RADARSAT-2 polarimetric SAR response to crop biomass for agricultural production monitoring. *IEEE J. Sel. Top. Appl. Earth Obs. Remote Sens.*, 7, 11, 4461–4471, 2014.
10. Ahmed, M., Aloshan, M.A., Mohammed, W., Mesbah, E., Alsaleh, N.A., Elghonaimy, I., Characterizing land surface temperature (LST) through remote sensing data for small-scale urban development projects in the gulf cooperation Council (GCC). *Sustainability*, 16, 9, 3873, 2024.
11. Wang, Y., Guo, E., Kang, Y., Ma, H., Assessment of Land Desertification and Its Drivers on the Mongolian Plateau Using Intensity Analysis and the Geographical Detector Technique. *Remote Sens. (basel)*, 14, 24, 6365, 2022.
12. MohanRajan, S.N., Loganathan, A., Manoharan, P., Survey on Land Use/Land Cover (LU/LC) change analysis in remote sensing and GIS environment: Techniques and Challenges. *Environ. Sci. Pollut. Res.*, 27, 24, 29900–29926, 2020.

13. Shehu, P., Rikko, L.S., Azi, M.B., Monitoring urban growth and changes in land use and land cover: A strategy for sustainable urban development. *Int. J. Hum. Capital Urban Manage.*, 8, 1, 111–126, 2023.
14. Zhang, C., Jiang, W., Zhang, Y., Wang, W., Zhao, Q., Wang, C., Transformer and CNN Hybrid Deep Neural Network for Semantic Segmentation of Very-High-Resolution Remote Sensing Imagery. *IEEE Trans. Geosci. Remote Sens.*, 60, 2022, doi:10.1109/TGRS.2022.3144894.
15. Ahmad, M.N., Shao, Z., Javed, A., Modelling land use/land cover (LULC) change dynamics, future prospects, and its environmental impacts based on geospatial data models and remote sensing data. *Environ. Sci. Pollut. Res.*, 30, 12, 32985–33001, 2023.
16. Baxter, W.B., *et al.*, A comprehensive approach to data collection, management, and visualization for terrain characterization in cold regions, 2024.
17. Zuo, Q., Zhou, Y., Wang, L., Li, Q., Liu, J., Impacts of future land use changes on land use conflicts based on multiple scenarios in the central mountain region, China. *Ecol. Indic.*, 137, 108743, 2022.
18. Shaikh, T.A., Mir, W.A., Rasool, T., Sofi, S., Machine learning for smart agriculture and precision farming: towards making the fields talk. *Arch. Comput. Methods Eng.*, 29, 7, 4557–4597, 2022.
19. Nedd, R., Light, K., Owens, M., James, N., Johnson, E., Anandhi, A., A synthesis of land use/land cover studies: Definitions, classification systems, meta-studies, challenges and knowledge gaps on a global landscape. *Land (basel)*, 10, 9, 994, 2021.
20. Regasa, M.S., Nones, M., Past and future land use/land cover changes in the Ethiopian Fincha Sub-Basin. *Land (basel)*, 11, 8, 1239, 2022.
21. Mehra, N. and Swain, J.B., Assessment of land use land cover change and its effects using artificial neural network-based cellular automation. *J. Eng. Appl. Sci.*, 71, 1, 70, 2024.
22. Xiong, J., *et al.*, Nominal 30-m cropland extent map of continental Africa by integrating pixel-based and object-based algorithms using Sentinel-2 and Landsat-8 data on Google Earth Engine. *Remote Sens. (basel)*, 9, 10, 1065, 2017.
23. Diek, S., Fornallaz, F., Schaepman, M.E., De Jong, R., Barest pixel composite for agricultural areas using landsat time series. *Remote Sens. (basel)*, 9, 12, 1245, 2017.
24. Talukdar, S., *et al.*, Land-use land-cover classification by machine learning classifiers for satellite observations-A review. *Remote Sens.*, MDPI AG, Apr. 01, 2020, doi:10.3390/rs12071135.
25. Luo, K., Lu, L., Xie, Y., Chen, F., Yin, F., Li, Q., Crop type mapping in the central part of the North China Plain using Sentinel-2 time series and machine learning. *Comput. Electron. Agric.*, 205, Feb. 2023, doi:10.1016/j.compag.2022.107577.
26. Ghassemi, B., Dujakovic, A., Żółtak, M., Immitzer, M., Atzberger, C., Vuolo, F., Designing a European-Wide Crop Type Mapping Approach Based on

Machine Learning Algorithms Using LUCAS Field Survey and Sentinel-2 Data. *Remote Sens. (basel)*, 14, 3, Feb. 2022, doi:10.3390/rs14030541.

27. Hudait, M. and Patel, P.P., Crop-type mapping and acreage estimation in smallholding plots using Sentinel-2 images and machine learning algorithms: Some comparisons. *Egypt. J. Remote Sens. Space. Sci.*, 25, 1, 147–156, Feb. 2022, doi:10.1016/j.ejrs.2022.01.004.
28. De Luca, G., Silva, J.M.N., Di Fazio, S., Modica, G., Integrated use of Sentinel-1 and Sentinel-2 data and open-source machine learning algorithms for land cover mapping in a Mediterranean region. *Eur. J. Remote Sens.*, 55, 1, 52–70, 2022, doi:10.1080/22797254.2021.2018667.
29. Xie, G. and Niculescu, S., Mapping and monitoring of land cover/land use (LCLU) changes in the crozon peninsula (Brittany, France) from 2007 to 2018 by machine learning algorithms (support vector machine, random forest, and convolutional neural network) and by post-classification comparison (PCC). *Remote Sens. (basel)*, 13, 19, Oct. 2021, doi:10.3390/rs13193899.
30. van Leeuwen, B., Tobak, Z., Kovács, F., Machine Learning Techniques for Land Use/Land Cover Classification of Medium Resolution Optical Satellite Imagery Focusing on Temporary Inundated Areas. *J. Environ. Geogr.*, 13, 1–2, 43–52, Apr. 2020, doi:10.2478/jengeo-2020–0005.
31. Huntingford, C., Jeffers, E.S., Bonsall, M.B., Christensen, H.M., Lees, T., Yang, H., Machine learning and artificial intelligence to aid climate change research and preparedness. *Environ. Res. Lett.*, 14, 12, 124007, 2019.
32. Jamali, A., Evaluation and comparison of eight machine learning models in land use/land cover mapping using Landsat 8 OLI: a case study of the northern region of Iran. *SN Appl. Sci.*, 1, 11, Nov. 2019, doi:10.1007/s42452–019-1527–8.
33. Nasiri, V., Deljouei, A., Moradi, F., Sadeghi, S.M.M., Borz, S.A., Land Use and Land Cover Mapping Using Sentinel-2, Landsat-8 Satellite Images, and Google Earth Engine: A Comparison of Two Composition Methods. *Remote Sens. (basel)*, 14, 9, May 2022, doi:10.3390/rs14091977.
34. Kafle, S., K.C., S., Poudyal, B., Devkota, S., Machine learning approach to detect Land Use Land Cover (LULC) change in Chure region of Sarlahi district, Nepal. *Arch. Agric. Environ. Sci.*, 8, 2, 168–174, Jun. 2023, doi:10.26832/24566632.2023.0802012.
35. Lohare, J., Nair, R., Sharma, S.K., Pandey, S.K., Ramakrishnan, S., Geospatial based Land Use Land Cover Change Detection in Jabalpur district, Madhya Pradesh. *Biol. Forum-an Int. J.*, 15, 10, 585, 2023.
36. Baig, M.F., Mustafa, M.R.U., Baig, I., Takaijudin, H.B., Zeshan, M.T., Assessment of Land Use Land Cover Changes and Future Predictions Using CA-ANN Simulation for Selangor, Malaysia. *Water (Switzerland)*, 14, 3, Feb. 2022, doi:10.3390/w14030402.
37. Alibabaei, K., *et al.*, A Review of the Challenges of Using Deep Learning Algorithms to Support Decision-Making in Agricultural Activities. *Remote Sens.*, MDPI, Feb. 01, 2022, doi:10.3390/rs14030638.

38. Helber, P., Bischke, B., Dengel, A., Borth, D., Eurosat: A novel dataset and deep learning benchmark for land use and land cover classification. *IEEE J. Sel. Top. Appl. Earth Obs. Remote Sens.*, 12, 7, 2217–2226, Jul. 2019, doi:10.1109/JSTARS.2019.2918242.
39. Yang, X., Chen, Z., Li, B., Peng, D., Chen, P., Zhang, B., A fast and precise method for large-scale land-use mapping based on deep learning, in: *IGARSS 2019–2019 IEEE International Geoscience and Remote Sensing Symposium*, pp. 5913–5916, 2019.
40. Bansal, C., *et al.*, IndiaSat: A Pixel-Level Dataset for Land-Cover Classification on Three Satellite Systems - Landsat-7, Landsat-8, and Sentinel-2, in: *Proceedings of 2021 4th ACM SIGCAS Conference on Computing and Sustainable Societies, COMPASS 2021*, Association for Computing Machinery, Inc., pp. 147–155, Jun. 2021, doi:10.1145/3460112.3471953.
41. Kussul, N., Lavreniuk, M., Skakun, S., Shelestov, A., Deep Learning Classification of Land Cover and Crop Types Using Remote Sensing Data. *IEEE Geosci. Remote Sens. Lett.*, 14, 5, 778–782, May 2017, doi:10.1109/LGRS.2017.2681128.
42. Kussul, N., Shelestov, A., Lavreniuk, M., Kolotii, A., Vasiliev, V., Land Cover and Land Use Monitoring Based on Satellite Data within World Bank Project, in: *2019 10th International Conference on Dependable Systems, Services and Technologies (DESSERT)*, pp. 127–130, 2019.
43. De Oliveira, J.P., Costa, M.G.F., Costa Filho, C.F.F., Methodology of data fusion using deep learning for semantic segmentation of land types in the amazon. *IEEE Access*, 8, 187864–187875, 2020, doi:10.1109/ACCESS.2020.3031533.
44. Ienco, Di., Gaetano, R., Dupaquier, C., Maurel, P., Land Cover Classification *via* Multitemporal Spatial Data by Deep Recurrent Neural Networks. *IEEE Geosci. Remote Sens. Lett.*, 14, 10, 1685–1689, Oct. 2017, doi:10.1109/LGRS.2017.2728698.
45. Zeferino, L.B., de Souza, L.F.T., do Amaral, C.H., Fernandes Filho, E.I., de Oliveira, T.S., Does environmental data increase the accuracy of land use and land cover classification? *Int. J. Appl. Earth Obs. Geoinf.*, 91, Sep. 2020, doi:10.1016/j.jag.2020.102128.
46. Naushad, R., Kaur, T., Ghaderpour, E., Deep transfer learning for land use and land cover classification: A comparative study. *Sensors*, 21, 23, Dec. 2021, doi:10.3390/s21238083.
47. Hindarto, D. and Santoso, H., Performance Comparison of Supervised Learning Using Non-Neural Network and Neural Network. *J. Nas. Pendidikan Tek. Inform.: JANAPATI*, 11, 1, 49–62, 2022.
48. Vaidya, N., *et al.*, Principal Component Analysis of Grain Yield and its Attributing Traits in Recombinant Inbred Lines of Rice (Oryza sativa L.) Developed for Submergence Tolerance. *Int. J. Plant Soil Sci.*, 36, 2, 143–148, 2024.
49. Guan, H., *et al.*, Improved Gaussian mixture model to map the flooded crops of VV and VH polarization data. *Remote Sens. Environ.*, 295, 113714, 2023.

50. Huang, A. and Chang, F.-J., Using a self-organizing map to explore local weather features for smart urban agriculture in northern Taiwan. *Water (basel)*, 13, 23, 3457, 2021.
51. Wu, K., Du, Q., Wang, Y., Yang, Y., Supervised sub-pixel mapping for change detection from remotely sensed images with different resolutions. *Remote Sens. (basel)*, 9, 3, Mar. 2017, doi:10.3390/rs9030284.

10

Remote Sensing and GIS-Based Approaches to LULC Analysis for Sustainable Agricultural Planning

Garima[1*], Ankita Kumari[2] and Soubhagya Ranjan Dhall[3]

[1]Department of Computer Science and Engineering, Vivekananda Global University, Jaipur, Rajasthan, India
[2]Department of Computer Science and Application, Vivekananda Global University, Jaipur, India
[3]New Product Launch, ABB Ltd., New Berlin, WI, USA

Abstract
Land use land cover (LULC) analysis is crucial for understanding the spatial and temporal dynamics of land systems, thereby supporting sustainable agricultural practices. In this chapter, the investigation focuses on the synergistic use of remote sensing, machine learning (ML), and cloud computing for effective and efficient LULC classification. Crop monitoring, resource optimization, and environmental planning can be enhanced using the Normalized Difference Vegetation Index (NDVI) and multi-resolution data from Sentinel, Landsat, MODIS, and UAVs. Classification accuracy can be enhanced by integrating advanced ML algorithms, and this process can be further improved by using the Google Earth Engine (GEE) for classification tasks. Various studies from India, Tunisia, and Europe are excellent examples of the use of geospatial methods for agricultural zoning, irrigation management, groundwater management, climate change strategies, and land health assessments. The following chapter focuses on the utilization of geospatial intelligence in food systems and environmental sustainability education, serving as a valuable resource.

Keywords: Land use land cover, sustainable agriculture, remote sensing, crop monitoring, land management, earth observation, precision farming

**Corresponding author*: garimalakhera04@gmail.com; ORCID: 0009-0001-4970-8638
Ankita Kumari: ORCID: 0009-0001-9048-7689

Gurwinder Singh, Vishakha Sood, Narayan Vyas, Surendra Yadav and Dankan Gowda V. (eds.) Geospatial Intelligence for Precision Agriculture, (197–212) © 2026 Scrivener Publishing LLC

10.1 Introduction

The global demand for sustainable agricultural landscape management has increased owing to population growth, urbanization, climate variability, and food security challenges. Spatial and temporal changes in farming ecosystems are broadly researched domains for both policymakers and researchers, and land use land cover (LULC) analysis is a useful tool for understanding these changes. Land changes over long periods are monitored using LULC mapping, which helps in understanding land resource dynamics [1]. Agricultural landscape monitoring using remote sensing along with geographic information systems (GIS) has made LULC analysis more effective, scalable, and data-driven. Satellite-based Earth Observation (EO) systems, such as Landsat and Sentinel-2, help gather essential temporal and geographically explicit data on variations in vegetation cover, cropping intensity, and soil health over time [2, 3].

The following chapter focuses on the use of the Normalized Difference Vegetation Index (NDVI) for the analysis of vegetation health and biomass. NDVI can be employed in agriculture to examine crop vigor, drought impacts, and yield forecasting [4]. Figure 10.1 depicts the different ranges associated with various vegetation conditions, such as vegetation density,

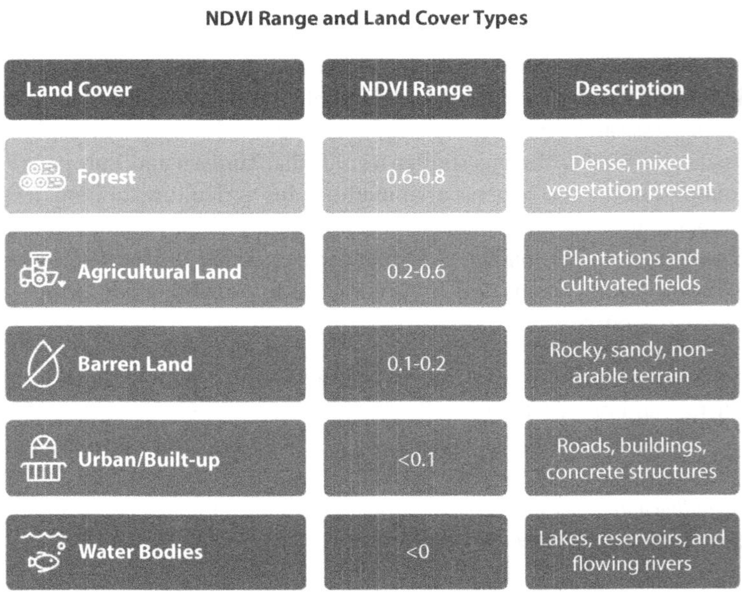
NDVI Range and Land Cover Types

Land Cover	NDVI Range	Description
Forest	0.6-0.8	Dense, mixed vegetation present
Agricultural Land	0.2-0.6	Plantations and cultivated fields
Barren Land	0.1-0.2	Rocky, sandy, non-arable terrain
Urban/Built-up	<0.1	Roads, buildings, concrete structures
Water Bodies	<0	Lakes, reservoirs, and flowing rivers

Figure 10.1 NDVI-based classification of land cover types.

sparse vegetation, or barren land, based on NDVI classifications. This framework enables accurate land-cover assessments and agricultural planning. The improvement in the accuracy of LULC classifications for agricultural zoning derived by NDVI is evident in this study and shows how machine learning (ML) algorithms, such as Random Forest (RF), Support Vector Machines (SVM), and k-Nearest Neighbors (k-NN), contribute to this enhancement [5, 6]. In geospatial settings, high-resolution data integration from hyperspectral sensors with unmanned aerial vehicles (UAVs) is beneficial for geospatial applications in LULC analysis. Farmers can make near-real-time decisions on irrigation, fertilization, and pest control based on centimeter-level resolution from UAVs, leading to improved productivity and reduced environmental effects [7].

Broader land management goals, such as evaluating urban sprawl, deforestation, and desertification, can also benefit from medium-resolution systems such as the Moderate Resolution Imaging Spectroradiometer (MODIS) and Landsat [8]. Cloud computing platforms, such as the Google Earth Engine (GEE), enhance the accessibility of geospatial processing tools, enabling the cost-effective analysis of large, multi-temporal satellite datasets in developing countries with limited computational infrastructure [9]. Local academics and institutions can support resilient agriculture and sustainable land-use planning by leveraging free satellite imagery and open-access platforms [10]. The use of cloud computing platforms, such as GEE, helps improve accessibility to tools for geospatial processing. This makes handling and computing huge, multi-temporal satellite datasets easy and cost-effective, which is highly challenging for developing countries with limited computational infrastructure. Local academics and institutions can support resilient agriculture and sustainable land-use planning by leveraging free satellite imagery and open-access platforms [11, 12]. The use of long-term Landsat imagery in assessing land degradation and shifts in cultivation patterns shows the critical relationship between nature and human interaction, as seen in studies conducted in Ethiopia and Tunisia, which show how remote sensing and GIS are important in agricultural planning. Similarly, research conducted in India and Vietnam has shown the usefulness of temporal LULC maps for controlling urban expansion into agricultural areas and adjusting zoning laws accordingly [13, 14]. These applications demonstrate the versatility of the framework across socioeconomic and ecological gradients.

However, the application of these technologies has its pros and cons. Image quality, seasonal fluctuation, and classification methodologies all affect the accuracy of LULC maps; to guarantee dependability, ground truthing and validations are necessary. Additionally, the combination of multi-source data (e.g., climate, topography, and socio-economic variables) into LULC models remains a methodological bottleneck that calls for the development of hybrid approaches that combine remote sensing with artificial intelligence (AI) and ML [15]. Thus, for modern agricultural planning, the combination of remote sensing and GIS for LULC analysis is necessary. These technologies not only provide mapping and tracking of land resources but also enable strategic interventions that focus on enhancing food security, climate resilience, and sustainable development. The following sections showcase how geospatial tools can change agricultural sustainability using real-world case studies and analytical frameworks.

10.2 Technological Advancements in LULC Mapping

Remote sensing is highly dependable for LULC mapping to detect and assess changes in land cover across broad geographic regions accurately. Recent breakthroughs in satellite photography, ML, remote sensing, and data integration technologies have further augmented the accuracy and utility of LULC products for resource management, urban planning, and environmental monitoring. High-resolution satellites, such as Landsat and IRS LISS-III, have improved the data available for analyzing temporal changes in LULC. Their ability to acquire detailed data corresponding to land cover types and their temporal dynamics helps track trends in urban growth and identify shifts in land use and how human activities affect the environment. The emphasis on high-resolution satellite imagery in this study is due to its importance in acquiring crucial data to identify changes in LULC and their effects on water supplies. In addition, the combination of various sensors and developments in data science analytics, such as the use of drones and web-based platforms, is more focused on important datasets that support the tracking and evaluation of the connections between LULC and water resources. In addition to this study, remote sensing faces several challenges associated with the use of remotely sensed data in assessing and monitoring land use. This study highlights the ongoing research and efforts to combine remote sensing data with ecosystem services, spatial patterns, and hydrological models. Table 10.1 presents various datasets utilized in different studies and how the choice of data influences the application of the study results in LULC classification analysis.

Table 10.1 Summary of remote sensing datasets used for LULC classification and agricultural applications.

Dataset	Resolution	Application	Source
Sentinel-2	30 m	Images are accessed by the Descartes Lab platform for processing and analysis.	[16]
Landsat 5	30 m	Identifies detailed observations of land cover types.	[17]
MODIS	250 m–1 km	Used to identify NDVI, essential for vegetative cover.	[18]
EuroSAT (Sentinel-2)	10 m	Determines crop classification.	[16]
L8-OLI	30 m	Provides multispectral imagery (with Sentinel-1 and Sentinel-2) showing seasonal variations.	[19]
LISS-IV	5.8 m	Used for land cover classification through ML models.	[20]

10.2.1 ML and AI

ML and AI techniques enhance the accuracy and efficiency of LULC mapping, monitoring, and change detection by enabling the automated analysis of complex spatial data. These technologies support urban planning, environmental management, and policy implementation. Algorithms, namely, RF, support vector machine (SVM), and artificial neural network (ANN), consistently achieve high classification accuracy. It is also used for urban LULC mapping, environmental impact assessment, deforestation monitoring, and forecasting future land cover changes. These technologies are typically used in diverse settings, including urban areas in Bangladesh and Australia. Many techniques for LULC mapping and dynamics have evolved from satellite-based observations, moving beyond traditional ground-based methods. However, these approaches require manual intervention, are tedious, and are time-consuming, especially when considering a larger landscape.

However, satellite- and aerial-photograph-based mapping of LULC is cost-saving, spatially extensive, multi-temporal, and time-effective. Utilizing advancements in remote sensing and microwave sensors, we can obtain satellite data at different spatial and temporal scales. To derive answers to the problem statements, the fetched satellite data were subjected to various ML techniques. These techniques are mainly bifurcated into supervised and unsupervised techniques. Supervised ML methods include a range of algorithms, such as support vector machines (SVM), random forest (RF), spectral angle mapper (SAM), fuzzy adaptive resonance theory-based supervised predictive mapping (fuzzy ARTMAP), Mahalanobis distance (MD), radial basis function (RBF), decision trees (DT), multilayer perceptron (MLP), Naive Bayes (NB), and maximum likelihood classifier (MLC). In contrast, unsupervised algorithms are used for pattern discovery, namely, the K-means algorithm and iterative self-organizing data (ISODATA), which also helps with classification without labeled data.

10.2.2 Cloud Computing Platforms

The efficiency and scalability of LULC mapping have been enhanced by cloud computing platforms, enabling dynamic planners and policymakers to conduct more thorough and expedient geographical analyses. GEE, which combines a vast collection of satellite imagery and geospatial datasets with potent cloud-based processing capabilities, is one of the most popular platforms for this purpose [21]. In addition to increasing computational efficiency, cloud-based processing has other advantages, such as providing facilities to collaborative researchers, automating workflows, and enabling faster decision-making by removing the load of data that are downloaded or stored in a computer system before the use of a cloud environment.

10.2.2.1 A Study of the Sonipat District, Haryana

This study focused on the Sonipat district in Haryana. The latitudes and longitudes of the Sonipat District are 280°48′15″ and 290°17′10″ north and 760°28′40″ and 770°12′45″ east, respectively. The city has strong connectivity *via* a broad-gauge railway that links it directly to major urban centers such as Delhi and Chandigarh. Climatically, the region experiences a continental weather pattern, characterized by extensive heat in the summer and extreme cold in the winter. In this area, the southwest monsoon season occurs from July to September. Landsat 7 (2011) and Sentinel 2 (2021) are used for the research, such as urban and built-in areas, bodies of water, and cropland [22].

10.2.2.2 Mekong Delta in Cambodia

This area is renowned for its variations in land usage, both within and between years. Talking about this area, the annual monsoon floods, which normally occur from July to November, affect the region. The floodplain is heavily farmed in one to three cropping cycles for the remainder of the year. Sentinel 1 and Sentinel 2 data were used to capture the images, which are maintained by the European Space Agency (ESA). High-resolution SPOT-6 and SPOT-7 imagery was also included to support fieldwork conducted in Cambodia. In November 2019, February 2020, and April 2020, images were used for classifier training [23].

10.3 Applications in Sustainable Agriculture

Precision farming is also well known for site-specific crop management, which is a modern practice for agriculture compared to traditional practices, which mainly focus on the basic resource utilization for specific conditions of different areas within a field. LULC mapping plays an important role in providing basic insights that inform data-driven agricultural decisions. High-resolution satellite imagery is used to detect variations in soil properties, vegetation health, moisture levels, and topography, enabling precise management actions such as seeding, fertilization, irrigation, and pesticide application. Farmers can use LULC data to prevent over-irrigation and under-fertilization in specific fields. Therefore, it allows for real-time monitoring and classification of crop types, growth stages, and field boundaries, which are generated through advanced ML and cloud-based platforms, such as GEE or Sentinel-2 data processed with CNN [24].

10.3.1 Climate Change Adaptation

Understanding how land-use changes affect local climate conditions, improve local resilience, and aid in climate change adaptation requires LULC mapping. By analyzing previous and current LULC data using satellite imagery and remote sensing tools, researchers can detect shifts in vegetation cover, urban expansion, deforestation, and changes in water bodies.

This forms the basis of environmental vulnerability. Furthermore, LULC trends help identify how land productivity and climate effects vary across regions. For example, areas that face more droughts or have decreased soil health can be recommended for drought-resistant crop varieties or sustainable water management techniques. Therefore, by combining LULC data with climatic variables, adaptive agricultural planning can be better informed, supporting improved food security, resource utilization, and community resilience in the face of climate change.

10.3.2 Soil Health and Carbon Sequestration

The close relationship between soil health and carbon sequestration is highlighted by the fact that effective management techniques improve both soil quality and carbon storage capacity. Most of the techniques explored aim to increase soil organic carbon (SOC) levels, including reduced tillage, the use of cover crops, and the incorporation of mycorrhizal fungi. These approaches promote microbial diversity and enhance soil structure, water retention, and nutrient cycling. Maintaining and assessing SOC is crucial for evaluating the sustainability of land management practices and their environmental impacts. SOC is a key indicator of soil fertility and long-term productivity, especially in agricultural landscapes where degradation risks are high. These tools allow researchers and land managers to identify trends, implement corrective actions, and optimize resource allocation. Thus, improving soil health through carbon-conscious land practices not only supports ecosystem functions but also plays a pivotal role in achieving sustainable agriculture and environmental resilience in the face of global changes.

10.4 Case Studies on LULC Mapping for Sustainable Agriculture

Developing sustainable agricultural planning strategies requires a clear understanding of LULC patterns and transformations. This does not limit the recognition of the existing patterns and distribution of topographical resources but rather the change in pattern over time by natural or unnatural interventions by humans. Several region-centric studies have been conducted using remote sensing and other geospatial tools to address these dynamics over the last couple of decades. However, the purpose of a study varies based on its geographic focus, temporal span, data sources, and intended application. Table 10.2 presents a comparative summary of

Table 10.2 Comparative analysis of recent LULC studies using remote sensing and ML techniques.

Study area	Data sources and platforms	Methodology	Application	Results	Limitations	Source
Söke/Aydın, Turkey	Sentinel-2, NDVI, GNDVI; GEE	PCA, RF classification	Agricultural land classification	High classification accuracy using vegetation indices and PCA. OA: 90%, Kappa: 0.87	Limited temporal scope	[24]
Vijayapura, Karnataka, India	Sentinel-2A, Landsat-5; ArcGIS, Google Earth Pro	NDVI, Supervised Classification (MLC)	LULC change over 25 years	Increased vegetation; good classification performance. OA: 88.16%	Mixed pixels	[25]
Shrirampur, Maharashtra, India	Landsat-8; GEE	RF (50/100 trees)	LULC classification and change detection	Accurate LULC maps. Training Accuracy: 99%, Validation Accuracy: 79–80%	Region-specific, less ground truth	[26]

(Continued)

Table 10.2 Comparative analysis of recent LULC studies using remote sensing and ML techniques. (*Continued*)

Study area	Data sources and platforms	Methodology	Application	Results	Limitations	Source
Kashmir Valley, India	Landsat 5, 7, 8; ERDAS Imagine, ArcGIS	Supervised Classification (MLC)	LULC trend analysis	Significant land cover changes detected. OA: 88.56%	Cloud cover issues	[27]
Pangari Watershed, India	LISS-III, Landsat-8; ERDAS Imagine, ArcGIS 10.3	NDVI, Supervised Classification	Watershed management	Transitions in land use and vegetation monitored *via* NDVI. Estimated OA: ~85%	Limited temporal resolution	[28]
European Union	Sentinel-1, LUCAS; Python, SNAP, ML pipeline	RF Classification	Crop type mapping (EU-scale)	First EU-wide crop type map (10 m). OA: 80.3%	Cross-region crop variability	[29]
Nord-Est Tunisia	Sentinel-2; QGIS, ENVI, Python	Supervised Classification (RF, KNN, MDC)	LU mapping in semi-arid zones	RF most effective. OA: 99.54%	Applicable only to similar climates	[30]

(*Continued*)

Table 10.2 Comparative analysis of recent LULC studies using remote sensing and ML techniques. (*Continued*)

Study area	Data sources and platforms	Methodology	Application	Results	Limitations	Source
Central North Italy	Sentinel-2; Python (PyTorch), SNAP	R-CNN (Recurrent-Convolutional Neural Network)	Land cover and crop classification	High performance deep learning on temporal features. OA: 96.5%	High computation cost	[31]
Landes Forest, France	GEDI, Sentinel-1, 2; Python (TensorFlow), SNAP, Google Colab	Deep Learning (U-Net), Multi-stream integration	Canopy height modeling	Created 10 m canopy map. MAE: 2.02 m	Complex data integration	[32]
Serralunga d'Alba, Italy	UAV multispectral, Sentinel-2; Python (PyTorch), QGIS, SNAP	CNN-based refinement of NDVI	Precision agriculture	Improved NDVI maps with UAV integration.	Limited spatial coverage	[33]
Sangrur, India	Sentinel-1, Sentinel-2, and ERDAS	Fusion of Sentinel-1 and Sentinel-2, then LULC classification	LULC Mapping	SVM achieved accuracy of 95.67% and a kappa of 0.9191	No temporal analysis	[34]

significant studies leveraging satellite images and ML approaches, along with their relevant key findings.

As seen in the table, the diversity of approaches reflects the complexity of land systems and the need for tailored solutions. For instance, while some studies focus on high-resolution classification for precision agriculture in specific districts, others attempt broader landscape-scale assessments to support regional planning. Similarly, classification accuracies, vegetation indices, and the degree of ground truthing vary, thereby influencing the reliability and applicability of the findings. Collectively, these case studies provide a rich evidence base for selecting appropriate methods, tools, and data sources for future LULC research. They also underscore the critical role of integrating geospatial data into agricultural policy formulation, particularly in the face of climate variability, land degradation, and food-security concerns. Readers are encouraged to explore the individual entries for nuanced insights into the methodology selection, regional adaptation, and challenges associated with data resolution and validation.

10.5 Future Directions

The use of drones and UAVs enhances LULC mapping by providing high-resolution and timely spatial data, particularly for localized and small-scale land-use analysis. This makes it possible to accurately identify patterns in agriculture, urbanization, vegetation health, and small-scale land-use changes. Accurate and speedy land feature classification is further improved by integrating AI-powered image analysis. The Internet of Things (IoT) can continuously monitor variables such as temperature, vegetation indices, soil moisture, and air quality, integrating with LULC data for enhanced analysis. These sensors provide dynamic inputs for adaptive land management techniques. When placed throughout metropolitan areas, forest regions, or agricultural fields. IoT enables prompt decision-making and quick reactions to changes in the environment or effects caused by humans when combined with edge computing and cloud platforms. Through public participation in data collection, such as geotagging photographs, reporting land changes, or categorizing satellite imagery, researchers can gain a variety of firsthand insights that might not otherwise be accessible.

These contributions serve as useful sources for validating AI-driven models and remotely sensed data, in addition to enriching datasets. Non-experts can now contribute meaningfully to such initiatives through platforms like OpenStreetMap and smartphone-based environmental monitoring applications. Participatory mapping, which directly involves local communities, helps create and analyze local maps. In a bottom-up approach, residents record land usage, cultural legacy, resource allocation, and environmental issues based on their personal experiences. This practice ensures that land management plans are in line with community needs and priorities, promotes ownership, and improves spatial literacy. Interestingly, this combination of knowledge from scientific and indigenous/local systems develops more sustainable and inclusive land-use planning through participatory GIS tools and workshops. Citizen and EO data are connected through this platform, which enables the analysis and sharing of data from various campaigns, allowing collaborative mapping and environmental monitoring through mobile and web applications [35].

10.5.1 Conclusion

By providing rapid, accurate, and scalable insights into land dynamics by leveraging the integration of cloud computing platforms, ML, AI, and remote sensing technologies, LULC mapping plays a significant role in advancing sustainable agricultural practices and land management, as it enables systematic monitoring of land dynamics and supports data-driven decision-making. These developments bring forward an effective strategy for monitoring crops, optimizing resources, managing soil health, and adapting to climate change. Cloud platforms, such as GEE, allow the processing, analysis, and extraction of insights from vast geospatial datasets, enabling near-real-time tracking of land cover changes across different spatial scales. This study presents the application of these methods across different geographic settings, illustrating their potential for LULC analysis under varying conditions. Ensemble learning techniques, such as RF, are highly consistent, accurate, and effective across numerous LULC applications. However, challenges such as limited temporal resolution, cloud cover interference, and dependency on ground truth data persist. Continued advancements in sensor technology and computational methods promise to overcome these limitations. Therefore, the convergence of remote sensing and AI-driven methodologies significantly enhances the capacity for sustainable land management and environmental conservation efforts. This synthesis offers a comprehensive reference for researchers and practitioners seeking to harness geospatial technologies for informed and sustainable decision making.

AI disclosure statement: The author acknowledges the use of AI-based tools for minor language editing or grammatical correction. However, all intellectual contributions, analyses, and conclusions presented in this chapter are entirely the author's own.

References

1. Mashala, M.J., Dube, T., Mudereri, B.T., Ayisi, K.K., Ramudzuli, M.R., A Systematic Review on Advancements in Remote Sensing for Assessing and Monitoring Land Use and Land Cover Changes Impacts on Surface Water Resources in Semi-Arid Tropical Environments. *Remote Sens.*, 15, 16, Jan. 2023, doi:10.3390/rs15163926.
2. Gupta, P., Singh, S.K., Gupta, P., Kanga, S., Mishra, V.N., Application of Remote Sensing and GIS Techniques for Identification of Changes in Land Use and Land Cover (LULC): A Case Study. *Indian J. Sci. Technol.*, 16, 46, 4456–4468, Dec. 2023, doi:10.17485/IJST/v16i46.2530.
3. Mariye, M., Jianhua, L., Maryo, M., Land use and land cover change, and analysis of its drivers in Ojoje watershed, Southern Ethiopia. *Heliyon*, 8, 4, e09267, Apr. 2022, doi:10.1016/j.heliyon.2022.e09267.
4. NDVI Ranges Explained: Boost Crop & Vegetation Health, Accessed: May 29, 2025, [Online] Available: https://farmonaut.com/remote-sensing/understanding-ndvi-ranges-from-agricultural-land-to-vegetation-health.
5. Linh, N.H.K., Pham, T.G., Pham, T.H., Tran, C.T.M., Nguyen, T.Q., Ha, N.T., Ngoc, N.B., Land-Use and Land-Cover Changes and Urban Expansion in Central Vietnam: A Case Study in Hue City. *Urban Science*, 8, 4, 242, 2024. https://doi.org/10.3390/urbansci8040242.
6. Basha, U.I., Suresh, U., Raju, G.S., Rajasekhar, M., Veeraswamy, G., Balaji, E., Landuse and Landcover Analysis Using Remote Sensing and GIS: A Case Study in Somavathi River, Anantapur District, Andhra Pradesh, India. *Nat. Environ. Pollut. Technol.*, 17, 3, 2018.
7. Tsouros, D.C., Bibi, S., Sarigiannidis, P.G., A Review on UAV-Based Applications for Precision Agriculture. *Information*, 10, 11, Nov. 2019, doi:10.3390/info10110349.
8. Braimoh, A.K., Vlek, P.L.G., Land-cover change trajectories in northern Ghana. *Environ. Manage.*, 36, 3, 356–373, Sep. 2005, doi:10.1007/s00267-004-0283-7.
9. Gorelick, N., Hancher, M., Dixon, M., Ilyushchenko, S., Thau, D., Moore, R., Google Earth Engine: Planetary-scale geospatial analysis for everyone. *Remote Sens. Environ.*, 202, 18–27, Dec. 2017, doi:10.1016/j.rse.2017.06.031.
10. Radočaj, D., Šiljeg, A., Marinović, R., Jurišić, M., State of Major Vegetation Indices in Precision *Agriculture* Studies Indexed in Web of Science: A Review. *Agriculture*, 13, 3, Mar. 2023, doi:10.3390/agriculture13030707.
11. Dagnachew, M., Kebede, A., Moges, A., Abebe, A., Land Use Land Cover Changes and Its drivers in Gojeb River Catchment, Omo Gibe Basin, Ethiopia. *J. Agric. Environ. Int. Dev. JAEID*, 114, 1, Jul. 2020, doi:10.12895/jaeid.20201.842.
12. Adesina, E.A., Muhammad, H.I., Assessment of land degradation using remote sensing approach. *Geod. Cartogr.*, 49, 3, Oct. 2023, doi:10.3846/gac.2023.16007.

13. Borana, S.L., Yadav, S.K., Parihar, S.K., Paturkar, R.T., Integration of remote sensing & GIS for urban land use/cover change analysis of the Jodhpur city. 33rd INCA International Congress, Jodhpur, Rajasthan, India, 2013.
14. Tuan, N.T., Urbanization and land use change: A study in Vietnam. *Environ. Socio-econ. Stud.*, 10, 2, 19–29, Jun. 2022, doi:10.2478/environ-2022–0008.
15. Joshi, A., Pradhan, B., Gite, S., Chakraborty, S., Remote-Sensing Data and Deep-Learning Techniques in Crop Mapping and Yield Prediction: A Systematic Review. *Remote Sens.*, 15, 8, Jan. 2023, doi:10.3390/rs15082014.
16. Kunwar, S., Ferdush, J., Mapping of Land Use and Land Cover (LULC) Using EuroSAT and Transfer Learning. *Rev. Int. Géomat.*, 33, 1–13, 2024, doi:10.32604/rig.2023.047627.
17. Phuong, V.T., Thien, B.B., Land Use Change Mapping and Analysis Using Remote Sensing and GIS: A Case Study in Tam Ky City, Quang Nam Province, Vietnam. *J. Multidiscip. Appl. Nat. Sci.*, 4, 2, 210–224, Apr. 2024, doi:10.47352/jmans.2774–3047.208.
18. Khaldi, R., *et al.*, TimeSpec4LULC: a global multispectral time series database for training LULC mapping models with machine learning. *Earth Syst. Sci. Data*, 14, 3, 1377–1411, Mar. 2022, doi:10.5194/essd-14–1377-2022.
19. Bhungeni, O., Ramjatan, A., Gebreslasie, M., Evaluating Machine-Learning Algorithms for Mapping LULC of the uMngeni Catchment Area, KwaZulu-Natal. *Remote Sens.*, 16, 12, Jan. 2024, doi:10.3390/rs16122219.
20. Prakash, S., Wijesingha, J., Nautiyal, S., Wachendorf, M., A Systematic Procedure for Land Use Land Cover Mapping Using Interpretable Machine Learning Model with Satellite Data, in: *2023 IEEE India Geoscience and Remote Sensing Symposium (InGARSS)*, pp. 1–3, Dec. 2023, doi:10.1109/InGARSS59135.2023.10490409.
21. Nasiri, V., Deljouei, A., Moradi, F., Sadeghi, S.M.M., Borz, S.A., Land Use and Land Cover Mapping Using Sentinel-2, Landsat-8 Satellite Images, and Google Earth Engine: A Comparison of Two Composition Methods. *Remote Sens.*, 14, 9, Jan. 2022, doi:10.3390/rs14091977.
22. Rana, D., Kumari, M., Kumari, R., Land Use and Land Coverage Analysis with Google Earth Engine and Change Detection in the Sonipat District of the Haryana State in India. *Eng. Proc.*, 27, 1, 2022, doi:10.3390/ecsa-9–13366.
23. Orieschnig, C.A., Belaud, G., Venot, J-P., Massuel, S., Ogilvie, A., Input imagery, classifiers, and cloud computing: Insights from multi-temporal LULC mapping in the Cambodian Mekong Delta. *Eur. J. Remote Sens.*, 54, 1, 398–416, Jan. 2021, doi:10.1080/22797254.2021.1948356.
24. İnalpulat, M., Civelek, N., Uşaklı, M., Genç, L., Agricultural Land Classification Using Vegetation Indices, PCA, and Google Earth Engine: Case Study of Söke/Aydın. *ÇOMÜ Ziraat Fak. Derg.*, 11, 1, Jul. 2023, doi:10.33202/comuagri.1295054.
25. Makandar, A. and Kaman, S., Land Use Land Cover Study of Sentinel-2A and Landsat-5 Images Using NDVI and Supervised Classification Techniques.

Comput. Intell. Mach. Learn., 2, 2, 20–29, Oct. 2021, doi:10.36647/CIML/02.02.A003.

26. Pande, C.B., *et al.*, Characterizing land use/land cover change dynamics by an enhanced random forest machine learning model: a Google Earth Engine implementation. *Environ. Sci. Eur.*, 36, 1, 84, Apr. 2024, doi:10.1186/s12302-024-00901-0.
27. Alam, A., Bhat, M.S., Maheen, M., Using Landsat satellite data for assessing the land use and land cover change in Kashmir valley. *Geojournal*, 85, 6, 1529–1543, Dec. 2020, doi:10.1007/s10708-019-10037-x.
28. Pande, C.B., Moharir, K.N., Khadri, S.F.R., Assessment of land-use and land-cover changes in Pangari watershed area (MS), India, based on the remote sensing and GIS techniques. *Appl. Water Sci.*, 11, 6, 96, May 2021, doi:10.1007/s13201-021-01425-1.
29. d'Andrimont, R., Verhegghen, A., Lemoine, G., Kempeneers, P., Meroni, M., van der Velde, M., From parcel to continental scale – A first European crop type map based on Sentinel-1 and LUCAS Copernicus *in-situ* observations. *Remote Sens. Environ.*, 266, 112708, Dec. 2021, doi:10.1016/j.rse.2021.112708.
30. Abida, K., *et al.*, Sentinel-2 Data for Land Use Mapping: Comparing Different Supervised Classifications in Semi-Arid Areas. *Agriculture*, 12, 9, Sep. 2022, doi:10.3390/agriculture12091429.
31. Mazzia, V., Khaliq, A., Chiaberge, M., Improvement in Land Cover and Crop Classification based on Temporal Features Learning from Sentinel-2 Data Using Recurrent-Convolutional Neural Network (R-CNN). *Appl. Sci.*, 10, 1, 238, Dec. 2019, doi:10.3390/app10010238.
32. Schwartz, M., *et al.*, High-resolution canopy height map in the Landes forest (France) based on GEDI, Sentinel-1, and Sentinel-2 data with a deep learning approach. *Int. J. Appl. Earth Obs. Geoinf.*, 128, 103711, Apr. 2024, doi:10.1016/j.jag.2024.103711.
33. Mazzia, V., Comba, L., Khaliq, A., Chiaberge, M., Gay, P., UAV and Machine Learning Based Refinement of a Satellite-Driven Vegetation Index for Precision Agriculture. *Sensors*, 20, 9, Jan. 2020, doi:10.3390/s20092530.
34. Vyas, N., Sethi, G.K., Singh, S., "Comparative Analysis of Different Classifiers on Fusion of Sentinel Datasets for Agriculture Land, in: *2024 4th International Conference on Technological Advancements in Computational Sciences (ICTACS)*, pp. 1173–1177, Nov. 2024, doi:10.1109/ICTACS62700.2024.10840995.
35. Moorthy, I., See, L., Batic, M., Matheus, A., The LandSense Engagement Platform: Connecting citizens with earth observation data to promote environmental monitoring, presented at the The ESA Earth Observation Week, Frascati, Italy, Nov. 2018, Accessed: May 30, 2025, [Online] Available: http://phiweek.esa.int/.

11

Image Classification Techniques for Precision Agriculture and Integrated Pest Management

Neetu Agrawal, Mehul Mahrishi* and Mukesh Kumar Gupta

Department of Computer Science & Engineering, Swami Keshvanand Institute of Technology, Management & Gramothan, Jaipur, Rajasthan, India

Abstract

The increasing demand for sustainable agricultural practices has accelerated the use of advanced technologies in pest management, mainly through precision agriculture (PA) and integrated pest management (IPM) systems. Uncontrolled pesticide uses negatively affects biodiversity, harming plants, animals, beneficial insects, and human health. The United Nations' Sustainable Development Goals highlight the need for environmental protection and biodiversity conservation, advocating restrictions on pesticide use. Correct and timely identification of pest species can substantially reduce the need for pesticides to control them. In this chapter, various AI-driven image classification techniques and a comprehensive framework for pest image classification are discussed. The chapter concludes with predictions for the future of pest management, emphasizing the importance of continuous innovation and the use of AI-powered precision technologies to address global food security concerns.

***Keywords*:** Artificial intelligence, deep learning, integrated pest management (IPM), precision agriculture, sustainable development goals

**Corresponding author*: mehul@skit.ac.in

Gurwinder Singh, Vishakha Sood, Narayan Vyas, Surendra Yadav and Dankan Gowda V. (eds.) Geospatial Intelligence for Precision Agriculture, (213–232) © 2026 Scrivener Publishing LLC

11.1 Introduction

Researchers have tried various pest control methods, including chemical and biological methods. In the 1940s, DDT was introduced for insect-borne disease control and agricultural pest management. Given the escalating demand for food due to population growth and the ongoing introduction of new pest species, pests require immediate attention. Chemical pesticide use is beneficial in one way, as it contributes to reducing the pest population and improving food production. However, the unregulated use of chemical pesticides harms biodiversity, pollutes soil, and contaminates water [1]. Precision agriculture (PA) promotes the selective application of pesticides, targeting only the affected areas instead of uniformly. This approach is essential for environmental conservation and advancing sustainable agricultural practices. Therefore, accurately identifying pest species is crucial to support more precise and limited pesticide use.

Integrated pest management (IPM) is an approach to keep pests away from crops. It integrates biological and cultural pest control methods, emphasizing the use of pesticides only as a final measure [2]. IPM is a crop protection strategy that prioritizes human health, friendly, non-targeted insects, and the environment [1]. The 1950s saw the introduction of IPM. In IPM, natural enemies, host–plant resistance, cultural control measures such as crop rotation, and chemical pesticides are used together to eliminate pests. Biological methods include the use of natural enemies of crop pests, such as pathogens, predators, and parasites [3, 4]. Cultural control of pests includes crop rotation (different crops are sown every season on the same field), multi-crops in a field, and plowing the field multiple times to expose the pests to pest enemies, such as birds and pathogens. The 1950s saw the introduction of IPM and educational programs for farmers [4]. Integrated pest control involves seeking the perfect combination of chemical and biological controls to minimize disruption to the agricultural ecosystem. Subsequently, the IPM concept expanded to encompass other methods, such as cultural control and host plant resistance. In 1972, the United States articulated IPM in its national policies.

The IPM pyramid, as depicted in Figure 11.1, prioritizes various methods for controlling pest attacks. Before implementing any of the pyramid's methods, one must identify the pests and assess the severity of pest attacks to determine the necessity of pest management. Threshold levels are available that ensure the need for pest control. Appropriate pest control methods can be adopted based on the severity of the attack. The goal of IPM is to manage pests and protect crops while considering the effects on

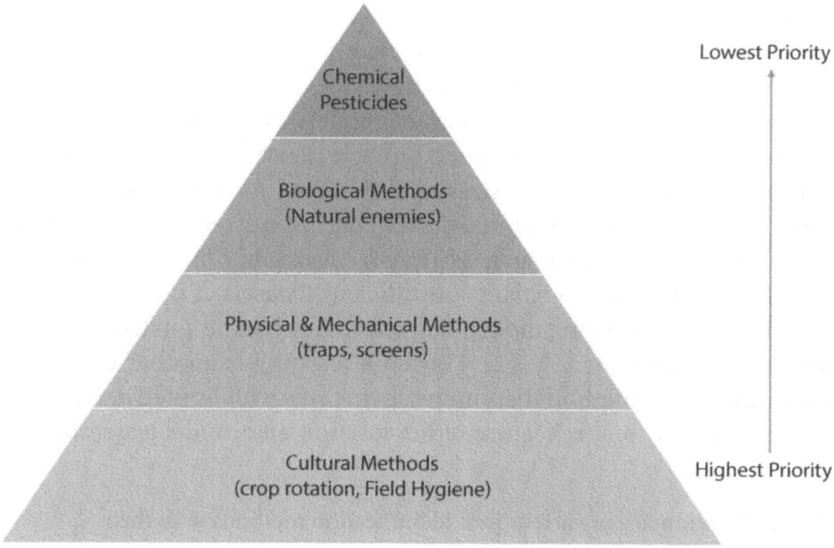

Figure 11.1 IPM pyramid showing the use of chemical pesticides as the last resort.

human health and biodiversity. However, the other side of the picture is that pesticides contribute to one-third of agricultural production, which is necessary to support the ever-increasing world population. Without its use, the world would face distressing losses in the production of various crops [5]. Accurate pest species identification enables targeted pesticide use and minimizes environmental harm. Manual identification of pest species requires detailed knowledge of insect taxonomy. Additionally, manual identification is a tedious and time-consuming task that requires field expertise. Deep learning techniques enable the precise classification and detection of pest-induced agricultural changes from field images, supporting smart pest management and change detection *via* remote sensing [6]. The constant emergence of new pest species makes it difficult to find an expert for accurate pest identification in agriculture.

PA leverages AI technology and data analytics to automate various crop and soil management tasks, helping to conserve the environment and support sustainable agriculture. AI plays a tremendous role in the automation of multiple subfields of agriculture, including crop disease recognition and detection, pest detection and identification, crop monitoring, and crop and soil quality assessment. Agricultural automation through AI began with machine learning techniques, which made the task easier, faster, and more efficient [7]. Deep learning techniques, which are capable of extracting features from complex images without human intervention and working with large numbers of classes more efficiently, have made further advancements [8, 9]. However, a significant drawback of deep learning techniques is the need for large datasets to train the models. Pest image data collection considerably hinders the pest detection and identification processes [10]. Natural obstacles, small pest sizes, and the need for sophisticated equipment to capture pest images make data collection strenuous. Researchers have adopted techniques such as transfer learning (TL) and fine-tuned deep learning models to balance insufficient datasets. TL combined with pre-trained deep learning models can effectively handle small datasets and yield improved results [11, 12]. Few-shot learning is another category of machine learning method that can perform tasks with limited data, but it is characterized by weaker feature representation and model generalization.

Table 11.1 Summary of various pest identification methods with their advantages and limitations.

Method of pest identification	Characteristics and limitations
Manual identification	Insect taxonomy experts observe the insect with their naked eyes. Thousands of pest species make the process intricate and time-consuming.
Machine learning	Automatically identify and classify the pest species. The process is both faster and more efficient. Image feature extraction is manual.
Deep learning (CNN)	Deep learning is more efficient than traditional machine learning, operating automatically without human intervention. Deep learning can extract features from complex images. However, it needs a large dataset for efficient training, and pest data collection has limitations.
Transfer learning	Small datasets with similar characteristics can utilize pre-trained CNN models with saved weights. It needs fewer parameter adjustments and saves training time, too.
Few-shot learning	This branch of machine learning algorithms can solve the problem of large datasets as it can work with only a few samples of images. However, feature representation is weaker and lacks model generalization.

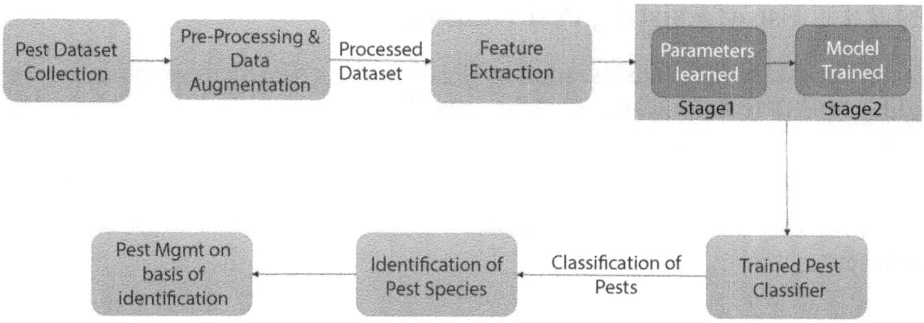

Figure 11.2 Complete pest classification process.

Table 11.1 summarizes the various techniques and their characteristics adopted for crop pest recognition, as well as the limitations that led to the search for a better method.

Figure 11.2 displays the complete pest classification process, including various steps: collecting the image dataset, preprocessing it, extracting features (only for machine learning algorithms), classifying it, and using it in real life. Pest image data are mostly unclear because pests are tiny and blend in with their surroundings. This makes the feature extraction task challenging. Thus, pre-processing the image dataset is crucial for obtaining the desired results from the deep learning models. Data augmentation is essential to compensate for small pest datasets because collecting large pest datasets is challenging.

This chapter highlights the urgent need to find a solution to the problem of crop pests in agriculture. This highlights the importance of automatic pest identification and the use of different AI-based methods for pest-related image-processing tasks. Machine learning and deep learning methods are discussed, which are frequently used by researchers for image classification tasks in general and are popular in pest image classification tasks. Various intermediate steps involved in the classification task, such as image data collection and image pre-processing methods, are also discussed.

11.2 Pest Image Acquisition

Pest image data collection is a significant hindrance to the fulfilment of automated pest detection and identification systems. Natural obstacles, small pest sizes, and the need for sophisticated equipment to capture pest images make data collection strenuous [13]. There are four ways to prepare a dataset for pest identification.

11.2.1 Use of Existing Pest Dataset

Several pest image datasets, such as IP102 and AgriPest, are publicly accessible. The IP102 dataset was created by [14] in 2019, the Xie1 dataset by [15] in 2015, and the Xie2 dataset by [16] in 2018, respectively. In 2018, [17] created a 10-class pest dataset, [18] created the MPD2018 dataset, and [19] created the AgriPest dataset, which is one of the most used pest datasets.

11.2.2 Images Taken from Search Engines

Pest images were collected from standard search engines such as Baidu, Google, Insert Images, IPM images, and Dave's Garden. Pictures obtained from online sources require careful annotation by experts.

11.2.3 Laboratory-Based Images

The pest image dataset prepared in the laboratory has the advantage of clarity and detail without background noise. Different insect postures were considered. To access images for lab-based dataset preparation, insect traps, such as sticky traps and pheromone traps, are placed in fields that work without human intervention. Digital cameras can capture insects stuck in traps from various angles.

11.2.4 On-Field Image Acquisition

Capturing pest images in uncontrolled fields is challenging because of environmental variability and pest size. Specialized acquisition equipment may be installed in the field, or handheld digital cameras can be used to capture pest images. The study in [20] used a wireless image monitoring system with an embedded system and camera. The monitoring system consists of a sticky trap board with a camera in front, which regularly sends images to a remote server. In [18], researchers designed task-specific image-acquisition equipment.

11.3 Image Pre-Processing

The image pre-processing task ensure that the image data were appropriately formatted for the machine learning model. Before feeding data to the classification model for training, the images must be prepared in terms of quality and quantity. This improves the chance of reducing the model training time and enhancing the model inference speed. In the presence of distortions, noise, or unclear images, image processing models based on CNN can be affected, leading to poor performance [21]. Three aspects require attention in image preprocessing: image quality, image quantity, and data imbalance, which are related to dataset quality. Improving image quality includes the following steps:

- Normalization scales pixel values to 0–1, thereby standardizing the image data. Variations in pixel intensity across images influence model processing. Min–max normalization, L1 and L2 normalization, and mean normalization are a few image normalization methods.
- Image resizing was performed according to the requirements of the CNN model used. Resizing makes all images consistent, resulting in enhanced model performance. Cropping the images is also part of preprocessing to enable the model to concentrate only on the relevant parts of the image.
- Contrast and brightness improvement enhances the features of the image, assisting the model in better recognizing images. Histogram equalization, gamma correction, and adaptive gamma correction are some of the methods used for contrast enhancement.
- Noise reduction: Noise or distortions in an image refer to objects other than the subject of interest, such as background elements. Removing noise is necessary for the model to learn meaningful features and to identify patterns. Gaussian, median, and morphological operations are basic image-denoising methods.
- The grayscale conversion of color images speeds up training if it preserves the representation of object features. Color space conversion is part of the pre-processing, highlighting different image features and enabling more efficient classification.

The image quantity refers to the number of images in the dataset used for model training. Deep learning models require vast amounts of data to learn features and operate efficiently. Data augmentation is a vital aspect of image data pre-processing, as it artificially increases the dataset size and assists models in generalizing effectively and reducing overfitting [22–24]. Sometimes, processes performed to highlight some features and remove distractions for better feature extraction as part of image pre-processing play the role of augmenting the dataset. The noise is removed to identify features more effectively, but noise is also added deliberately to increase the number of image samples and make the model robust [25–28]. Similarly, color space conversion is performed to enhance feature representation and, simultaneously, is used to add more samples to the dataset. Some standard techniques of data augmentation for image processing tasks include the following:

- Geometric transformations include rotation, horizontal and vertical flipping, scaling, translation, and shear.
- Noise insertion, including Gaussian noise and salt-and-pepper noise.
- Random and center-cropping
- Brightness and contrast adjustment
- GANs generate realistic pest images from noisy data, augmenting small datasets for robust model training.

11.4 AI-Driven Pest Identification

Various AI-based methods have proved fruitful in automating tasks in diverse fields, namely agriculture, medicine, education, finance, e-commerce, social networking websites, and security, to name a few. A subset of AI-based methods has been effective in computer vision tasks. Machine learning and deep learning methods are used for the image processing of agricultural crop pests and diseases for detection and classification. Situations with data scarcity also involve the use of few-shot learning algorithms. Next, we discuss these methods and their advantages and limitations in detail.

11.4.1 Machine Learning Models for Pest Image Classification

Traditional machine-learning techniques were the first step in the automation of agriculture. Different machine learning algorithms, like SVM and Artificial Neural Networks, have been implemented for tasks such as pest detection and classification, as well as crop disease detection and classification [14, 29]. Handcrafted features, selected manually using various algorithms, are a prerequisite for ML algorithms [30]. Features are selected based on the domain and characteristics deemed relevant to the task. Irrelevant feature selection would hinder the classification or detection task.

The most used feature descriptors include histogram-oriented gradients (HOG) for capturing edge gradients, color histograms (CH) for color features, local binary patterns (LBP) for capturing texture features, scale-invariant feature transforms (SIFT) for local features that are invariant to scale and rotation, Gabor filters, shape features, and a few more [15, 17]. The model then receives these features for training.

11.4.1.1 Support Vector Machine

The machine learning algorithm Support Vector Machine (SVM) is capable of performing classification tasks as well as regression tasks. Primarily used for binary classification, SVM can adapt to multiclass classification by implementing specific strategies [31]. The objective of the SVM algorithm is to find the most appropriate decision boundary, called a hyperplane. The hyperplane isolates the n-dimensional space into classes or categories with maximum distance between the data points. A few extreme points in each space, called support vectors, help find the hyperplane. The extreme points on negative and positive hyperplanes allow for the maximum margin between the two classes and are the support vectors. These support vectors establish a maximum margin between the classes, resulting in the optimal decision boundary known as a hyperplane, as shown in Figure 11.3. SVM can prove to be a robust model if the margin between the classes

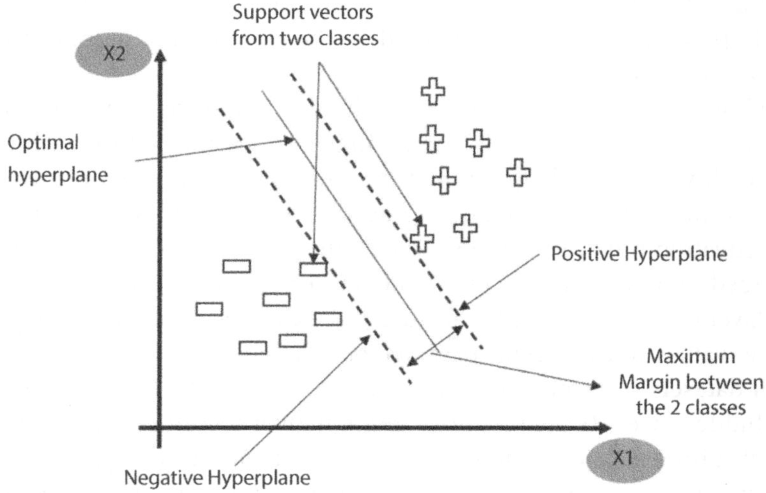

Figure 11.3 SVM algorithm depicting optimal hyperplane with the help of support vectors from two classes.

is evident. However, in the case of extensive datasets, SVM performs well with small datasets when features are carefully engineered [31]. Task and dataset characteristics affect how an algorithm performs.

11.4.1.2 Artificial Neural Network

The human brain is the inspiration for designing a neural network. Similar to neurons in the human brain, neurons or nodes in ANN are the central computational units. In its basic form, an ANN consists of an input layer, one or two hidden layers, and an output layer. The input layer receives the input, with the number of neurons in the layer matching the number of features in the input data. This layer does not perform any computations. Each neuron in the hidden layers receives input from all neurons in the previous layer and performs summation using input data, weight, and bias parameters. The summation is fed to the activation function which then determines whether and to what extent the neuron should be activated. The activation function adds nonlinearity to the network, enabling the model to learn complex and nonlinear image patterns. A linear transformation from input to output can only capture simple and linear image patterns, whereas real-world data cannot be simple and linear. Feature extraction and significant computations are performed in the hidden layers. The number of neurons in the output layer is equal to the number of classes in the classification task. In pest classification, it would be the number of pest species chosen for classification. The results (labels) from the output layer were matched with the actual values, and the error was computed. The weight parameters are updated based on the error values to minimize the loss or error values.

In Figure 11.4, neuron N_1 in the first hidden layer receives input from all neurons in the input layer. Computations were performed for each input data with weight and bias parameters. The summation is performed, and an activation function is applied to decide the activation of neuron N_1. If the activation function outputs zero for the neuron, its output will not be provided as input to the second hidden layer. Consequently, the next hidden layer does not receive the features N1 extracted. This mainly depends on the values of the weight parameters. Machine learning is suitable for small datasets with manually extractable features [32]. ANN with one or two hidden layers (called shallow neural networks) can efficiently classify complex images, and features can be extracted with fewer layers. Machine learning models are also favorable when computational resources are

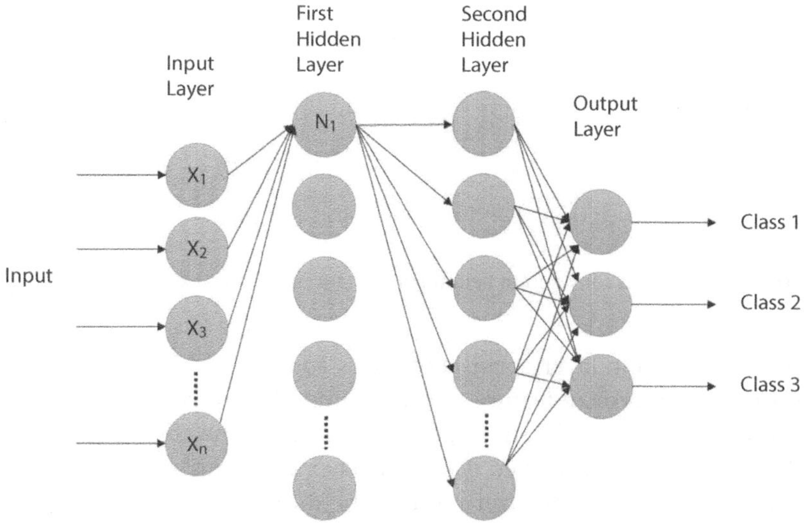

Figure 11.4 Example of a neural network for classification of three classes.

limited. However, when it comes to complex patterns in the data, extracting features is a challenge that leads to erroneous results for the model.

11.4.2 Deep Learning Models

Deep learning models use ANN with more than two hidden layers and require large amounts of data and many computations. However, DL models can represent complex image patterns owing to their multiple layers [3]. Deep learning continues to evolve and is impressive in various fields, particularly regarding the extent to which machines can automate tasks. DL models include many networks, such as CNNs, LSTMs, and RNNs. However, none of these fits all types of functions. Some algorithms are suited for specific tasks, such as CNN, image classification, object recognition, and detection tasks. A typical CNN model comprises input, convolutional, pooling, and fully connected layers, as shown in Figure 11.5. The CNN model operates in two phases: feature extraction and classification. The convolutional and pooling layers are part of the feature extraction phase, whereas the fully connected layer is part of the classification phase. Feature extraction processes raw images to extract information or features from the image, such as shape, color, texture, and edges. This simplifies the task of analyzing images for classification.

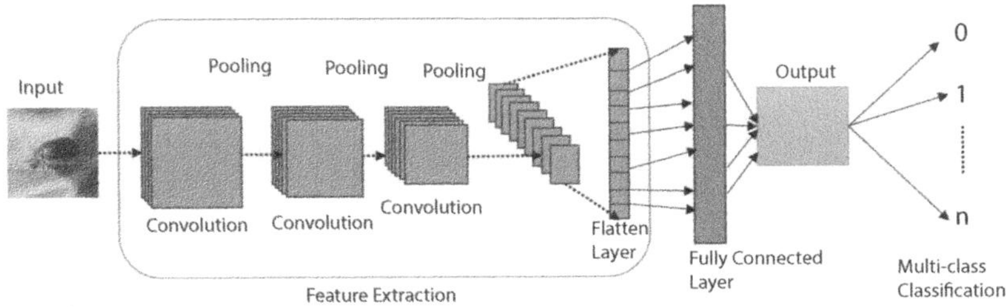

Figure 11.5 A typical convolution neural network for multi-class classification.

The convolution layer consists of filters applied to the image to extract features in the form of feature maps. Multiple filters were used to extract different features at different stages. The output of the convolutional layer is the input for the next layer. A pooling layer is present after every convolution layer. The pooling layer reduces the dimensionality and, therefore, reduces the computation time while simultaneously preserving important image information. There can be many convolution layers in a network that extract more complex image features. A fully connected layer, which consists of dense layers, does the task of connecting the convolutional layers to the output layer. The output from the convolution layer is a multi-dimensional array, and the dense layer requires a one-dimensional array as input. The two layers employ the flatten method to transform a multidimensional array into a one-dimensional array.

Different CNN models have been proposed for image processing, proving their worth in various tasks [33]. The earliest CNN networks were Neocognitron in 1979, ConvNet in 1989, and LeNet-5, constructed in 1998. With changes in requirements and improvements in computing resources, new and improved CNN networks have been presented, namely AlexNet, ZFNet, GoogleNet, VGGNet, and ResNet. AlexNet leverages GPUs to accelerate CNN training. It consists of five convolutional and three fully connected layers. It consists of two parallel CNN lines trained on two GPUs with cross connections. This was followed by GoogleNet and various Inception models, which were based on the concept of the inception module and included 22 layers. The Inception module provides an advantage in terms of filter size. Different filter sizes can be used with the other layers. Although these models were efficient at the same time, their complexity also increased. VGG-16 and VGG-19 were introduced in 2014, which are 16 and 19 layers deep, and they lowered the computing complexity by reducing the filter size to 3×3. In 2016, ResNet, an extensive network consisting of 152 layers that operates on residual blocks, underwent numerous transformations. ResNet uses shortcut and skip connections. Below is an explanation of the architectures of a few CNN models that are most used for pest image classification tasks in agriculture.

11.4.2.1 InceptionV3

Inception-based CNN models have evolved through various versions, beginning WITH Inception V1 to Inception-ResNet, which was proposed recently. The Inception V3 model has been implemented in several image classification tasks [34, 35]. Here, we delve into the Inception V3 model, which comprises a set of parallel convolutional branches with varying filter sizes that facilitate enhanced feature representation. The main components of Inception V3 are parallel branches, known as inception modules. Parallel convolutions allow for multi-scale feature extraction, making them resistant to changes in object size and orientation. Typically, these convolutions are 1 × 1, 3 × 3, and 5 × 5, but Inception V3 introduced the factorization of convolutions to reduce computational complexity. Therefore, this model used two consecutive 3 × 3 convolutions instead of 5 × 5 convolutions, as shown in Figure 11.6. However, this change did not reduce the feature representation efficiency. We used reduction blocks, such as 3 × 3 convolutions and max pooling layers, between the inception modules to downsample the feature maps. The reduction blocks also help reduce the spatial dimensions. Intermediate layers use auxiliary classifiers to address the vanishing gradient problem. However, they did not play any role in the final classification stage.

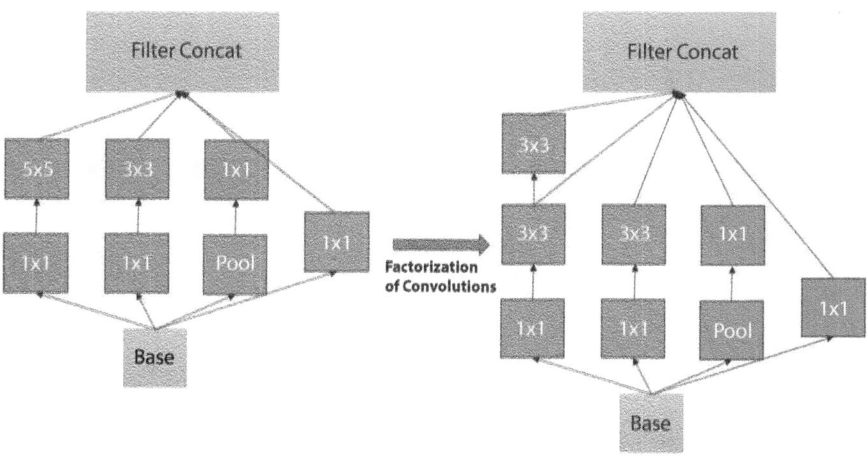

Figure 11.6 Inception V3 model consisting of Inception modules with convolutions of 5 × 5 factorized to two convolutions of 3 × 3.

11.4.2.2 ResNet

ResNet, or residual networks, are residual blocks designed to solve the vanishing gradient problem. As the number of layers in the network increases, the gradient decreases, eventually leading to zero. The residual blocks consist of shortcut connections that can skip one or more connections, as shown in Figure 11.7. Skip connections, which form identity mappings, bypass layers and connect the activation of one layer to the subsequent layers. Each residual block included a skip connection. The ResNet architecture comprises four stages, each containing a specific number of residual blocks. There are various versions of ResNet, such as ResNet18, ResNet34, ResNet50, ResNet101, and ResNet152, with differences in the number of residual blocks at each stage. The skip connections of ResNet allow deeper networks, enabling the model to learn features effectively. Researchers are interested in the ResNet architecture because it works well in image processing, even in agriculture. They want to learn more about it and use it for tasks such as image classification and detection [36–38].

11.4.2.3 DenseNet

The DenseNet architecture primarily includes dense blocks, transition blocks, and a classification head. Each dense block consists of layers, and every layer receives feature maps directly from all preceding layers. This allows for feature reuse and better gradient flow through the network, which is the highlight of the DenseNet models. A dense connection allows better feature reuse with fewer parameters [39]. However, because of the

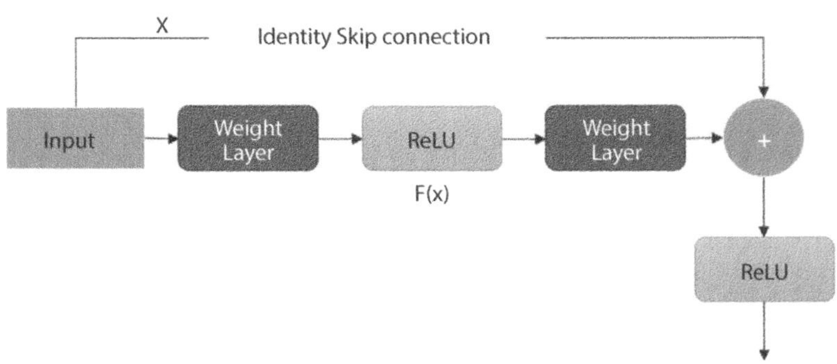

Figure 11.7 One residual block with single skip connection.

dense connections, it can be computationally expensive. Between each dense block, there is a transition block that performs downsampling and reduces the feature maps. A transition block consists of a 1 × 1 convolutional layer and an average 2 × 2 pooling layer. After the last dense block, there is a global average 7 × 7 pooling layer. The resultant feature maps from this pooling layer were flattened and connected to a fully connected layer for classification.

Large publicly available datasets, such as ImageNet, serve as training grounds for many pre-trained CNN-based models (such as VGGNet, AlexNet, ResNet, DenseNet, and MobileNet). Researchers have used such pre-trained models with their saved weights and trained only task-specific layers on their dataset, known as transfer learning. This enables deep learning models to be trained on smaller datasets [40, 41]. Fine-tuned models are pre-trained models that are re-trained on the dataset for the task. New training takes advantage of the previously acquired knowledge of the model, thus requiring fewer parameter adjustments [42]. This allows it to work with smaller datasets and low resource requirements. Using pre-trained models is beneficial when the dataset available for research is small, with very few images [34].

11.4.3 Few-Shot Learning

The requirement for a large dataset hinders the high performance of deep learning models. Transfer learning techniques and data augmentation partially overcome this challenge. Few-shot learning (FSL) is a specialized machine learning technique developed to solve the problem of small datasets. FSL is advantageous for limited data problems and helps manage class imbalances in datasets [43]. FSL finds its application in domains with the possibility of rare cases, such as the medical domain for rare diseases and agriculture for detecting or identifying tiny pest insects. Rare cases can provide only a few samples for training the models. FSL requires 1–10 samples per class for simpler tasks, but complex tasks may require more [43]. This set of algorithms is based on prior knowledge of data and learning. It works on the concept of meta-learning, that is, learning to learn. This is achieved by identifying similarities between various objects. FSL is primarily used for tasks related to supervised learning, such as image classification and object recognition [44]. FSL for image processing came into the limelight with a study [45] in which matching networks were introduced. However, the shortcomings of FSL lie in its feature representation and generalization capabilities, necessitating further research. The feature representation weakens because the technique relies on limited data, making it harder for the FSL model to differentiate patterns from noise.

11.4.4 Conclusion

Machine learning, deep learning, and few-shot learning methods are widely employed in image-processing tasks to automate processes across different domains while delivering accurate results. Among these, convolutional neural networks (CNNs) are dominant in image classification tasks. Their widespread adoption extends to numerous agricultural applications, such as pest classification and detection. Despite their effectiveness, achieving robust results with deep learning models often requires a substantial dataset of pest images, which can be challenging to obtain. Image augmentation, transfer learning with CNN models that have already been trained, and the new Few-Shot Learning (FSL) method have all made the problem of not having enough image data much easier to deal with. However, FSL cannot be used in many situations because it is not very good at representing and generalizing features. This highlights the need for further research to enhance its effectiveness and to broaden its usability.

AI disclosure statement: The author acknowledges the use of AI-based tools for minor language editing or grammatical correction. However, all intellectual contributions, analyses, and conclusions presented in this chapter are entirely the author's own.

References

1. Teixeira, A.C., Ribeiro, J., Morais, R., Sousa, J.J., Cunha, A., A systematic review on automatic insect detection using deep learning. *Agriculture*, 13, 3, 2023, ISSN: 2077–0472, doi:10.3390/agriculture13030713.
2. Isman, M.B., Challenges of pest management in the twenty first century: New tools and strategies to combat old and new foes alike. *Front. Agron.*, 1, 2019, ISSN: 2673–3218, doi:10.3389/fagro.2019.00002.
3. Sawicka, B. and Egbuna, C., Chapter 1-Pests of agricultural crops and control measures, in: *Natural Remedies for Pest, Disease and Weed Control*, C., Egbuna and B., Sawicka (Eds.), pp. 1–16, Academic Press, 2020, ISBN: 978–0-12–819304-4, doi:https://doi.org/10.1016/B978-0-12-819304-4.00001-4.
4. Deguine, J-P., Aubertot, J-N., Flor, R.J., Lescourret, F., Wyckhuys, K.A., Ratnadass, A., Integrated pest management: Good intentions, hard realities. a review. *Agron. Sustainable Dev.*, 41, 3, 38, May 2021, ISSN: 1773–0155, doi:10.1007/s13593–021-00689-w.

5. Tudi, M., Ruan, H.D., Wang, L., Lyu, J., Sadler, R., Connell, D., Chu, C., Phung, D.T., Agriculture development, pesticide application and its impact on the environment. *Int. J. Environ. Res. Public Health*, 18, 3, 1112, 2021, doi:10.3390/ijerph18031112.1.
6. Singh, G., *et al.*, A novel pixel-based deep neural network in posterior probability space for the detection of agriculture changes using remote sensing data. *Remote Sens. Appl.: Soc. Environ.*, 38, 101591, Apr. 2025, doi:10.1016/j.rsase.2025.101591.
7. Wei, D., Chen, J., Luo, T., Long, T., Wang, H., Classification of crop pests based on multi-scale feature fusion. *Comput. Electron. Agric.*, 194, 106 736, Mar. 2022, doi:10.1016/j.compag.2022.106736.
8. Tiwari, V., Joshi, R.C., Dutta, M.K., Dense convolutional neural networks based multiclass plant disease detection and classification using leaf images. *Ecol. Inf.*, 63, 101 289, 2021, ISSN: 1574–9541, doi:https://doi.org/10.1016/j.ecoinf.2021.101289.
9. Ali, F., Qayyum, H., Iqbal, M.J., Faster-pestnet: A lightweight deep learning framework for crop pest detection and classification. *IEEE Access*, 11, 104016–104027, 2023, doi:10.1109/ACCESS.2023.3317506.
10. Pattnaik, V.K.S.G. and Parvathi, K., Transfer learning-based framework for classification of pest in tomato plants. *Appl. Artif. Intell.*, 34, 13, 981–993, 2020, doi:10.1080/08839514.2020.1792034.
11. Ayan, E., Erbay, H., Var¸cın, F., Crop pest classification with a genetic algorithm-based weighted ensemble of deep convolutional neural networks. *Comput. Electron. Agric.*, 179, 105809, 2020, ISSN: 0168–1699, doi:10.1016/j.compag.2020.105809.
12. Jia, X., Jiang, X., Li, Z., Mu, J., Wang, Y., Niu, Y., Application of deep learning in image recognition of citrus pests. *Agriculture*, 13, 5, 2023, ISSN: 2077–0472, doi:10.3390/agriculture13051023.
13. Zhang, X., Bu, J., Zhou, X., Wang, X., Automatic pest identification system in the greenhouse based on deep learning and machine vision. *Front. Plant Sci.*, 14, 2023, ISSN: 1664–462X, doi:10.3389/fpls.2023.1255719.
14. Wu, X., Zhan, C., Lai, Y.-K., Cheng, M-M., Yang, J., Ip102: A largescale benchmark dataset for insect pest recognition, in: *2019 IEEE/CVF Conference on Computer Vision and Pattern Recognition (CVPR)*, pp. 8779–8788, 2019, doi:10.1109/CVPR.2019.00899.2.
15. Xie, C., Zhang, J., Li, R., Li, J., Hong, P., Xia, J., Chen, P., Automatic classification for field crop insects *via* multiple-task sparse representation and multiple-kernel learning. *Comput. Electron. Agric.*, 119, 123–132, 2015, ISSN: 0168–1699, doi:10.1016/j.compag.2015.10.015.
16. Xie, C., Wang, R., Zhang, J., Chen, P., Dong, W., Li, R., Chen, T., Chen, H., Multi-level learning features for automatic classification of field crop pests. *Comput. Electron. Agric.*, 152, 233–241, 2018, ISSN: 0168–1699, doi:10.1016/j.compag.2018.07.014.

17. Deng, L., Wang, Y., Han, Z., Yu, R., Research on insect pest image detection and recognition based on bio-inspired methods. *Biosyst. Eng.*, 169, 139–148, 2018, ISSN: 1537–5110, doi:10.1016/j.biosystemseng.2018.02.008.
18. L., L., *et al.*, Pestnet: An end-to-end deep learning approach for large-scale multi-class pest detection and classification. *IEEE Access*, 7, 45301–45312, 2019, doi:10.1109/ACCESS.2019.2909522.
19. Wang, R., Liu, L., Xie, C., Yang, P., Li, R., Zhou, M., Agripest: A large-scale domain-specific benchmark dataset for practical agricultural pest detection in the wild. *Sensors*, 21, 5, 2021, ISSN: 1424–8220.
20. Rustia, D.J.A., Chao, J.-J., Chiu, L.-Y., Wu, Y.-F., Chung, J.-Y., Hsu, J.-C., Lin, T.-T., Automatic greenhouse insect pest detection and recognition based on a cascaded deep learning classification method. *J. Appl. Entomol.*, 145, 3, 206–222, 2021, doi:https://doi.org/10.1111/jen.12834.
21. Li, Y., Wang, H., Dang, L.M., Sadeghi-Niaraki, A., Moon, H., Crop pest recognition in natural scenes using convolutional neural networks. *Comput. Electron. Agric.*, 169, 105174, 2020.
22. Nanni, L., Manf`e, A., Maguolo, G., Lumini, A., Brahnam, S., High performing ensemble of convolutional neural networks for insect pest image detection. *Ecol. Inf.*, 67, 101515, 2022, ISSN: 1574–9541, doi:10.1016/j.ecoinf.2021.101515.
23. Wang, J., Li, Y., Feng, H., Ren, L., Du, X., Wu, J., Common pests image recognition based on deep convolutional neural network. *Comput. Electron. Agric.*, 179, 105834, Dec. 2020, doi:10.1016/j.compag.2020.105834.
24. Zheng, T., Yang, X., Lv, J., Li, M., Wang, S., Li, W., An efficient mobile model for insect image classification in the field pest management. *Eng. Sci. Technol.*, 39, 101 335, Mar. 2023, doi:10.1016/j.jestch.2023.101335.3.
25. Xia, D., Chen, P., Wang, B., Zhang, J., Xie, C., Insect detection and classification based on an improved convolutional neural network. *Sensors*, 18, 12, 2018, ISSN: 1424–8220.
26. Khanramaki, M., Askari Asli-Ardeh, E., Kozegar, E., Citrus pests classification using an ensemble of deep learning models. *Comput. Electron. Agric.*, 186, 106192, 2021, ISSN: 0168–1699, doi:10.1016/j.compag.2021.106192.
27. Kusrini, K., Suputa, S., Setyanto, A., Agastya, I.M.A., Priantoro, H., Chandramouli, K., Izquierdo, E., Data augmentation for automated pest classification in mango farms. *Comput. Electron. Agric.*, 179, 105842, 2020, ISSN: 0168–1699, doi:10.1016/j.compag.2020.105842.
28. Xin, D., Chen, Y.-W., Li, J., Fine-grained butterfly classification in ecological images using squeeze-and-excitation and spatial attention modules. *Appl. Sci.*, 10, 5, 2020, ISSN: 2076–3417, doi:10.3390/app10051681.
29. Kasinathan, T., Singaraju, D., Uyyala, S.R., Insect classification and detection in field crops using modern machine learning techniques. *Inf. Process. Agric.*, 8, 3, 446–457, 2021, ISSN: 2214–3173, doi:10.1016/j.inpa.2020.09.006.
30. Amarathunga, D.C., Grundy, J., Parry, H., Dorin, A., Methods of insect image capture and classification: A systematic literature review. *Smart Agric.*

Technol., 1, 100023, 2021, ISSN: 2772–3755, doi:https://doi.org/10.1016/j.atech.2021.100023.
31. Ebrahimi, M., Khoshtaghaza, M., Minaei, S., Jamshidi, B., Vision-based pest detection based on svm classification method. *Comput. Electron. Agric.*, 137, C, 52–58, May 2017, ISSN: 0168–1699, doi:10.1016/j.compag.2017.03.016.
32. Thenmozhi, K. and Reddy, U.S., Crop pest classification based on deep convolutional neural network and transfer learning. *Comput. Electron. Agric.*, 164, 104906, 2019, ISSN: 0168–1699, doi:10.1016/j.compag.2019.104906.
33. Almryad, A. and Kutucu, H., Automatic identification for field butterflies by convolutional neural networks. *Eng. Sci. Technol.*, 23, Feb. 2020, doi:10.1016/j.jestch.2020.01.006.4.
34. Chulu, F., Phiri, J., Nkunika, P., Nyirenda, M., Kabemba, M., Sohati, P., A convolutional neural network for automatic identification and classification of fall army worm moth. *Int. J. Adv. Comput. Sci. Appl.*, 10, 112, Aug. 2019, doi:10.14569/IJACSA.2019.0100717.
35. Mahrishi, M., Morwal, S., Dahiya, N., *et al.*, A framework for index point detection using effective title extraction from video thumbnails. *Int. J. Syst. Assur. Eng. Manage.*, 2021, https://doi.org/10.1007/s13198-021-01166-z.
36. Chen, Y., Zhang, X., Chen, Z., Song, M., Wang, J., Fine-grained classification of fly species in the natural environment based on deep convolutional neural network. *Comput. Biol. Med.*, 135, 104 655, Jul. 2021, doi:10.1016/j.compbiomed.2021.104655.
37. Alanazi, A., Shakeabubakor, A., Abdel-Khalek, S., Alkhalaf, S., Iot enhanced metaheuristics with deep transfer learning based robust crop pest recognition and classification. *Alexandria Eng. J.*, 84, 100–111, Dec. 2023, doi:10.1016/j.aej.2023.11.008.
38. Hassan, S. and Maji, A., Pest identification based on fusion of self-attention with resnet. *IEEE Access*, 1–1, Jan. 2024, doi:10.1109/ACCESS.2024.3351003.
39. Peng, H., Xu, H., Gao, Z., Zhou, Z., Tian, X., Deng, Q., He, H., Xian, C., Crop pest image classification based on improved densely connected convolutional network. *Front. Plant Sci.*, 14, 2023, ISSN: 1664–462X, doi:10.3389/fpls.2023.1133060.
40. Alzubaidi, L., Zhang, J., Humaidi, A., Al-Dujaili, A., Duan, Y., Al-Shamma, O., Santamar´ıa, J., Fadhel, M., Al-Amidie, M., Farhan, L., Review of deep learning: Concepts, cnn architectures, challenges, applications, future directions. *J. Big Data*, 8, Mar. 2021, doi:10.1186/s40537–021-00444–8.
41. Tetila, E., Brandoli, B., Astolfi, G., Belete, N.A., Amorim, W., Roel, A., Pistori, H., Detection and classification of soybean pests using deep learning with uav images. *Comput. Electron. Agric.*, 179, 105 836, Dec. 2020, doi:10.1016/j.compag.2020.105836.
42. Alves, A., Souza, W., Borges, D., Cotton pests classification in field-based images using deep residual networks. *Comput. Electron. Agric.*, 174, 105 488, Jul. 2020, doi:10.1016/j.compag.2020.105488.

43. Sun, J., Cao, W., Fu, X., Ochi, S., Yamanaka, T., Few-shot learning for plant disease recognition: A review. *Agron. J.*, 116, Feb. 2023, doi:10.1002/agj2.21285.5.
44. Tassis, L. and Krohling, R., Few-shot learning for biotic stress classification of coffee leaves. *Artif. Intell. Agric.*, 6, Apr. 2022, doi:10.1016/j.aiia.2022.04.001.
45. Vinyals, O., Blundell, C., Lillicrap, T., Koray, K.K., Wierstra, D., Matching networks for one shot learning, in: *Advances in Neural Information Processing Systems*, vol. 29, D., Lee, M., Sugiyama, U., Luxburg, I., Guyon, and R., Garnett (Eds.), Curran Associates, Inc., 2016.

12

Spatio-Temporal Analysis of Tropospheric NO_2 Over Haryana Using Sentinel-5P and Google Earth Engine

Narayan Vyas[1,2], Neha Bhati[3], Rajesh Yadav[4*] and Abdullah Alzahrani[5]

[1]*Department of Computer Science and Engineering, Vivekananda Global University, Jaipur, Rajasthan, India*
[2]*Centre of Excellence in Geospatial Intelligence for Climate Studies, Vivekananda Global University, Jaipur, Rajasthan, India*
[3]*Department of Computer Science and IT, IIS (Deemed to be University), Jaipur, Rajasthan, India*
[4]*Department of Zoology, JECRC University, Jaipur, Rajasthan, India*
[5]*Department of Electrical and Computer Engineering, Oakland University, Rochester, MI, USA*

Abstract
This study investigated the temporal variation in tropospheric nitrogen dioxide (NO_2) concentrations over the Indian state of Haryana using satellite-derived data from the Sentinel-5P Tropospheric Monitoring Instrument (TROPOMI) sensor. The monthly NO_2 values were extracted and analyzed for 2020 and 2024 using the Google Earth Engine (GEE) platform. The year 2020, significantly influenced by COVID-19 lockdown measures, exhibited lower NO_2 levels across most months, whereas 2024 reflected a notable increase, indicating a rebound in anthropogenic emissions from industrial, vehicular, and agricultural sources. The analysis revealed peak NO_2 concentrations during May–June and October–December, corresponding to the periods of active agricultural operations and stubble burning of stubble. Spatial maps confirmed the widening of pollution hotspots in 2024, particularly in urban-industrial regions such as Gurugram, Faridabad, and Panipat. Monthly mean values ranged from 78.83 μg/m^3 to 99.91 μg/m^3 in 2020 and from 76.81 μg/m^3 to

**Corresponding author*: ryadav70@yahoo.co.in; ORCID: 0000-0002-4909-4744
Narayan Vyas: ORCID: 0000-0003-4225-0764
Neha Bhati: ORCID: 0009-0008-0171-2786

Gurwinder Singh, Vishakha Sood, Narayan Vyas, Surendra Yadav and Dankan Gowda V. (eds.) Geospatial Intelligence for Precision Agriculture, (233–250) © 2026 Scrivener Publishing LLC

121 μg/m^3 in 2024. This study demonstrates the value of satellite remote sensing for air quality monitoring in regions with limited ground-based infrastructure. It also highlights GEE's ability to efficiently process and visualize large-scale air quality data for environmental monitoring. Overall, this study provides a comprehensive assessment of NO_2 dynamics over Haryana and offers a replicable methodology for regional air quality monitoring using open-access satellite data.

***Keywords*:** Nitrogen dioxide (NO_2), Sentinel-5P, Google Earth Engine (GEE), air quality monitoring, remote sensing, Haryana

12.1 Introduction

Nitrogen dioxide (NO_2) is one of the most significant tropospheric trace gases, primarily produced by anthropogenic sources, including fossil fuel combustion, industrial emissions, and biomass burning [1, 2]. It serves as an important precursor for the formation of ground-level ozone (O_3) and secondary particulate matter ($PM_{2.5}$) [3]. Long-term elevated NO_2 levels can cause respiratory diseases, such as asthma, bronchitis, and decreased lung function, particularly in children and the elderly [4]. NO_2 can harm the environment by causing soil acidification, reducing vegetation growth, and contributing to photochemical smog, which exacerbates climate-related impacts [5]. Transportation, thermal power plants, and stubble burning contribute to increased NO_2 concentrations in urban and industrial areas of Haryana [6]. Conventional ground-based monitoring stations, while reliable, provide only localized measures of pollution and often miss the wider distributions of pollutants that affect larger areas, particularly in rural and remote regions [7]. Thus, a regional and temporal understanding of NO_2 variability is key to formulating appropriate air quality management policies and environmental health risk assessments while achieving sustainable development and climate resilience goals.

Conventional air quality monitoring systems, although important, offer limited spatial coverage at higher temporal granularities because of rapid industrialization and urbanization. On contrast, satellite remote sensing provides a comprehensive, low-cost, and scalable solution for the long-term monitoring of atmospheric pollutants over vast geographic regions at a high temporal frequency [8]. Hyperspectral and multispectral sensors onboard satellites detect trace gases such as NO_2, SO_2, and CO at high spectral resolutions, enabling the retrieval of tropospheric concentrations with high accuracy [9]. When combined with geospatial tools such as Google Earth Engine (GEE), satellite data enable researchers to process and visualize large-scale environmental datasets in hours or days rather than weeks or months [10]. Satellites provide regional data that are essential for mitigation planning during public health crises, crop burning, and industrial surges [11]. In this sense, satellite-derived datasets, such as those from the Sentinel-5P mission, provide a practical approach for solid environmental observation, policymaking, and outreach.

The Sentinel-5 Precursor (Sentinel-5P) is a satellite launched by the European Space Agency as part of the Copernicus Programme, providing high-temporal and spatial resolution measurements of atmospheric trace gases [12]. TROPOMI, the primary instrument on this satellite, provides measurements of a wide range of atmospheric constituents (e.g., NO_2, O_3, SO_2, CO, CH_4, and aerosols). TROPOMI also produces global daily coverage, with a spatial resolution of 7 $km^2 \times 3.5$ km^2 (originally 5.5 $km^2 \times 3.5$ km^2 in later phases), making TROPOMI viable for more localized air quality measurements [13]. Sentinel-5P Level-3 data products representing vertically integrated tropospheric NO_2 column densities (indicated as NO_2 columns in this study) can be utilized to retrieve regional pollutant concentrations concerning NO_2 using the Sentinel-5P satellite based on the technique demonstrated in this study [14]. Retrieved using sophisticated algorithms and made available through tools such as GEE, they provide prescriptive access to temporal and spatial information on atmospheric pollution. These data are particularly useful in countries such as India, where there are multiple sources of pollution and regulating pollution through traditional means is challenging [15]. The combination of consistent measurements, cloud filtering, and free data access makes Sentinel-5P a vital tool for monitoring the Earth, conducting environmental research, and supporting public health surveillance [16]. Using this satellite, researchers can monitor trends, identify pollution outliers, and provide evidence to decision-makers for data-driven air quality policies in the future.

This study aimed to investigate the temporal variation in tropospheric NO_2 concentrations over the Indian state of Haryana using satellite-derived data from Sentinel-5P. The core objective was to compare the monthly NO_2 trends between 2020 and 2024, thereby capturing the evolving dynamics of atmospheric pollution in a critical industrial and agricultural region. The selection of these two years is motivated by contrasting socio-economic activities: 2020, marked by pandemic-induced lockdowns, and 2024, representing a return to typical or even elevated emission levels. Specifically, this study focused on generating monthly NO_2 composites, quantifying pollutant levels, visualizing spatial variations, and identifying seasonal peaks. The study also utilized GEE for data preprocessing, visualization, and export of georeferenced outputs. Another goal of this study was to assess the implications of changing NO_2 levels in terms of public health and environmental sustainability. By establishing a comparative baseline, these findings contribute to the growing body of evidence required for regional air quality management. Ultimately, this study supports the broader agenda of integrating satellite-based environmental monitoring into policymaking, planning, and community awareness, with an emphasis

Table 12.1 Comparative analysis of air quality studies.

Dataset	Study area	Methodology	Limitations	Applications	Source
Sentinel-5P (CO, PM2.5)	Nile Delta, Egypt	Satellite-based pollution mapping, vulnerability index, *in situ* validation	Only CO and PM2.5 analyzed; limited spatial ground validation	Urban air quality assessment and health risk evaluation	[17]
Sentinel-5P (NO_2)	Guayaquil, Ecuador	Interpolation (EBK, IDW, Kriging), COVID-based temporal analysis	Cloud cover, coarse resolution, limited *in situ* validation	City-scale NO_2 monitoring where ground stations are absent	[18]
Sentinel-5P, ERA5	Milan, Italy	ML models (MLP + SVR) for ground-level NO_2 estimation	Focused on urban NO_2; excludes rural/other pollutants	Ground-level NO_2 estimation for urban health monitoring	[19]
Sentinel-5P, AIS	Canary Islands	NO_2 mapping, ship tracking with AIS, spatial correlation analysis	Multitemporal approach needed due to atmospheric noise	Maritime traffic pollution monitoring	[20]
Sentinel-5P, GEE	India (multi-city)	Country-level pollutant analysis using GEE (NO_2, CO, AOD)	Broad scale; limited city-specific inference	Nationwide air quality trend analysis and policy insights	[21]
Sentinel-5P, GEE, SRTM, ArcGIS	Caspian Sea Basin	Spatiotemporal analysis of seven pollutants, correlation with elevation/population	Complex pollutant behavior not fully modeled	Regional pollution dynamics and environmental planning	[22]

on sustainability and resilience. A comparative analysis of Sentinel-5P-based air quality studies, covering methodologies, applications, and limitations across diverse geographical contexts is presented in Table 12.1.

12.2 Study Area

Haryana, situated in northern India, is a landlocked state with a geographical area of approximately 44,212 km^2 [23]. It was chosen as the study area in this research, as shown in Figure 12.1. It lies between latitudes 27.39°N and 30.35°N and longitudes 74.28°E and 77.36°E, bordered by Punjab and Himachal Pradesh to the north, Rajasthan to the west and south, and Uttar Pradesh to the east. The state surrounds the national capital, Delhi, on three sides, making it an integral part of the National Capital Region (NCR). Haryana comprises 22 districts, featuring significant urban centers such as Gurugram, Faridabad, and Panipat and vast agricultural zones. Its topography is primarily composed of plains, with the Shivalik Hills touching the northern edge and semi-arid regions in the southwest. The economy of the state is largely driven by agriculture and industry. Haryana is one of India's leading producers of wheat and rice and plays a crucial role in food grain production through practices such as paddy stubble burning, which contributes to air pollution during the post-harvest season [24]. Industrial clusters, especially those in districts adjacent to Delhi, contribute significantly to the emission load, particularly of NO_2. Additionally,

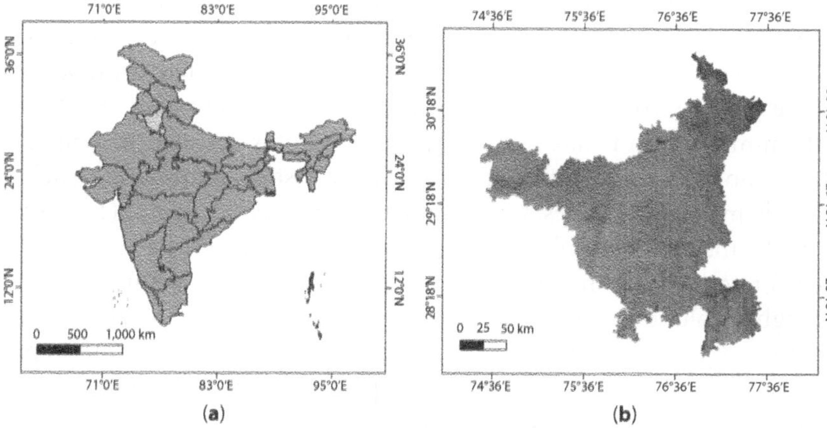

Figure 12.1 Study area where (a) shows a map of India highlighting Haryana state, and (b) Landsat-8 satellite imagery of Haryana state (June 2024).

the high density of vehicular traffic and the presence of coal-fired thermal power plants make Haryana a critical area for studying atmospheric pollution [25]. The spatial diversity of Haryana, with both rural and urban zones, makes it a representative case for analyzing the seasonal variability in tropospheric NO_2 concentrations using satellite-based remote sensing [26].

12.3 Dataset and Tools

This study utilized Sentinel-5P's Level-3 NO_2 data, processed on GEE, a cloud-based platform for efficient analysis of Earth observations. In particular, the dataset used was COPERNICUS/S5P/NRTI/L3_NO2, which provides global daily NO_2 observations with a spatial resolution of approximately 7 $km^2 \times 3.5$ km^2 [27]. The study period spanned two full calendar years (1 January–31 December) for both 2020 and 2024, allowing for a comparison of NO_2 trends during and after the COVID-19 pandemic. NO_2 column densities are provided in the dataset units of mol/m^2 and are vertically integrated from the surface to the tropopause. Finally, using a scaling factor of 1×10^6, the values were scaled and expressed in micrograms per cubic meter (μg/m^3) for improved interpretability and comparative analysis with air quality thresholds, without implying direct physical conversion. The GEE script then masked the image collection for the dates that fell within the given ranges, in addition to the spatial extent of the Haryana region, which was generated using external data from the FAO GAUL (2015) Level-2 administrative boundaries.

Monthly mean composites were generated for each month of both years by averaging the daily images over the respective one-month periods. For visualization, false-color NO_2 maps were created using a color palette ranging from black (lowest values) to red (highest concentrations), with minimum and maximum display thresholds set at 0 mol/m^2 and 0.0002 mol/m^2, respectively. Additionally, statistical summaries, such as monthly mean NO_2 values, were extracted using .reduceRegion() at a scale of 10,000 m, and the results were exported as CSV files. These were further used for temporal comparison through column charts within GEE and for high-resolution map exports (scale: 100 m, CRS: EPSG:4326) for each month and year. The datasets and GEE tools enable an efficient and reproducible methodology for assessing air quality that is applicable to other regions with similar emission monitoring needs.

12.4 Methodology

12.4.1 Region of Interest Definition

The first step in the methodology involved spatially delineating the study area, Haryana, using administrative boundary data, as shown in Figure 12.2. The region of interest (ROI) was extracted from the FAO's Global Administrative Unit Layers (GAUL) Level-2 dataset, which provides precise political boundaries for global subnational divisions. Haryana, located in northern India, was identified by filtering for the country name "India" and the first-level administrative division name "Haryana." This spatial geometry was fundamental for later operations, as it specified the space in which all possible Sentinel-5P data would be filtered, clipped, and analyzed. Using Level-2 boundaries from GAUL ensures spatial consistency, and more importantly, in a country like India, where many environmental governance decisions are made using administrative boundaries, the boundaries are the same for environmental and social processes that occur simultaneously. Additionally, using a pre-validated, globally-accepted dataset enhances reproducibility and dependability across diverse geospatial platforms. The study defined this region of interest (ROI) *a priori* to ensure that each monthly NO_2 composite and statistical extraction equated to the same surface area, thereby eliminating geometric discrepancies and providing the ability to compare air quality variability on a year-on-year basis with fidelity within the bounds of Haryana.

12.4.2 Temporal Scope and Parameter Selection

The temporal framework of this study was selective, based on the calendar year; two strategic years, 2020 and 2024, were chosen to investigate the spatial and temporal variability of tropospheric NO_2 concentrations over the Haryana region. We aimed to compare emissions between two contrasting scenarios: 2020, characterized by reduced industrial and vehicular activity due to pandemic-driven restrictions, and 2024, defined as a time when human emissions returned to pre-pandemic levels or higher, following the resumption of economic activity. The year was divided into 12 discrete monthly intervals, allowing within- and between-year standardization every month. The study used the Level-3 NO_2 product retrieved from Sentinel-5P satellite data and its tropospheric column number density variable for each of month from January to December. This parameter measures the column-integrated vertical Tropospheric NO_2 content,

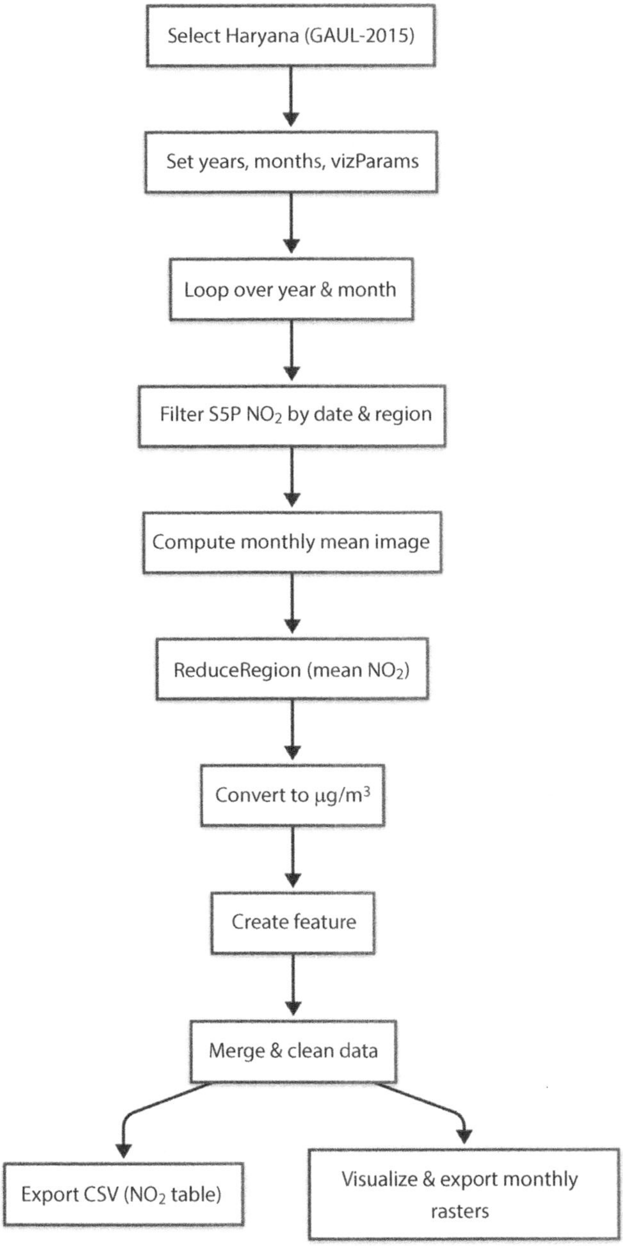

Figure 12.2 Methodology of the proposed framework.

which is crucial for assessing near-surface air pollution. The satellite images were filtered based on temporal (limiting the date range to when each month falls) and spatial (defining the Boundary of the Haryana state). Daily images for each month were averaged to generate monthly composite images. These composites mitigate the impact of daily noise associated with meteorological conditions, yielding a stable regional signal for NO_2 that is consistent with the expected trends. Level-3 products were used to ensure spatial continuity and reduce noise levels, thereby facilitating integration with visual and statistical tools. The methodology created a regularized time series for each year, allowing for comparisons between months and the target month, identifying pollution spikes, and examining seasonal aspects of pollution. This frequency is crucial for identifying key risk hotspots, aligning with agricultural seasons, and implementing targeted policy measures.

12.4.3 Feature Generation and Unit Conversion

Following the generation of monthly NO_2 composites, statistical summaries were extracted to quantify pollution levels across the Haryana region. For each monthly image, the mean tropospheric NO_2 concentration was calculated using a region-based reducer function within the specified administrative boundary. This process involved computing the average of all pixel values within the region, resulting in a single representative value for each month. The original data obtained from Sentinel-5P are provided in units of moles per square meter (mol/m^2), which are not directly interpretable in conventional air quality metrics. To enhance interpretability and align the output with commonly used standards, the values were scaled and expressed in micrograms per cubic meter (μg/m^3) using a multiplication factor of 1×10^6, without implying a direct physical conversion. This unit standardization is critical for comparing the results with national air quality thresholds and existing ground-based measurements. The monthly NO_2 values, along with the corresponding metadata, such as year, numeric month, and abbreviated month name, were structured into geospatial features. These were stored in a single unified feature collection, which allowed the entire time-series dataset to be handled as a tabular structure for further analysis and export. By converting satellite raster outputs into structured feature collections, this methodology enables detailed temporal analyses, statistical visualizations, and downstream integration into CSV files and charts. This data structuring step was pivotal for summarizing NO_2 patterns, ensuring the traceability of monthly values, and supporting reproducible workflows for future temporal expansions or geographic scaling of the analysis.

12.4.4 Visualization and Export

The final step in the methodology involved visualizing and exporting both spatial and temporal outputs. For each monthly NO_2 composite, a color-coded visualization layer was created using a defined palette ranging from black to red, representing the gradation from low-to-high NO_2 concentrations. The visualization parameters were standardized across all images using a fixed value range between 0 mol/m^2 and 0.0002 mol/m^2 to ensure consistent interpretation and comparison. These monthly maps were overlaid on the interactive GEE map, enabling researchers to assess the spread and fluctuations in pollution intensity throughout the year. The integration of geospatial visualization and data export ensures that both quantitative and qualitative assessments of air quality are accessible, interpretable, and ready for dissemination to policymakers and researchers.

12.5 Results

12.5.1 Visual Analysis

Monthly NO_2 maps for 2020 (Figure 12.3) and 2024 (Figure 12.4) revealed higher concentrations and expanded hotspots in 2024, particularly in June, July, October, and November. Each subplot (a–l) in both figures corresponds to the months from January (a) to December (l), and the observed patterns strongly align with the tabular data presented. In 2020, most maps showed dominant blue and green tones, indicating relatively low NO_2 concentrations. Higher NO_2 intensities were localized and primarily visible in districts such as Faridabad, Panipat, and Gurugram, all of which are major industrial and vehicular emission hubs. The summer months (May–June) exhibited a slight uptick, with small red patches (e.g., subplots e and f), which aligned with the reported peaks of 99.91 μg/m^3 and 98.42 μg/m^3. The monsoon months (July–September) showed the most diluted NO_2 levels (subplots g–i), consistent with the rain-induced dispersion observed in the table. Notably, December 2020 (subplot l) showed one of the highest NO_2 concentrations, corresponding to the post-harvest burning season. In contrast, the 2024 visuals exhibited significantly more intense green, yellow, and red zones across nearly all months. The spatial extent and intensity of the high NO_2 areas have increased notably. May and June (e, f) were characterized by saturated yellow-red tones, consistent with the peak tabular values of 121 μg/m^3 and 112.2 μg/m^3. Post-monsoon months, such as October and November (j, k), also displayed widespread pollution plumes,

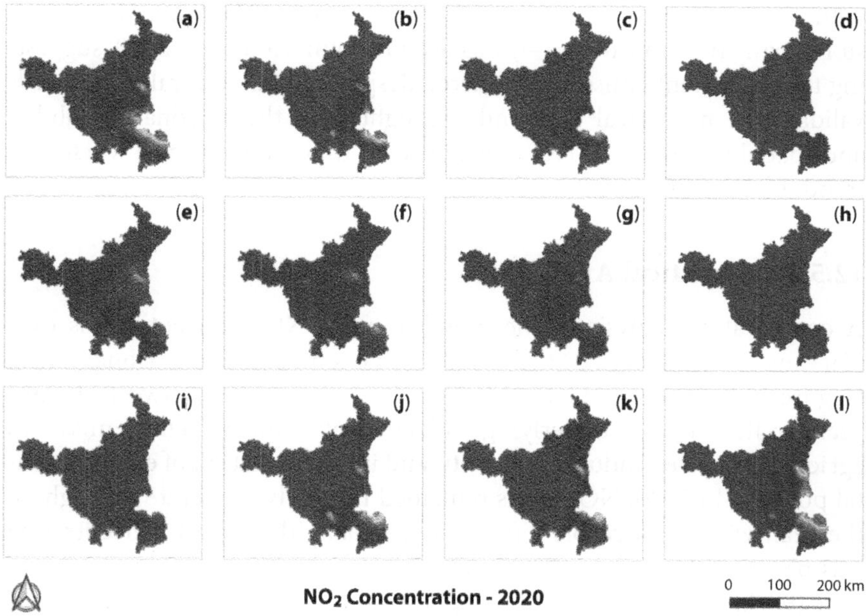

Figure 12.3 NO_2 concentration of the study area for the year 2020.

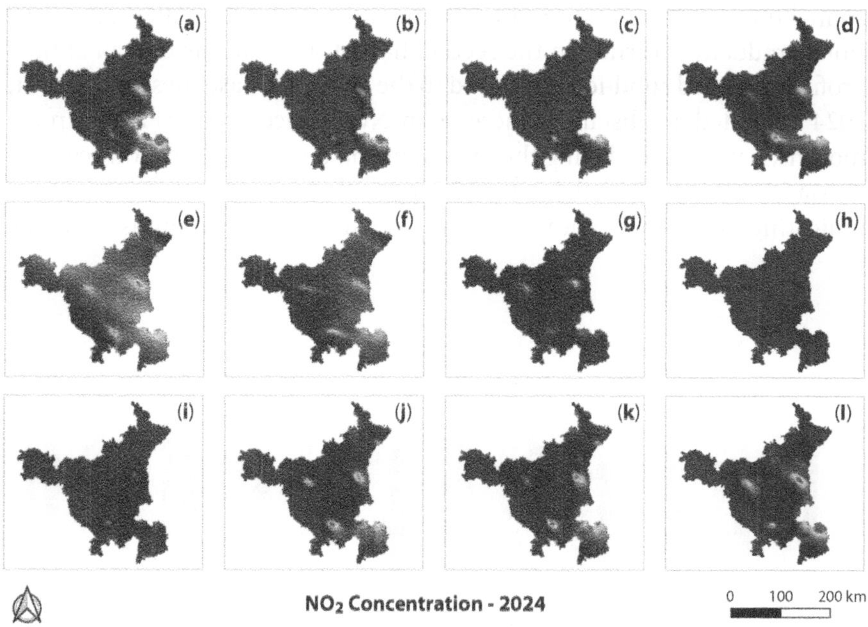

Figure 12.4 NO_2 concentration of the study area for the year 2024.

with multiple hotspots across southern and central Haryana. The dispersion during the monsoon (g–i) was less pronounced than in 2020, suggesting that persistent emissions occurred despite rainfall. Overall, the visuals validate the numerical data and highlight both the seasonal variability and spatial expansion of NO_2 hotspots over time, confirming a worsening trend in air quality in 2024.

12.5.2 Statistical Analysis

A comparative analysis of the monthly mean NO_2 concentrations over Haryana for 2020 and 2024 revealed critical temporal and seasonal trends (Figure 12.5). The data highlighted significant variations between the two years, reflecting the underlying influence of anthropogenic emissions, agricultural cycles, industrial activity, and implementation of environmental policies. In 2020, NO_2 levels remained relatively moderate throughout the year, with a noticeable dip in the first half. The lowest concentration was observed in August (78.83 $\mu g/m^3$), whereas May (99.91 $\mu g/m^3$) and June (98.42 $\mu g/m^3$) recorded the highest values. This pattern closely aligns with the impacts of the COVID-19 pandemic, which led to nationwide lockdowns during the initial months of 2020. The restrictions on vehicular traffic, closure of industries, and reduced power generation collectively contributed to the temporary improvement in air quality. Although emissions gradually returned in the second half of the year, the overall annual profile remained subdued compared to the historical baselines. In contrast, 2024 exhibited a substantial increase in NO_2 concentrations across most months, particularly during the pre-monsoon and post-monsoon seasons.

May (121 $\mu g/m^3$) and June (112.2 $\mu g/m^3$) showed the highest values, indicating intensified anthropogenic activities. These months coincide

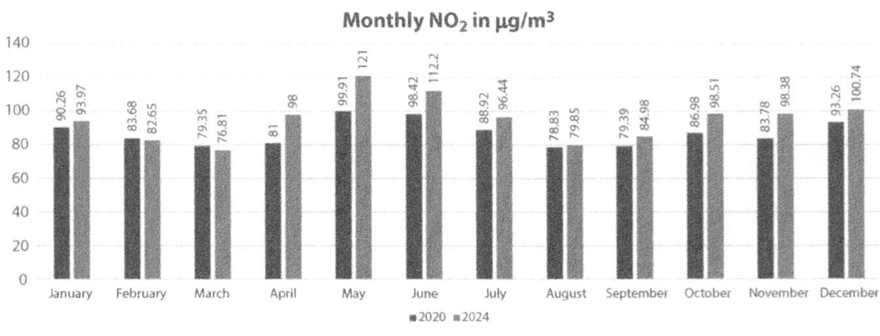

Figure 12.5 Monthly analysis of NO_2 concentration.

with the harvest and pre-sowing periods in Haryana's agrarian calendar, which involves significant use of diesel-powered machinery and transportation, contributing to elevated emissions. Furthermore, rising temperatures during this period promote photochemical reactions, increasing NO_2 formation. The monsoon months (July to September) showed a stabilizing effect due to rainfall and atmospheric cleansing. However, even in these months, 2024 values remained consistently higher than those in 2020, with September 2024 reaching 84.98 μg/m^3, compared to 79.39 μg/m^3 in 2020. This persistence suggests that emission sources have intensified, or that meteorological dispersion conditions were less favorable in 2024. The post-monsoon period (October to December) experiences a sharp resurgence of NO_2, likely due to the prevalence of paddy stubble burning practices in Haryana and the neighboring Punjab. October (98.51 μg/m^3), November (98.38 μg/m^3), and December (100.74 μg/m^3) showed significantly higher NO_2 levels in 2024 than in the corresponding months in 2020.

During these months, atmospheric inversion layers tend to form, trapping pollutants close to the ground, which later become significant contributors to air quality problems. The rise in December NO_2 concentrations confirmed this assumption, with a significant contribution from emissions associated with residential heating and seasonal industrial activities. However, the NO_2 values in February and March 2024 were slightly below those observed in 2020. The recorded level in February 2024 was 82.65 μg/m^3, compared to 83.68 μg/m^3 in 2020. In March, the levels were 76.81 μg/m^3, compared to 79.35 μg/m^3. Although small, these differences could be the result of short-term meteorological changes or the success of local efforts to reduce emissions. In general, the results show that NO_2 concentrations were substantially higher in 2024 than those observed in 2020, suggesting a post-COVID bounce or resurgence in emissions. This highlights the importance of implementing air quality management strategies, especially during high-risk periods. Specific interventions in this regard will require the right set of policies that emphasize controlled stubble burning, climate-friendly energy technologies, emission control policies, and other overarching policies to reduce emissions.

12.6 Discussion

Sentinel-5P-derived tropospheric NO_2 concentrations over Haryana in two characteristically different years, 2020 and 2024, provide insights into the evolving air pollution dynamics of a rapidly urbanizing and industrializing region. The reduced levels of NO_2 observed in 2020, following the closure of many human activities to control COVID-19, served as a unique real-life control experiment, directly demonstrating that human activity is a major cause of NO_2 pollution [28]. However, the record-high levels and widespread pollution hotspots in 2024 are consistent with a major resurgence in emissions, and in some months, return to above-background emissions in several regions compared to pre-pandemic baselines. The periodic peaks in May–June and October–December coincide with the agricultural calendar and pertinent climatology, including enhanced pollution during thermal inversions. The sustained rise in NO_2 levels during the monsoon months of 2024, which are typically characterized by washout, suggests an increase in emission sources or a change in the meteorological conditions. GEE enables near-real-time analysis and visualization, leading to the rapid generation of monthly composites and effective spatial mapping. These results demonstrate the potential of cloud-based geospatial platforms for environmental monitoring in sparsely populated regions where ground data are limited. Ultimately, this study highlights the importance of policy actions to address the major seasonal contributions and sources of emissions that control NO_2 pollution. Frequent satellite-based evaluations can inform policy implementation, strengthen health-centered interventions, and design sustainable development trajectories.

12.6.1 Conclusion

Using data from the Sentinel-5P satellite, this study presents a detailed spatiotemporal analysis of tropospheric NO_2 concentrations over the Indian state of Haryana during 2020 and 2024. For NO_2, the results indicated a notable increase in seasonal variation in 2024 compared to 2020, highlighting the impact of anthropogenic events, including industrial activity, vehicular emissions, and post-harvest stubble burning. Significant peaks in emissions in May–June and October–December 2024 were notably pronounced, resulting from the concurrent influence of crop-based contributions, summer photochemical reactions, and winter atmospheric inversion environments. In the visual maps of satellite NO_2 for 2024, however, this trend has also continued, with larger and hotter pollution hotspots appearing geographically over the heavily urban and industrial areas of Gurugram, Faridabad, and Panipat. Conversely, 2020, characterized by COVID-19 lockdowns associated with decreased emissions, had lower NO_2 values and fewer zones with high NO_2 concentrations. These observations corroborate the success of emission reductions while highlighting the need for robust regulations to ensure sustainability in the post-COVID era.

The synthesis of Sentinel-5P data using GEE provided a rapid, reproducible, and scalable analysis pipeline, highlighting the potential of satellite-based monitoring for air quality evaluation. This study highlights the urgent need for preemptive policy interventions, including more stringent emission standards, the promotion of clean technologies, and the enforcement of a stubble burning ban. With the ongoing urbanization and industrialization of Haryana, continuous satellite monitoring can support long-term environmental sustainability and public health by providing the necessary information for decision-making.

AI disclosure statement: The author acknowledges the use of AI-based tools for minor language editing or grammatical correction. However, all intellectual contributions, analyses, and conclusions presented in this chapter are entirely the author's own.

References

1. Most., M.M.F.J., A critical review and prospect of NO_2 and SO_2 pollution over Asia: Hotspots, trends, and sources. *Sci. Total Environ.*, 876, 162851, Jun. 2023, doi:10.1016/j.scitotenv.2023.162851.
2. Ojeda-Castillo, V., *et al.*, Tropospheric NO_2: Anthropogenic Influence, Global Trends, Satellite Data, and Machine Learning Application. *Remote Sens.*, 17, 1, Jan. 2025, doi:10.3390/rs17010049.
3. Baruah, U.D., Robeson, S.M., Saikia, A., Mili, N., Sung, K., Chand, P., Spatio-temporal characterization of tropospheric ozone and its precursor pollutants NO_2 and HCHO over South Asia. *Sci. Total Environ.*, 809, 151135, Feb. 2022, doi:10.1016/j.scitotenv.2021.151135.
4. Chen, J., *et al.*, Chronic exposure to ambient PM2.5/NO_2 and respiratory health in school children: A prospective cohort study in Hong Kong. *Ecotoxicol. Environ. Saf.*, 252, 114558, Mar. 2023, doi:10.1016/j.ecoenv.2023.114558.
5. Manisalidis, I., Stavropoulou, E., Stavropoulos, A., Bezirtzoglou, E., Environmental and Health Impacts of Air Pollution: A Review. *Front. Public Health*, 8, 14, Feb. 2020, doi:10.3389/fpubh.2020.00014.
6. Singh, J., Paddy and wheat stubble blazing in Haryana and Punjab states of India: A menace for environmental health. *Environ. Qual. Manage.*, 28, 2, 47–53, 2018, doi:10.1002/tqem.21598.
7. Relvas, H., Lopes, D., Armengol, J.M., Empowering communities: Advancements in air quality monitoring and citizen engagement. *Urban Clim.*, 60, 102344, Mar. 2025, doi:10.1016/j.uclim.2025.102344.
8. Agate, J., Ballinger, R., Ward, R.D., Satellite remote sensing can provide semi-automated monitoring to aid coastal decision-making. *Estuarine Coastal Shelf Sci.*, 298, 108639, Mar. 2024, doi:10.1016/j.ecss.2024.108639.

9. Reshi, A.R., Pichuka, S., Tripathi, A., Applications of Sentinel-5P TROPOMI Satellite Sensor: A Review. *IEEE Sens. J.*, 24, 13, 20312–20321, Jul. 2024, doi:10.1109/JSEN.2024.3355714.
10. Yang, L., Driscol, J., Sarigai, S., Wu, Q., Chen, H., Lippitt, C.D., Google Earth Engine and Artificial Intelligence (AI): A Comprehensive Review. *Remote Sens.*, 14, 14, Jan. 2022, doi:10.3390/rs14143253.
11. Chakrabarti, S., Khan, M.T., Kishore, A., Roy, D., Scott, S.P., Risk of acute respiratory infection from crop burning in India: estimating disease burden and economic welfare from satellite and national health survey data for 250 000 persons. *Int. J. Epidemiol.*, 48, 4, 1113–1124, Aug. 2019, doi:10.1093/ije/dyz022.
12. Compernolle, S., *et al.*, Validation of the Sentinel-5 Precursor TROPOMI cloud data with Cloudnet, Aura OMI O_2–O_2, MODIS, and Suomi-NPP VIIRS. *Atmos. Meas. Tech.*, 14, 3, 2451–2476, Mar. 2021, doi:10.5194/amt-14–2451-2021.
13. Veefkind, J.P., *et al.*, TROPOMI on the ESA Sentinel-5 Precursor: A GMES mission for global observations of the atmospheric composition for climate, air quality and ozone layer applications. *Remote Sens. Environ.*, 120, 70–83, May 2012, doi:10.1016/j.rse.2011.09.027.
14. Martín, C., Amigo, J.M., Castro, K., Environmental science through orbiters. An example of Sentinel-5P on how the Covid-19 pandemic impacted the air quality. *Chemom. Intell. Lab. Syst.*, 240, 104927, Sep. 2023, doi:10.1016/j.chemolab.2023.104927.
15. Singh, S., Avtar, R., Jain, A., Alsulamy, S., Ouda, M.M., Kharrazi, A., Identifying Micro-Level Pollution Hotspots Using Sentinel-5P for the Spatial Analysis of Air Quality Degradation in the National Capital Region, India. *Sustainability*, 17, 5, Jan. 2025, doi:10.3390/su17052241.
16. Vîrghileanu, M., Săvulescu, I., Mihai, B.-A., Nistor, C., Dobre, R., Nitrogen Dioxide (NO_2) Pollution Monitoring with Sentinel-5P Satellite Imagery over Europe during the Coronavirus Pandemic Outbreak. *Remote Sens.*, 12, 21, Jan. 2020, doi:10.3390/rs12213575.
17. Hassaan, M.A., Abdallah, S.M., Shalaby, E.-S.A., Ibrahim, A.A., Assessing vulnerability of densely populated areas to air pollution using Sentinel-5P imageries: a case study of the Nile Delta, Egypt. *Sci Rep*, 13, 1, 17406, Oct. 2023, doi:10.1038/s41598–023-44186–4.
18. Mejía C., D., Alvarez, H., Zalakeviciute, R., Macancela, D., Sanchez, C., Bonilla, S., Sentinel satellite data monitoring of air pollutants with interpolation methods in Guayaquil, Ecuador. *Remote Sens. Appl. Soc. Environ.*, 31, 100990, Aug. 2023, doi:10.1016/j.rsase.2023.100990.
19. Cedeno Jimenez, J.R. and Brovelli, M.A., NO_2 Concentration Estimation at Urban Ground Level by Integrating Sentinel 5P Data and ERA5 Using Machine Learning: The Milan (Italy) Case Study. *Remote Sens.*, 15, 22, Jan. 2023, doi:10.3390/rs15225400.

20. Rodriguez Valido, M., Perez Marrero, J., Mauro González, A., Fabiani Bendicho, P., Efrem Mora, C., Evaluation of the Potential of Sentinel-5P TROPOMI and AIS Marine Traffic Data for the Monitoring of Anthropogenic Activity and Maritime Transport NOx-Emissions in Canary Islands Waters. *Sustainability*, 15, 5, Jan. 2023, doi:10.3390/su15054632.
21. Halder, B., *et al.*, Machine learning-based country-level annual air pollutants exploration using Sentinel-5P and Google Earth Engine. *Sci. Rep.*, 13, 1, 7968, May 2023, doi:10.1038/s41598-023-34774-9.
22. Tabunshchik, V., *et al.*, The Dynamics of Air Pollution in the Southwestern Part of the Caspian Sea Basin (Based on the Analysis of Sentinel-5 Satellite Data Utilizing the Google Earth Engine Cloud-Computing Platform). *Atmosphere*, 15, 11, Nov. 2024, doi:10.3390/atmos15111371.
23. Raheja, H., Goel, A., Pal, M., Prediction of groundwater quality indices using machine learning algorithms. *Water Pract. Technol.*, 17, 1, 336–351, Dec. 2021, doi:10.2166/wpt.2021.120.
24. Saxena, P., *et al.*, Impact of crop residue burning in Haryana on the air quality of Delhi, India. *Heliyon*, 7, 5, e06973, May 2021, doi:10.1016/j.heliyon.2021.e06973.
25. Sahu, S.K., Mangaraj, P., Beig, G., Decadal growth in emission load of major air pollutants in Delhi. *Earth Syst. Sci. Data*, 15, 7, 3183–3202, Jul. 2023, doi:10.5194/essd-15-3183-2023.
26. Misra, P., *et al.*, Nitrogen oxides concentration and emission change detection during COVID-19 restrictions in North India. *Sci. Rep.*, 11, 1, 9800, May 2021, doi:10.1038/s41598-021-87673-2.
27. Crosman, E., Meteorological Drivers of Permian Basin Methane Anomalies Derived from TROPOMI. *Remote Sens.*, 13, 5, Jan. 2021, doi:10.3390/rs13050896.
28. Singh, D., Nanda, C., Dahiya, M., State of air pollutants and related health risk over Haryana India as viewed from satellite platform in COVID-19 lockdown scenario. *Spatial Inf. Res.*, 30, 1, 47–62, Feb. 2022, doi:10.1007/s41324-021-00410-9.

13

Agriculture Crop Pest Monitoring and Management Using LiDAR Technology

Satyabrata Sarangi[1,2*], Tarini Prasad Das[3], Ekamra Kanan Padhan[2], Jyoti Prakash Das[2] and Harmohan Pradhan[2]

[1]*Department of Entomology, College of Agriculture, OUAT, Bhubaneswar, Odisha, India*
[2]*Crop Protection Division, ICAR-CRRI, Cuttack, Odisha, India*
[3]*Department of Agronomy, College of Agriculture, OUAT, Bhubaneswar, Odisha, India*

Abstract

Precision farming has advanced significantly with the application of Light Detection and Ranging (LiDAR) technology for agricultural crop pest monitoring and control. The majority of current studies have focused on more general topics like crop growth evaluation, yield forecasts, and overall plant health assessments, so even though there has been a lot of progress, there is still a significant research gap regarding the direct use of LiDAR for the early identification and assessment of pest-related damage to crops. The ability of LiDAR technology to provide high-resolution three-dimensional (3D) maps of crop canopies has been leveraged in recent developments to identify structural abnormalities and biomass declines associated with pest outbreaks. LiDAR devices on unmanned aerial vehicles (UAVs) provide detailed canopy structure data, which, when analyzed using machine learning (ML) algorithms, can detect regions with pest-induced structural changes, such as reduced canopy density. The combination of these developments has significantly improved the temporal and spatial resolution of pest surveillance, enabling the planning of more effective pest management responses. The high costs and complex data processing requirements of LiDAR limit its adoption in agricultural research and among smallholder farmers. To further increase its adoption, cost reduction and LiDAR sensor miniaturization are critical. If LiDAR-based pest monitoring is to be useful for diverse crop types, pest species, and environmental conditions, it must be developed over the long term. In the

**Corresponding author*: satyasarangi42478@gmail.com

Gurwinder Singh, Vishakha Sood, Narayan Vyas, Surendra Yadav and Dankan Gowda V. (eds.) Geospatial Intelligence for Precision Agriculture, (251–274) © 2026 Scrivener Publishing LLC

long run, addressing these research gaps will pave the way for holistic real-time pest monitoring and management systems, leading to enhanced and sustainable farming practices.

Keywords: LiDAR, hyperspectral remote sensing, precision agriculture, crop pests, monitoring, sensors

13.1 Introduction

Precision farming relies heavily on the collection, management, and analysis of fast and precise data to make informed decisions [1, 2]. The fundamental components of this approach are artificial intelligence (AI), big data analytics, the Internet of Things (IoT), and smart sensors, which enable farmers to improve crop management, make better use of resources, and promote sustainability [3]. Precision agriculture uses various technologies, including global positioning systems (GPS), sensors, and Big Data, to boost crop production. Decision support systems based on information and communication technology (ICT) provide insights into every facet of farming with a level of detail previously unattainable [4]. In precision agriculture, the use of ICT has allowed farmers to increase crop yields, monitor pests, administer pesticides more efficiently, reduce expenses, and lessen their impact on the environment [5]. The role of remote sensing technology through the inclusion of hyperspectral imaging (HI) and Light Detection and Ranging (LiDAR) sensors has boosted crop production, which is discussed in the subsequent subsections.

By using techniques such as remote sensing, which has become popular in recent decades owing to the availability of detailed satellite imagery for crop pest monitoring and disease detection, smart agriculture targets to increase crop production while reducing harm to the environment and to develop effective management strategies [6, 7]. The use of AI in image analysis enables farmers to make critical decisions regarding the management of pests and diseases, and precision agriculture is a prime example of how technology might revolutionize farming in the modern era [8]. The potential for worldwide smartphone-assisted biotic stress diagnosis has been highlighted by the capacity of deep learning (DL), machine learning (ML), and artificial neural network (ANN) models to reliably identify plant pests and diseases from images [9]. When used in conjunction with geographic information systems (GIS), multispectral imaging, and HI is especially useful for field crop health monitoring, which allows for targeted interventions, helps map the distribution of pests and diseases, and identifies high-risk areas by integrating spatial data with other relevant information [6, 10].

In precision agriculture, prompt and precise detection of plant diseases is essential for preventing the waste of money and other resources, addressing the problem of long-term pathogen resistance, and reducing the impacts of climate change while producing healthier crops [11]. With AI algorithms crucial for image analysis to identify pests and diseases, machine vision technologies are increasingly being used to identify plant diseases automatically through leaf images (e.g., grapevine leaf diseases), which provides comprehensive support for precision irrigation by incorporating data on soil texture, water retention, moisture levels, and crop water needs at different growth phases [12, 13]. The unification of smartphone applications (such as *riceXpert*, PLANTIX, Agra Scout, Ag PHD, Field Pest, IPM Toolkit, PestXpert, and others) [14, 15], ecological simulation models (DYMEX—for rice YSB; and EPIRICE and EPIBLA—blast disease) [16, 17], and decision support systems (DSS) (such as "CropPest DSS" for rice pests and "DSSAT" for cotton pests) [18], empowered by DL and extensive data analysis through AI, has transformed the diagnosis of crop pests and diseases, particularly in areas with limited resources [9].

LiDAR technology is emerging as a valuable precision tool for agricultural pest monitoring and management when paired with precision farming methods [20–22]. As the agricultural industry increasingly uses smart technologies, LiDAR's ability to provide quick, non-invasive, and extremely accurate assessments helps improve crop health assessments, lower costs, and track pests (pine shoot beetle, *Tomicus* spp.), all of which contribute to sustainable agricultural practices and essential for precision farming systems and modern IPM programs [23, 24]. The working principles of LiDAR technology for pest management and monitoring are described in the following sections. Farmers can improve crop security by using AI to make better judgments regarding pest and disease management. Based on factual knowledge, farmers can make decisions using techniques such as DL and real-time data [8, 19].

13.2 Principles of Lidar Technology in Pest Monitoring and Management

13.2.1 LiDAR Technology Operating Principle in Smart Farming

The fundamental idea underlying LiDAR technology is to send laser pulses and measure the time taken for the echoed signals to return to the sensor, which enables extremely precise distance measurements of objects. This enables the production of three-dimensional (3D) point clouds that map environmental disturbances and swift data collection [25, 26]. LiDAR improves perceptual abilities when incorporated into autonomous vehicles, allowing for accurate object detection, self-localization, and mapping in complex and dynamic environments (by capturing reflected light and using the time-of-flight method to calculate the distance to the object [27, 28]. LiDAR's ability to create intricate 3D point clouds, which provide a wealth of geometric data necessary for autonomous navigation and environmental awareness (uses nanometer (nm) wavelengths to record backscattered light, providing both distance and intensity details), is the reason for its growing significance [29].

According to [30] and [31], LiDAR technology has many advantages for monitoring insect pests, including the ability to detect disease-carrying mosquitoes, characterize agricultural pests (migratory insects, sap suckers, and leaf chewing insects), and detect high insect populations of cereal-based ecosystems in the atmosphere, with precisely observed patterns of everyday activity as well as the size, movement, and density of insects, and it can also measure the optical cross-section of flying insects aiding in group identification and offers behavioral insights. The following subsection provides a detailed description of several types of LiDAR systems-mediated pest identification, monitoring, and management scheduling in Table 13.1.

13.2.2 LiDAR Technology Used for Insect–Pest Monitoring and Management

The use of LiDAR technology in precision farming offers an advanced method for pest management, providing detailed 3D information about crops and their environments [41]. These detailed data can be used to identify variations in plant health, structure, and density, which are indicators of pest presence or damage [42]. LiDAR technology facilitates the

Table 13.1 Types of LiDAR systems used for monitoring several crop pest behaviours, as well as crop protection aspects.

Pest behavior/pattern	LiDAR types	Source
Migratory behavior pattern of insect monitoring	Fluorescent LiDAR	[20]
Detects biotic stress in the ground areas	Terrestrial (Ground-based) LiDAR	[32]
Visible multiple pest or disease symptoms on the plant and leaf surface	Multi-sensor LiDAR	[33]
Aerial moving migratory insects and spore dispersal from pathogens	Mobile LiDAR system	[34]
Aerial and migratory insect pests' behavior patterns	Aerial (Drone/UAV) LiDAR	[35]
Leaf chewing and sap sucking insects (insect flight velocity measurements)	(CW) Scheimpflug LiDAR	[36]
Abiotic stress effects on insect phenology (emigratory flights of moths)	Doppler weather LiDAR	[37]
Quantification of insect activity and dispersal	Passive kHz LiDAR	[38]
Chlorophyll damage status by leaf chewing insects (caterpillars, locusts)	Green laser LiDAR	[39]
Terrestrial insects' movement patterns	Amplitude modulated continuous wave (AMCW) LiDAR	[29]
Migratory behavior pattern of flying insects	Pulsed laser-based LiDAR	[40]

prompt and precise identification of plant diseases, which is important for preventing financial losses and promoting sustainable farming practices, especially with the increased ease of global disease transfer and potential for pesticide resistance [43]. Early pest and disease identification, whether achieved through conventional agricultural extension services or modern online resources such as mobile applications, LiDAR sensors, remote sensing, and AI, is essential for effective disease management [9].

LiDAR sensors can detect the physiological changes that pest infestations often cause in plants, such as reduced photosynthetic activity, altered leaf reflectance, reduced nutrient absorption, and stunted growth in both shoots and roots, using laser light reflectance at several wavelength detection points in the damaged area [44]. The detailed functioning of LiDAR-mediated insect monitoring in the agricultural ecosystem, generating a 3D image, is depicted in Figure 13.1. Several LiDAR sensor systems have been adapted for crop pest surveillance, each with its own advantages and disadvantages [32–34]. Airborne LiDAR systems are suitable for regional pest surveillance because they can quickly cover large regions and can be mounted on unmanned aircraft or drones, which use powerful lasers to penetrate crop canopies and produce detailed ground elevation models that can be used to assess crop biomass and identify areas experiencing biotic stress [35]. However, the data from these airborne LiDAR systems can be affected by weather conditions, which require sophisticated processing techniques to remove noise and artifacts [45].

Although, terrestrial LiDAR systems, also known as ground-based LiDAR, have limitations in terms of spatial coverage, they provide data with even higher resolution than aerial systems by scanning individual plants or rows of crops. This is because they are frequently mounted on mobile platforms, such as tractors or unmanned ground vehicles (UGVs),

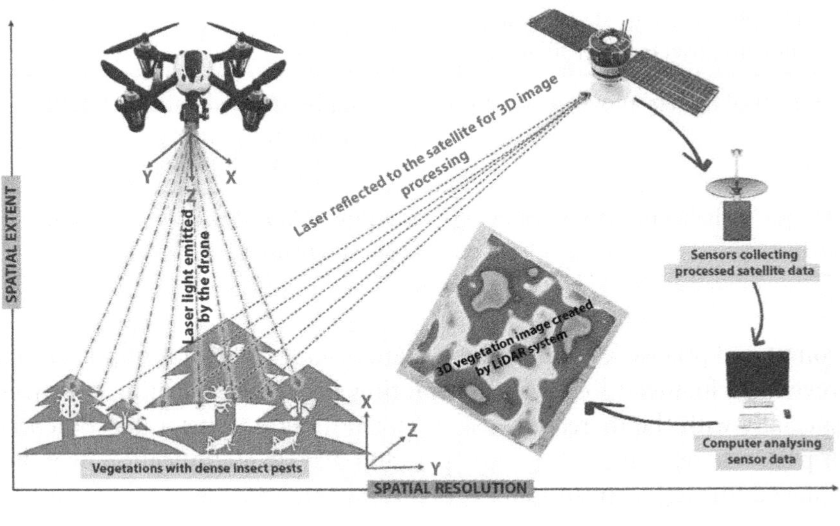

Figure 13.1 LiDAR technology used in agricultural crop pest monitoring.

which are especially useful for in-depth analysis of plant architecture, such as stem diameter, leaf area index, and canopy volume, which help assess plant health and spot minute indications of pest damage [46]. Mobile LiDAR devices enable the rapid evaluation of crop conditions over comparatively broad areas, offering a balanced choice between spatial coverage and data resolution. The high spatial and temporal resolution offered by continuous-wave LiDAR systems improves insect monitoring by enabling the characterization of insects in terms of their size, number, and preferred habitats [74]. Airborne LiDAR measurements of depth can be used to evaluate insect-induced defoliation, where the miniaturization of topographical LiDAR technology has opened new ecological applications [75]. Although the laser beam and canopy structure interaction can be complex, requiring the development of specific interpretation algorithms to achieve the desired level of accuracy and precision [47], these systems use real-time kinematic GPS to ensure accurate location information, which is typically mounted on vehicles, causing efficient monitoring of pest and disease infestations in orchards, vineyards (grapevine, betel, etc.), and other high-value crops.

13.3 Integration of Lidar with Other Digital Technologies for Precision Pest Monitoring and Control

Furthermore, the integration of AI into agricultural practices has spurred the development of precision farming, a data-intensive approach that uses advanced tools and techniques to inform farming decisions and add value to agriculture through extensive data analysis [48]. By combining LiDAR data with HI data, farmers can learn more about crop health and the effects of pests. Once collected, processed, and analyzed, these data help determine the state of the agroenvironment, and when linked with agroclimatic and economic models, technical interventions can be implemented at the field level using either conventional means or automated/robotized solutions [49]. DL techniques, along with LiDAR principles, have made it possible to identify crop pests and diseases with high accuracy using image data [9]. Furthermore, by providing additional information about the biochemical composition and physiological status of plants through wavelength detection (in nm), combining LiDAR with other sensing technologies, such as HI, enhances the accuracy and reliability of pest detection in damaged areas [50]. The higher spatial resolution of LiDAR makes it easier to identify the patterns of pest distribution (both aerial and terrestrial), enabling more focused interventions and minimizing the use of manual application of pesticides [6].

Self-operating robots with LiDAR sensors using AI can traverse fields, detect pest infestations, and administer treatments with minimal human assistance, guaranteeing that pesticides are applied only where necessary. This combination of LiDAR technology and robotic platforms, guided by DL and ML, offers promising opportunities for autonomous pest monitoring and targeted actions [51]. ML algorithms are increasingly used to analyze LiDAR data and automate pest detection because they can be trained to recognize certain pest signatures in LiDAR point clouds, allowing for early identification and quick responses [7]. Combining direct and indirect techniques, such as multi-sensor systems of different sizes, offers a promising method for detecting insect infestations. Combining LiDAR technology with hyperspectral push-broom imaging and remote dark-field observation (observing low-light or scattered signals to identify specific features under dark back ground) enhances the potential for LiDAR entomological research, particularly when used at telecom wavelengths [33, 76]. Both the temporally and spectrally resolved tracking of sugar water-tagged mosquitoes and the remote measurement of auto-powder-tagged honeybees are made possible using a sophisticated LiDAR technique, such as the hyperspectral Scheimpflug LiDAR (HSL) method, which simultaneously determines the range, timing, and photon energy from tagged insects flying freely in full daylight [75]. According to [52], LiDAR data can be integrated with other geospatial datasets, such as satellite imaging and meteorological data, to obtain a more thorough picture of crop conditions and pest-related dynamics.

13.4 Successful Case Studies of Pest Identification and Forecasting Using Lidar System

By identifying flying insects and assessing plant vitality and structural changes caused by pest infestations, LiDAR technology has had a significant impact on the direct and indirect detection and monitoring of pests. The important case studies and applications listed below demonstrate how different LiDAR systems have been used effectively for pest detection and monitoring, as shown in Table 13.2.

Table 13.2 Important case studies illustrating the role of LiDAR technology in major agricultural crop protection practices.

Crop/target	LiDAR application	Key findings/ outcomes	Source
Planthopper (*Hemiptera*) and moth (*Lepidoptera*) identification	Real-time detection and identification of flying pest species using fluorescent LiDAR	Migration and movement of pest insects were detected up to 50 m, identified at a peak of 450 nm.	[20]
Monitoring of pesticide clouds in the apple orchard	Drone-based terrestrial LiDAR (TLSs) for precision pesticide spray in undulated terrain for pest management	At the maximum pulse repetition frequency, the LiDAR system captured the mid-range pesticide spray drift with a temporal resolution of 100 ms and a distance resolution of 2.4 m.	[21]
Airborne and terrestrial insects	Doppler weather radars and harmonic radars in LiDAR systems are used for pest monitoring of migration and weather impacts on flight behavior.	Observing high-altitude insect migration pathways and spotting dense groups of flying and ground-dwelling insects at night and during the day.	[30]

(*Continued*)

Table 13.2 Important case studies illustrating the role of LiDAR technology in major agricultural crop protection practices. (*Continued*)

Crop/Target	LiDAR application	Key findings/ Outcomes	Source
Rice insects in paddy fields	LiDAR remote sensing used in the Chinese scene through continuous-wave (CW) Scheimpflug LiDAR.	By analyzing the frequencies of wingbeats, insect species were identified, and their attraction to an ultraviolet (UV) lamp placed nearer to the horizontal laser sampling route at night was examined.	[22]
General crops (pest impact on yield)	Remote sensing fusion (LiDAR + AI) for pest-induced crop yield prediction after pest damage	Improved yield forecasts and pest impact assessment using LiDAR-derived structural data and AI models	[53]
Crop fields (migratory insect pests)	Passive kHz LiDAR (dark-field methods that measure scattered sunlight) is used for the quantification of insect activity and dispersal due to less power consumption and ease of use.	Time-lag correlations of segments from the quadrant detector are used to determine insect motions, and the results show that during the research period, insects tended to fly more in the direction of the wind than against it.	[38]

(*Continued*)

Table 13.2 Important case studies illustrating the role of LiDAR technology in major agricultural crop protection practices. (*Continued*)

Crop/Target	LiDAR application	Key findings/ Outcomes	Source
Basal stem rot (*Ganoderma boninense*)	Ground-based LiDAR for tree-level morphometric trait analysis for basal stem rot	Using JMP software (SAS Institute, North Carolina, USA), the accuracy of a severity level classification was 80%, and the accuracy of the healthy-unhealthy classification was 86.67% among the five categories of disease severity.	[54]
Crop canopies (nitrogen and pest stress)	Green laser LiDAR for chlorophyll and nitrogen status in leaves after pest damage	Detected physiological stress possibly due to pests (loss in photosynthesis, reduced energy production, increased vulnerability to abiotic stresses), supporting precision pest management	[55]

(*Continued*)

Table 13.2 Important case studies illustrating the role of LiDAR technology in major agricultural crop protection practices. (*Continued*)

Crop/Target	LiDAR application	Key findings/ Outcomes	Source
Pine shoot beetle (PSB), *Tomicus* spp.	Monitoring the shoot damage ratio (SDR) of PSB through HI and LiDAR	The Lidar method (R^2 = 0.69, RMSE = 12.28%) fared better than the hyperspectral method (R^2 = 0.67, RMSE = 15.87%) for more accurate SDR estimation of individual trees, demonstrating how well HI and Lidar data integration can predict SDRs at the individual tree level.	[24]
Autonomous pesticide spraying	A thorough and reliable technique that uses LiDAR technology to enable autonomous robot navigation in agricultural fields.	Achieved accurate (location-specific) pesticide spray using autonomous 2D models using a retractable sprayer boom	[56]

(*Continued*)

Table 13.2 Important case studies illustrating the role of LiDAR technology in major agricultural crop protection practices. (*Continued*)

Crop/Target	LiDAR application	Key findings/ Outcomes	Source
Wheat ecosystem (pest migration)	Pattern observation with remote sensing (LiDAR-integrated) of daily flying insects at low altitudes	Low-altitude movement observations (150 m) and detailed information on insects in southern Sweden during the annual northward migration phase, which lasted from May to July, helped scientists better understand insect migration patterns and improve pest control techniques.	[31]
Migratory pattern of flying insects	A wing-beat-modulated scattered laser light operated Lidar device was used to investigate the geographic distributions of flying insects.	Position and densities of flying insects were observed through three types of mechanisms (wings up—full signal, wings down—no signal, and wings beating—modulated signal). With a maximum distance of 87.6 m, 662 insects were considered to be found, while 76 insects were categorized as extremely likely to be detected.	[57]

(*Continued*)

Table 13.2 Important case studies illustrating the role of LiDAR technology in major agricultural crop protection practices. (*Continued*)

Crop/Target	LiDAR application	Key findings/ Outcomes	Source
Vineyard pesticide spray	Canopy characterization for improved pesticide spray through ultrasonic and LiDAR sensors	LIDAR sensors provide precise data about the canopy, with a strong relationship between ultrasonic sensors and the leaf area index, or LAI (R2 = 0.51), and between the canopy volume (CVU) measured with ultrasonic and LIDAR sensors (R2 = 0.52), which seem to be ideal supplementary tools to increase pesticide efficiency use in vineyard.	[58]
Wood-boring forest insects	Echo-intensity of tree trunks damaged by borers was measured by hyperspectral LiDAR (HSL) based on an acousto-optic tunable filter (AOTF) system.	A 100% quick detection accuracy in terms of the echo intensity of the damaged area was achieved when the reflectivity from five spectral channels was used as input parameters.	[59]

(*Continued*)

Table 13.2 Important case studies illustrating the role of LiDAR technology in major agricultural crop protection practices. (*Continued*)

Crop/Target	LiDAR application	Key findings/ Outcomes	Source
Forest insects and diseases	Potential of airborne HSL (AHSL) checked for monitoring forest biotic stresses using 3D radiative transfer modelling.	As a promising and trustworthy tool to monitor forest disturbances, AHSL was found to have significantly higher overall accuracies (OA: 65.95%-89.45%) than HI (OA: 33.99%-57.02%) for predicting insect and disease stress levels, particularly for structural and spectral changes in the lower and middle parts of canopy.	[60]

13.5 Challenges and Future Studies

LiDAR is an essential technology in many industries because of its capacity to produce precise 3D environmental maps by measuring distances with laser beam reflectance from specific objects [61]. However, despite its potential, there are several obstacles to overcome before LiDAR can be used for agricultural monitoring, particularly in insect pest management, such as the intricate interactions between laser beams and plant canopies, which require the development of specialized algorithms to ensure accurate interpretation [47]. Its frequent application in crop monitoring is further hampered by the high expenses of data collection and the significant processing power required to process entire datasets [62], which must be overcome to fully utilize LiDAR's potential for accurate and successful pest management. There are possible advantages and disadvantages to using LiDAR technology for insect pest monitoring and control, especially regarding data interpretation and strategy formulation.

While LiDAR provides 3D canopy structure and precise localisation, improved YOLOv8 algorithms process 2D image data to identify, classify, and segment disease (Grapevine leaf disease) or pest characteristics, enabling real-time, targeted precision agriculture [63, 64]. Future research should focus on creating advanced analytical tools that can measure pest populations, identify particular pest species, and assess the extent of agricultural damage (by assessing specific parameters such as plant height in wheat pest damage), among other useful insights that can be extracted from LiDAR data. This would lessen the need for pesticides and promote sustainable farming methods by making it easier to conduct timely and targeted interventions [65].

A more complete picture of plant health can be obtained by integrating LiDAR data with thermal imaging, which can identify temperature variations in plants associated with pest infestations. Integrating LiDAR with other technologies can enable early and precise diagnosis of plant biotic stresses, addressing the challenges posed by emerging pests in regions lacking local expertise [66, 67]. Future studies should focus on developing robust algorithms that can adapt to variants in crops, growth stages, and environmental conditions to address the challenges associated with LiDAR-based pest management. These algorithms will enable the consistent and accurate identification and tracking of pests across diverse agricultural landscapes by efficiently removing noisy data and isolating core pest information from the outputs of the LiDAR system. Furthermore, efforts should be directed towards developing user-friendly graphical interfaces and decision-support tools that enable farmers and other agricultural professionals to easily retrieve and interpret LiDAR data [68]. Furthermore, expanding the use of ecological models by designing them to identify plant pests and diseases across broader geographic areas and using drone-captured aerial imagery could significantly affect crop quality and sustainable development.

As an effective and noninvasive method of evaluating the health of plants, image processing techniques have grown in significance in the identification of plant diseases [69]. With its capacity to produce intricate 3D geographical maps, LiDAR technology creates new avenues for early disease identification in crop leaves using a convolutional neural network (CNN) model, and prompt action may be able to stop the spread of disease to nearby crops [70]. Exploring novel technologies, such as cloud computing platforms and data-sharing initiatives, is necessary to address data limits and computational hurdles in LiDAR-centered pest management, which offers scalable and affordable resources, eliminating the need for expensive on-site infrastructure to store, process, and analyze large LiDAR datasets [71].

Furthermore, the ability of technology to transform global agriculture and food security has been demonstrated by the remarkable accuracy of sensor-captured image-based pest and disease diagnosis using ML models [72]. Working together can also help develop open-source tools and software, which will increase the number of people who can utilize LiDAR technology, including low-resource communities and smallholder farmers in developing countries. Industry stakeholders can greatly aid the development and marketing of LiDAR-based pest monitoring technologies. Through incentive schemes and instructional campaigns, government organizations can improve research and development efforts and promote farmers' adoption of LiDAR technology. These stakeholders, including farmers, researchers, NGOs, students, and industry representatives, can accelerate the development and application of LiDAR-based pest monitoring technologies by working together to promote a more resilient and sustainable agricultural ecosystem [12, 73].

13.6 Conclusion

The need for advanced solutions is highlighted by the drawbacks of traditional pest surveillance methods, such as delayed detection and inadequate temporal resolution. Because various biotic and abiotic pressures make plant stress indicators inherently complicated, it is important to be careful when identifying causal agents based only on visual inspection. Advanced remote sensing tools, such as LiDAR and HI, open new possibilities for measuring the structure and function of vegetation and identifying even minute signs of pest infestation. LiDAR complements traditional sensors and e-traps by detecting insects at low altitudes with low to moderate frequency, whereas radar is used for high-altitude insect migration monitoring. While acoustic techniques offer real-time insect detection through sound analysis, HI can distinguish between insect species based on their spectral characteristics. While LiDAR technology is an effective active detection technology for insects, increasing detection certainty and specificity and allowing discrimination between individual insects and observation of their behaviors in various monitoring situations, the combination of ML, the IoT, and ANN is revolutionizing agriculture to enable more efficient and sustainable production.

Additionally, greater detection sensitivity and specificity provide informative information about insect populations, weather, and time of day.

Owing to the rapid potential for pest infestations, sensor-based detection requires excellent spatial and temporal resolution to detect even small patches of contaminated crops. Because inexpensive LiDAR units are portable, they can be mounted on cars, drones, or even backpacks, allowing for *in situ* applications in remote areas and comparatively large-scale studies using many units. The importance of LiDAR in sustainable agriculture will increase as technology develops further, achieving higher sensor sensitivity and integration with predictive analytics. LiDAR's ability to monitor and manage pests is a significant step toward more environmentally friendly, efficient, and sustainable agriculture. It also marks the beginning of a new era in precision agriculture which balances ecological stewardship with productivity.

AI disclosure statement: The author acknowledges the use of AI-based tools for minor language editing or grammatical correction. However, all intellectual contributions, analyses, and conclusions presented in this chapter are entirely the author's own.

References

1. Karunathilake, E.M.B.M., Le, A.T., Heo, S., Chung, Y.S., Mansoor, S., The Path to Smart Farming: Innovations and Opportunities in Precision Agriculture. *Agriculture*, 13, 8, 1593, 2023.
2. Piccoli, F., Locatelli, S.G., Schettini, R., Napoletano, P., An Open-Source Platform for GIS Data Management and Analytics. *Sensors*, 23, 8, 3788, 2023.
3. Soussi, A., Zero, E., Sacile, R., Trinchero, D., Fossa, M., Smart Sensors and Smart Data for Precision Agriculture: A Review. *Sensors*, 24, 8, 2647, 2024.
4. Kumar, L., Ahlawat, P., Rajput, P., Navsare, R.I., Singh, P.K., Internet of Things (IoT) for Smart Precision Farming and Agricultural Systems Productivity: A Review. *IJEAST*, 5, 141–146, 2021.
5. Abdullahi, H.S., Mahieddine, F., Sheriff, R.E., Technology Impact on Agricultural Productivity: A Review of Precision Agriculture Using Unmanned Aerial Vehicles, in: *Wireless and Satellite Systems*, 7, 388–400, Springer Nature, Cham, 2015.
6. Roberts, D.P., Short, N.M., Sill, J., Lakshman, D.K., Hu, X., Buser, M., Precision Agriculture and Geospatial Techniques for Sustainable Disease Control. *Indian Phytopathol.*, 74, 2, 287, 2021.
7. Atalla, S., Tarapiah, S., Gawanmeh, A., Daradkeh, M., Mukhtar, H., Himeur, Y., Daadoo, M., IoT-Enabled Precision Agriculture: Developing an Ecosystem for Optimized Crop Management. *Information*, 14, 4, 205, 2023.

8. Javidan, S.M., Ampatzidis, Y., Banakar, A., Vakilian, K.A., Rahnama, K., An Intelligent Group Learning Framework for Detecting Common Tomato Diseases Using Simple and Weighted Majority Voting with Deep Learning Models. *Agriengineering*, 7, 2, 31, 2025.
9. Mohanty, S.P., Hughes, D.P., Salathé, M., Using Deep Learning for Image-Based Plant Disease Detection. *Front. Plant Sci.*, 7, 215232, 2016.
10. Omia, E., Bae, H., Park, E., Kim, M.S., Baek, I., Kabenge, I., Cho, B.K., Remote Sensing in Field Crop Monitoring: A Comprehensive Review of Sensor Systems, Data Analyses, and Recent Advances. *Remote Sens.*, 15, 2, 354, 2023.
11. Toselli, M., Baldi, E., Ferro, F., Rossi, S., Cillis, D., Smart Farming Tool for Monitoring Nutrients in Soil and Plants for Precise Fertilization. *Horticulturae*, 9, 9, 1011, 2023.
12. Sladojevic, S., Arsenovic, M., Anderla, A., Culibrk, D., Stefanovic, D., Deep Neural Networks-Based Recognition of Plant Diseases by Leaf Image Classification. *Comput. Intell. Neurosci.*, 2016, 1, 3289801, 2016.
13. Kapetas, D., Kalogeropoulou, E., Christakakis, P., Klaridopoulos, C., Pechlivani, E.M., Comparative Evaluation of AI-Based Multi-Spectral Imaging and PCR-Based Assays for Early Detection of Botrytis cinerea Infection on Pepper Plants. *Agriculture*, 15, 2, 164, 2025.
14. Mohapatra, S.D., Tripathi, R., Kumar, A., Kar, S., Mohapatra, M., Shahid, M., Pathak, H., Eco-Smart Pest Management in Rice Farming: Prospects and Challenges. *Oryza*, 56, 143–155, 2019.
15. Hampf, A.C., Nendel, C., Strey, S., Strey, R., Biotic Yield Losses in the Southern Amazon, Brazil: Making Use of Smartphone-Assisted Plant Disease Diagnosis Data. *Front. Plant Sci.*, 12, 621168, 2021.
16. Savary, S., Nelson, A., Willocquet, L., Pangga, I., Aunario, J., Modeling and Mapping Potential Epidemics of Rice Diseases Globally. *Crop Prot.*, 34, 6–17, 2012.
17. Nurhayati, E. and Koesmaryono, Y., Predictive Modeling of Rice Yellow Stem Borer Population Dynamics under Climate Change Scenarios in Indramayu, in: *IOP Conference Series: Earth and Environmental Science*, IOP Publishing, vol. 58, no. 1, pp. 012054, 2017.
18. Patel, A. and Kadam, P., An Ample Study of Numerous Intelligent Systems Used for Agriculture: Review. *Int. J. Comput. Appl.*, 143, 1, 20–24, 2016.
19. Atila, Ü., Uçar, M., Akyol, K., Uçar, E., Plant Leaf Disease Classification Using the Efficient-Net Deep Learning Model. *Ecol. Inf.*, 61, 101182, 2021.
20. Mei, L., Guan, Z.G., Zhou, H.J., Lv, J., Zhu, Z.R., Cheng, J.A., Somesfalean, G., Agricultural Pest Monitoring Using Fluorescence Lidar Techniques: Feasibility Study. *Appl. Phys. B*, 106, 733–740, 2012.
21. Gregorio, E., Rocadenbosch, F., Sanz, R., Rosell-Polo, J.R., Eye-Safe Lidar System for Pesticide Spray Drift Measurement. *Sensors*, 15, 2, 3650–3670, 2015.

22. Song, Z., Zhang, B., Feng, H., Zhu, S., Hu, L., Brydegaard, M., Li, D., Application of Lidar Remote Sensing of Insects in Agricultural Entomology on the Chinese Scene. *J. Appl. Entomol.*, 144, 3, 161–169, 2020.
23. Dong, P. and Chen, Q., *LiDAR Remote Sensing and Applications*, 1st ed., CRC Press, 2017.
24. Lin, Q., Huang, H., Wang, J., Huang, K., Liu, Y., Detection of Pine Shoot Beetle (PSB) Stress on Pine Forests at the Individual Tree Level Using UAV-Based Hyperspectral Imagery and Lidar. *Remote Sens.*, 11, 21, 2540, 2019.
25. Polat, N., Lidar-Derived 3D City Modelling. *Int. Arch. Photogramm. Remote Sens. Spat. Inf. Sci.*, 44, 339–342, 2020.
26. Pagrotra, A., Vyas, N., Dhiman, H., Dutt, V., Enhancing Land-Based Robotics Through the Development of Remotely Operated Vehicles for Military, Rescue and Industrial Applications, in: *2023 3rd Asian Conference on Innovation in Technology (ASIANCON)*, Ravet IN, India, pp. 1–6, 2023, doi:10.1109/ASIANCON58793.2023.10270756.
27. Chen, S., Niu, S., Lan, T., Liu, B., PCT: Large-Scale 3D Point Cloud Representations *via* Graph Inception Networks with Applications to Autonomous Driving, in: *2019 IEEE International Conference on Image Processing (ICIP)*, IEEE, pp. 4395–4399, 2019.
28. Ai, L., Zhao, Y., Huang, X., Enabling 3-D object detection with a low-resolution LiDAR. *IEEE Embedded Syst. Lett.*, 14, 4, 163–166, 2022.
29. Zhang, X., Huo, L., Liu, Y., Zhuang, Z., Yang, Y., Gou, B., Research on 3D phenotypic reconstruction and micro-defect detection of green plum based on multi-view images. *Forests*, 14, 2, 218, 2023.
30. Dwivedi, M., Shadab, M.H., Santosh, V.R., Insect pest detection, migration, and monitoring using radar and LiDAR systems, in: *Innovative Pest Management Approaches for the 21st Century: Harnessing Automated Unmanned Technologies*, pp. 61–76, Springer, 2020.
31. Chen, H., Li, M., Månefjord, H., Travers, P., Salvador, J., Müller, L., *et al.*, Lidar as a potential tool for monitoring migratory insects. *Iscience*, 27, 5, 2024.
32. Peteinatos, G.G., Weis, M., Andújar, D., Rueda Ayala, V., Gerhards, R., Potential use of ground-based sensor technologies for weed detection. *Pest. Manage. Sci.*, 70, 2, 190–199, 2014.
33. Bieganowski, A., Dammer, K.H., Siedliska, A., Bzowska-Bakalarz, M., Bereś, P.K., Dąbrowska-Zielińska, K., *et al.*, Sensor-based outdoor monitoring of insects in arable crops for their precise control. *Pest. Manage. Sci.*, 77, 3, 1109–1114, 2021.
34. Farhan, S.M., Yin, J., Chen, Z., Memon, M.S., A comprehensive review of LiDAR applications in crop management for precision agriculture. *Sensors (basel, Switzerland)*, 24, 16, 5409, 2024.
35. Kashyap, B. and Kumar, R., Sensing methodologies in agriculture for monitoring biotic stress in plants due to pathogens and pests. *Inventions*, 6, 2, 29, 2021.

36. Li, Y., Wang, K., Quintero-Torres, R., Brick, R., Sokolov, A.V., Scully, M.O., Insect flight velocity measurement with a CW near-IR Scheimpflug LiDAR system. *Opt. Express*, 28, 15, 21891–21902, 2020.
37. Westbrook, J.K., Eyster, R.S., Doppler weather radar detects emigratory flights of noctuids during a major pest outbreak. *Remote Sens. Appl.: Soc. Environ.*, 8, 64–70, 2017.
38. Jansson, S. and Brydegaard, M., Passive kHz LiDAR for the quantification of insect activity and dispersal. *Anim. Biotelem.*, 6, 1–10, 2018.
39. Nadimi, M., Sun, D.W., Paliwal, J., Recent applications of novel laser techniques for enhancing agricultural production. *Laser Phys.*, 31, 5, 053001, 2021.
40. Wang, Y., Zhao, C., Dong, D., Wang, K., Real-time monitoring of insects based on laser remote sensing. *Ecol. Indic.*, 151, 110302, 2023.
41. Cope, J.S., Corney, D., Clark, J.Y., Remagnino, P., Wilkin, P., Plant species identification using digital morphometrics: A review. *Expert Syst. Appl.*, 39, 8, 7562–7573, 2012.
42. Sishodia, R.P., Ray, R.L., Singh, S.K., Applications of remote sensing in precision agriculture: A review. *Remote Sens.*, 12, 19, 3136, 2020.
43. Wang, Z. and Menenti, M., Challenges and opportunities in Lidar remote sensing. *Front. Remote Sens.*, 2, 641723, 2021.
44. Radoglou-Grammatikis, P., Sarigiannidis, P., Lagkas, T., Moscholios, I., A compilation of UAV applications for precision agriculture. *Comput. Networks*, 172, 107148, 2020.
45. Li, W., Niu, Z., Huang, N., Wang, C., Gao, S., Wu, C., Airborne LiDAR technique for estimating biomass components of maize: A case study in Zhangye City, Northwest China. *Ecol. Indic.*, 57, 486, 2015.
46. Liu, S., Baret, F., Abichou, M., Boudon, F., Thomas, S., Zhao, K., Fournier, C., Andrieu, B., Irfan, K., Hemmerlé, M., Solan, B. de, Estimating wheat green area index from ground-based LiDAR measurement using a 3D canopy structure model. *Agric. For. Meteorol.*, 247, 12, 2017.
47. Mateo-Aroca, A., García-Mateos, G., Ruiz-Canales, A., Molina-García-Pardo, J., Molina-Martínez, J.M., Remote image capture system to improve aerial supervision for precision irrigation in agriculture. *Water*, 11, 2, 255, 2019.
48. Ganeshkumar, C., Sankar, J.G., David, A., Impact of artificial intelligence on agriculture value chain performance: Agritech perspective, in: *Blockchain, IoT, and AI Technologies for Supply Chain Management*, pp. 71–98, CRC Press, 2023.
49. Guevara, I., Ryan, S., Singh, A., Brandon, C., Margaria, T., Edge IoT prototyping using model-driven representations: A use case for smart agriculture. *Sensors*, 24, 2, 495, 2024.
50. Caballero, D., Calvini, R., Amigo, J.M., Hyperspectral imaging in crop fields: Precision agriculture, in: *Data Handling in Science and Technology*, vol. 32, pp. 453–473, Elsevier, BV, 2019.

51. Tang, Y., Chen, C., Leite, A.C., Xiong, Y., Precision control technology and application in agricultural pest and disease control. *Front. Plant Sci.*, 14, 1163839, 2023.
52. Qiu, Q., Sun, N., Bai, H., Wang, N., Fan, Z., Wang, Y., Cong, Y., Field-based high-throughput phenotyping for maize plants using 3D LiDAR point cloud generated with a "Phenomobile". *Front. Plant Sci.*, 10, 554, 2019.
53. Noskov, A., Bendix, J., Friess, N., A review of insect monitoring approaches with special References to radar techniques. *Sensors*, 21, 4, 1474, 2021.
54. Husin, N.A., Khairunniza-Bejo, S., Abdullah, A.F., Kassim, M.S., Ahmad, D., Azmi, A.N., Application of ground-based LiDAR for analysing oil palm canopy properties on the occurrence of basal stem rot (BSR) disease. *Sci. Rep.*, 10, 1, 6464, 2020.
55. Eitel, J.U., Magney, T.S., Vierling, L.A., Dittmar, G., Assessment of crop foliar nitrogen using a novel dual-wavelength laser system. *ISPRS J. Photogramm. Remote Sens.*, 97, 229–240, 2014.
56. Thamaraiselvan, S., Vivin, M.I., Ronald, K., Ray, R., Bini, D., Lidar-based navigation rover for fields with a smart pest sprayer using machine vision, in: *Proceedings of the 2023 9th International Conference on Advanced Computing and Communication Systems (ICACCS)*, vol. 1, pp. 1889–1894, 2023.
57. Tauc, M.J., Fristrup, K.M., Shaw, J.A., Development of a wing-beat-modulation scanning lidar system for insect studies, in: *Lidar Remote Sensing for Environmental Monitoring*, vol. 10406, pp. 94–103, SPIE Optical Engineering, San Diego, California, 2017.
58. Llorens, J., Gil, E., Llop, J., Escolà, A., Ultrasonic and LIDAR sensors for electronic canopy characterization in vineyards. *Sensors*, 11, 2, 2177–2194, 2011.
59. Shao, H., Zhang, H., Hu, P., Chen, Y., Li, W., Sun, L., Hu, Y., Feasibility study of forest wood-boring pest detection based on hyperspectral LiDAR. *Int. J. Remote Sens.*, 45, 6, 1926–1948, 2024.
60. Zhao, X., Qi, J., Xu, H., Yu, Z., Yuan, L., Chen, Y., Huang, H., Evaluating the potential of airborne hyperspectral LiDAR for assessing forest insects and diseases. *Remote Sens. Environ.*, 297, 113759, 2023.
61. Alsadik, B., Ellsäßer, F.J., Awawdeh, M., Al-Rawabdeh, A., Almahasneh, L., Oude Elberink, S., Al Asmar, Y., Remote sensing technologies using UAVs for pest and disease monitoring. *Remote Sens.*, 16, 23, 4371, 2024.
62. Chávez, J.D.Y. and Tullis, J.A., Deciduous forest structure estimated with LIDAR-optimized spectral remote sensing. *Remote Sens.*, 5, 1, 155, 2013.
63. Omasa, K., Hosoi, F., Konishi, A., 3D lidar imaging for detecting and understanding plant responses and canopy structure. *J. Exp. Bot.*, 58, 4, 881–898, 2007.
64. Hu, Q. and Liu, T., GCS-YOLO: A lightweight detection algorithm for grape leaf diseases based on improved YOLOv8. *Appl. Sci.*, 15, 7, 3910, 2025.
65. Guo, T., Fang, Y., Cheng, T., Tian, Y., Zhu, Y., Chen, Q., Qiu, X., Yao, X., Detection of wheat height using optimized multi-scan mode of LiDAR. *Comput. Electron. Agric.*, 165, 104959, 2019.

66. Ennouri, K., Triki, M.A., Kallel, A., Applications of remote sensing in pest monitoring and crop management, in: *Bioeconomy for Sustainable Development*, pp. 65–77, Springer Nature, Singapore, 2019.
67. Abd El-Ghany, N.M., Abd El-Aziz, S.E., Marei, S.S., A review: application of remote sensing for insect pests and diseases management. *Environ. Sci. Pollut. Res.*, 27, 27, 33503–33515, 2020.
68. Chlingaryan, A., Sukkarieh, S., Whelan, B., Machine learning approaches for crop yield prediction and nitrogen status estimation. *Comput. Electron. Agric.*, 151, 61–69, 2018.
69. Ngugi, L.C., Abelwahab, M., Abo-Zahhad, M., Recent advances in image processing techniques for automated leaf pest and disease recognition–A review. *Inf. Process. Agric.*, 8, 1, 27–51, 2021.
70. Kim, T.H., Shahroz, M., Alabdullah, B., Innab, N., Baili, J., Umer, M., Ashraf, I., ANFIS fuzzy convolutional neural network model for leaf disease detection. *Front. Plant Sci.*, 15, 1465960, 2024.
71. Ouhami, M., Hafiane, A., Es-Saady, Y., Hajji, M.E., Canals, R., Computer vision, IoT and data fusion for crop disease detection using machine learning: A survey and ongoing research. *Remote Sens.*, 13, 13, 2486, 2021.
72. Fedoroff, N.V., Food in a future of 10 billion. *Agric. Food Secur.*, 4, 1–10, 2015.
73. Singh, K., Yadav, M., Barak, D., Bansal, S., Moreira, F., Machine-learning-based frameworks for reliable and sustainable crop forecasting. *Sustainability*, 17, 10, 4711, 2025.
74. Jansson, S., Brydegaard, M., Mei, L., Li, T., Larsson, J., Malmqvist, E., Svanberg, S., Spatial monitoring of flying insects over a Swedish lake using a continuous-wave lidar system. *R. Soc. Open Sci.*, 10, 5, 221557, 2023.
75. Månefjord, H., Müller, L., Li, M., Salvador, J., Blomqvist, S., Runemark, A., Brydegaard, M., 3D-printed fluorescence hyperspectral lidar for monitoring tagged insects. *IEEE J. Sel. Top. Quantum Electron.*, 28, 5, 1–9, 2022.
76. Brydegaard, M., Advantages of shortwave infrared LIDAR entomology, in: *Laser Applications to Chemical, Security and Environmental Analysis 2014*, LW2D-6.

14

Role of Google Earth Engine (GEE) in Natural Hazard Monitoring and Management

Vijaya Ilango[1]* and Bhavadev Gajendiran[2]

[1]*Department of Chemistry, Birla Institute of Technology & Science Pilani, Dubai Campus, Dubai, UAE*

[2]*Department of Electrical and Electronics Engineering, Birla Institute of Technology & Science Pilani, Dubai Campus, Dubai, UAE*

Abstract

Millions of people are affected by disasters every year, resulting in thousands of deaths, significant economic losses, and community displacement. Google Earth Engine (GEE) is a robust geospatial platform that enables researchers, disaster managers, and policymakers to enhance their understanding, monitoring, and response to natural hazards, thereby facilitating more efficient disaster risk reduction and management. Globally, the frequency and intensity of climate-related disasters, such as hurricanes and floods, are increasing owing to the impacts of climate change, such as rising sea levels and altered precipitation patterns. The GEE is a powerful tool for disaster analysis. The Google Earth Engine offers free access to vast volumes of remote sensing datasets, including Landsat and Sentinel, as well as a robust cloud-based computing environment. In terms of technology, tools such as geographical information systems (GIS) and remote sensing can be used to manage catastrophes. Owing to its numerous applications, GEE, a cloud platform for big data and planetary-scale geospatial research, has recently gained recognition. This chapter aims to review the role of GEE, a cloud-based platform for large-scale geospatial data analysis in natural hazard monitoring and management. Initially, various techniques adopted for natural hazard assessment are explored, along with their specific applications. The potential use of GEE to track and forecast natural disasters is discussed, which will help individuals to pursue further research on GEE disaster management.

**Corresponding author*: srisaiviji2@gmail.com

Gurwinder Singh, Vishakha Sood, Narayan Vyas, Surendra Yadav and Dankan Gowda V. (eds.) Geospatial Intelligence for Precision Agriculture, (275–294) © 2026 Scrivener Publishing LLC

Keywords: Natural disasters, Google Earth Engine (GEE), hazards, monitoring, cloud platform, remote sensing

14.1 Introduction

Natural disasters have long posed a serious risk to infrastructure and human life, resulting in significant losses and destruction. The need for more effective and efficient disaster management systems has been brought to light in recent years due to the rising frequency and severity of natural disasters [1]. Natural events such as landslides, sinkholes, volcanic activity, and land subsidence can cause surface changes that have a substantial negative impact on the economy and society. Significant losses of natural and agricultural regions, harm to tourist attractions and cultural heritage, salinization of coastal soils, coastal retreat, and structural damage to infrastructure are all possible outcomes. Science and technology help us understand the mechanisms of atmospheric, geological, and hydrological natural hazards and analyze how these hazards become disasters. Over the past few years, significant advancements in the integrated approach to addressing natural disasters have been made possible owing to developments in the science and technology of natural hazards and related coping mechanisms [2].

This harmful effect can be exacerbated by climate change (e.g., increasing sea levels, changing rainfall patterns and intensities), particularly on unstable slopes and in low-lying coastal areas. Because it offers affordable options for risk-reduction techniques, ground deformation monitoring is essential for the management of these natural hazards, both during and after a disaster. The vulnerability of the agricultural sector to climate change has become increasingly evident in recent times. Hailstorms are only one example of the many ways in which the complex web of weather patterns and atmospheric dynamics is changing as the planet warms. Climate change manifests in several ways, including increased atmospheric instability, altered wind patterns, rising temperatures, and shifting precipitation patterns. These factors mutually influence the intensity, spatial distribution, and frequency of the hailstorms.

Hailstorms can cause havoc by tearing apart leaves, destroying fragile fruits, and puncturing plant tissues [3]. Hailstorm-related crop losses result in lower yields, poorer quality products, and economic hardships for farmers [4]. It is critical to understand the differences between hazards and disasters. When a potentially harmful phenomenon (hazard), such as an earthquake, occurs in a region that is not populated, it is not considered a disaster. When an event occurs in a crowded area and causes harm, loss, or devastation, it is referred to as a catastrophe. Natural catastrophes can have unforeseen effects, but they can be lessened by being prepared and acting quickly. It is vital for the community to prevent or alleviate the harm caused by natural disasters [5, 6]. Understanding the distinction between the phrases "hazard" and "disaster" is crucial. A hazard is a phenomenon that has the potential to cause harm or injury. For instance, an earthquake occurring in a deserted area is not considered a tragedy. However, if it occurs in a populated region and causes significant loss, harm, or devastation, it is referred to as a disaster. Various forms of disasters are categorized in several ways. The classification of disasters [7] is shown in Figure 14.1.

Natural disasters, including hurricanes, earthquakes, and volcanic eruptions, are catastrophic events that occur naturally and harm human society. Events caused by human activity (such as oil spills, nuclear catastrophes, major military wars, industrial chemical accidents, and air pollution) are referred to as human-made disasters. Natural catastrophes exacerbated or hastened by human activity are referred to as human-induced disasters. A sizable portion of the global population is affected by natural disasters, which are becoming increasingly frequent and severe. This hazard is expected to grow as climate issues worsen and the world's population becomes increasingly urbanized. A significant portion of urban populations is exposed to natural disasters, often with greater socioeconomic impacts than those in rural areas [8].

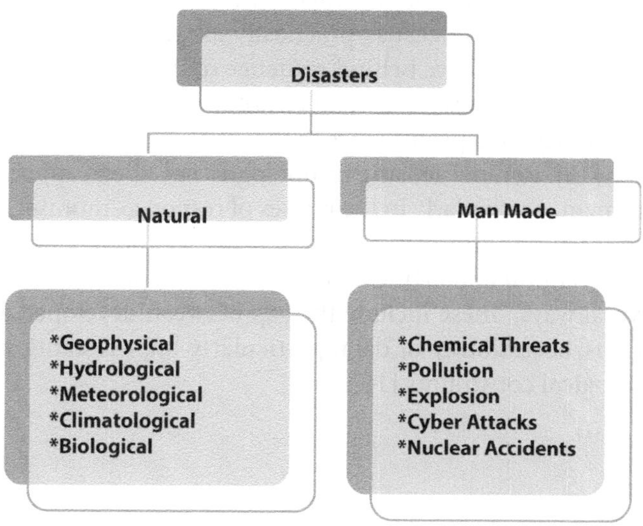

Figure 14.1 Classification of disasters.

14.2 Tools and Techniques for Natural Hazard Assessment

14.2.1 Geographical Information Systems Technology Method

A Geographic Information System (GIS) is a computer-based tool for geographical studies that can process and analyze spatial data, which is typically easy to comprehend. The most effective way to prevent or mitigate natural disasters is to implement risk zoning and elevate natural disaster management to the level of risk management. GIS has been extensively utilized in the risk analysis of geological catastrophes and is the most effective technique for risk zoning [9, 10]. Geographic studies integrate social and physical sciences, applying them to the spatial domain, as well as concepts and applications of GIS [11]. GIS supports a wide range of data in various forms in a user-friendly modeling environment. Governments have been assisted in making well-informed decisions regarding the establishment of emergency response strategies through mapping, modeling, and related assessments [12].

14.2.2 Remote Sensing

Remote sensing is a technique that enables the collection of environmental data from a distance. Natural disasters, including hurricanes, wildfires, and floods, can be tracked and monitored using remote sensing technology that utilizes satellites or airplanes. This method can be applied to the management of natural hazards because practically all geologic, hydrologic, and atmospheric processes leave behind evidence of their previous occurrence [13, 14]. The use of drones for civilian remote sensing emerged in the early 21st century and gained prominence in the 2010s for disaster management applications [15]. Remote sensing technologies have been applied to disaster management, particularly in the phases of response, monitoring, preparation, and warning. Although remote sensing technology can be used to manage both natural and artificial disasters, its adoption is currently limited in several ways. These include the gap between developed and developing nations, accessibility of data, particularly high-resolution imaging, and technological constraints [16].

14.2.3 Satellite Imaging

Satellites provide a means to gather data on the severity and extent of a natural disaster, as well as its impact on the environment and infrastructure [17]. While data from various sensors can be used for debris mapping, satellites with higher spatial resolution, such as SPOT, Landsat, and ASTER, are preferred [18]. The utilization of satellite remote sensing technology offers both quantitative and qualitative benefits for a range of applications, including risk reduction, operational aid, and post-disaster damage assessment. The post-disaster damage assessment process, which utilizes change detection, is a notable application of remote-sensing imagery. Disaster management applications utilize a range of remote sensing methods, including integrated, diverse, and high-resolution methods [19]. Several other tools and data sources used in disaster management are presented in Table 14.1.

14.2.4 Google Earth Engine

The Google Earth Engine (GEE) is a cloud-based geospatial processing tool used for extensive environmental monitoring and analysis. With the help of an explorer web app, users of the free GEE platform can access the following:

- Petabytes of remotely sensed pictures made accessible to the public
- Fast parallel processing and machine learning methods were implemented using Google's computer infrastructure.
- A collection of Application Programming Interfaces (APIs) that enable popular programming languages, such as Python and JavaScript, in development environments.

Without the need for supercomputers or specialized coding skills, people can search, evaluate, and show huge geographic data in practical ways by combining these basic operations. Google launched GEE in 2010 as a free cloud computing platform that allows anyone to utilize Google's infrastructure to undertake global geospatial research [20]. Earth and planetary sciences, environmental science, and agricultural and biological sciences are the top three fields in which GEE has been widely applied. GEE is based on a cloud-computing architecture that effectively manages satellite imagery and monitors data sources. Scholars from China and the US wrote the majority of GEE-related publications (58% of the total) [21], with US scholars having the most influence on the community (more than 50% of citations) [21]. Typically, disasters follow various phases, as depicted in Figure 14.2.

Regardless of the magnitude of the event, management entails all actions taken before, during, and after an event to minimize financial, social, and bodily harm. Both planned and spontaneous acts must be incorporated [22].

Table 14.1 Tools and data sources used in disaster management with their specific applications.

Disasters	Data sources used	Applications	Source
Earthquake	On board the NOAA-15 satellite, the Medium Energy Proton and Electron Detector (MEPED) instrument data	Vibration, seismology, tremor, seismic, and soil mechanics and Earthquake Disaster prevention, evacuation and rescue efforts.	[23–25]
Wildfires	Spatial multi-criteria modeling and evaluation (SMCME) of fire risk	Forest fire risk and hazard to enable the sustainability of the natural resources.	[26, 27]
Landslides/volcanic activities and the effects of earthquakes, wildfires, and floods.	Remotely piloted aircraft systems	Ultra-high-resolution images that can be used for the identification of active processes	[28, 29]
Floods	Machine learning and data analytics	Flood disaster management systems can benefit from the integration of machine learning and data analytics approaches to enhance flood forecasting, risk assessment, and response planning.	[30, 31]

(*Continued*)

Table 14.1 Tools and data sources used in disaster management with their specific applications. (*Continued*)

Disasters	Data sources used	Applications	Source
Floods	Spark and storm tools are used for machine learning	An efficient early warning system for flood management	[32]
Hurricanes	Dropsondes Multi-criteria Decision Making (MCDM), numerical models, and machine learning models.	Used to monitor hurricane conditions during climate events. Helps meteorologists create weather models and more accurate forecasts	[33, 34]
Volcanic eruptions	AI techniques Volcano geodesy	An AI model can accurately forecast the exact timing, magnitude, or specific details of volcanic eruptions and tracking the evolution of volcanic eruptions	[35]

Figure 14.2 Various phases involved in disaster management.

14.3 Role of GEE in Disaster Monitoring and Prediction

While Google Earth Engine (GEE) officially launched in 2010, it provides a 40-to-50-year historical record by hosting massive, pre-existing archives from international satellite missions. With its user-friendly interface, it addresses the challenges posed by traditional satellite data analysis methods and reduces the processing time. Users can interact with libraries, a code editor, and an explorer. The code editor enables users to implement the necessary coding steps in two languages: JavaScript and Python. It can handle data processing for large geospatial regions and enables users to easily input and implement commands to analyze, derive, or process satellite images. The GEE cloud computing interface can process data from large geospatial datasets, allowing users to analyze a wide study area over a longer duration. The GEE has been adapted to study the potential for disasters, such as predicting flooding due to water level rise or drought, by analyzing trends in precipitation indices and soil temperatures over the years. The damage caused by disasters, such as floods, avalanches, or other natural calamities, can be measured by analyzing the land cover area or changes in vegetation patterns over the affected area. It is easier to export the index records of regions over a period to examine them using complex and hybrid methodologies. The processing capabilities of GEE complement those of machine learning and artificial intelligence. Figure 14.3 illustrates the diverse applications of the GEE in natural hazard monitoring. This highlights how the GEE supports the real-time analysis and long-term assessment of various hazards.

14.3.1 Vegetation and Wildfires

The area covered by vegetation is essential for maintaining environmental balance among various organisms. It is necessary to monitor areas of vegetation and types of vegetation to study biodiversity, which also helps predict the likelihood of wildfire occurrence. Vegetation helps in understanding the trends of climate change cycles. GEE contains petabytes of data and image sets of vegetation areas over the years, allowing researchers to process them through the code editor or export them from the open-access Earth Engine for further study [36]. GEE has been utilized for monitoring vegetation mapping, deforestation, forest health assessment, and other applications [37]. Sentinel-2 has a high temporal resolution and is openly available on GEE [38]. The damage caused by wildfires in the forest areas

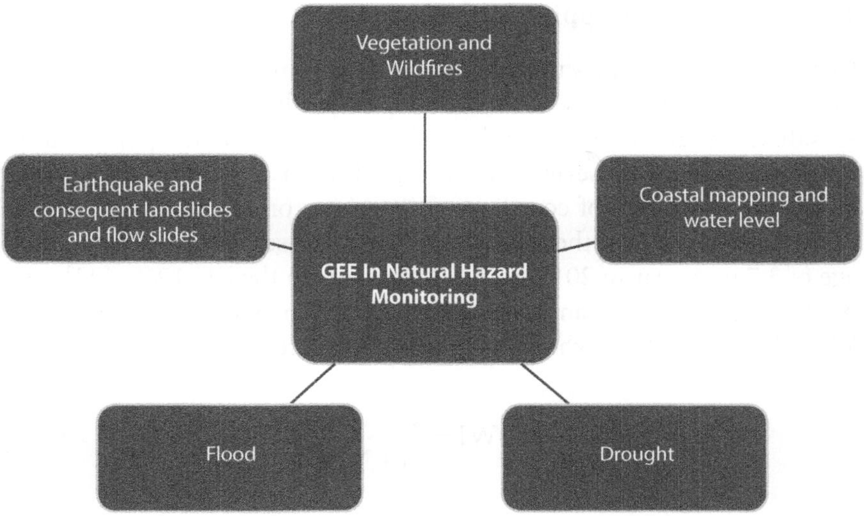

Figure 14.3 Use of GEE in various natural hazard monitoring.

of Uljin and Gangneung was estimated using integrated sample images from the Sentinel-2 satellite through GEE [39]. The Normalized Difference Vegetation Index (NDVI) is a remote sensing technique that helps assess the density or greenness of vegetation in an area. It is the ratio between the reflectance of red and near-infrared values, as shown in Equation (14.1) [40].

$$NDVI = \frac{NIR - R}{NIR + R} \tag{14.1}$$

GEE provides a straightforward calculation of the NDVI time series, eliminating time-consuming calculations. Studying changes in forest cover area is made easier by grouping regions according to the index. The CART-CVAPS method was implemented along with a Savitzky–Golay filter to eliminate the effect of cloud coverage over the landmass of China, providing a descriptive view of the change in land use. It detected the reduction of the flood bank into a desert area with an accuracy of 72% [41]. GEE open-source image sets to process and predict tree coverage using machine learning models such as the Global Forest Change (GFCD) tree cover and Clearing Assessment System Lite (CLASlite) tree cover [42]. The use of satellite imagery through GEE played a crucial role in this study, as emphasized by the authors.

14.3.2 Coastal Mapping and Water Level

Coastal mapping is the study of coastlines over time and their monitoring to inform improvements or further steps regarding land-use patterns in seaside towns and cities. With the melting of glaciers at the poles and the caution expressed by several global organizations regarding rising water levels, the possibility of coastlines encroaching on land areas may result in the submergence of these coastal regions. The sea level rose by an average of 3.7 mm/year in 2018 from 2.3 mm/year in the late 1960s [43]. Like NDVI, the shoreline can be classified based on Normalized Difference Water Index (NDWI), where the green reflectance replaces the red [44].

$$\mathrm{NDWI} = \frac{\mathrm{G} - \mathrm{NIR}}{\mathrm{G} + \mathrm{NIR}} \tag{14.2}$$

The End Point Rate (EPR) and Linear Regression Rate (LRR) are used for both short- and long-term observations of shoreline changes [45]. The GEE JavaScript editor is frequently used to analyze and visualize coastal changes. The coastlines of Algeria were studied using a dataset from the LANDSAT and Pleiades satellites, and the rate of erosion of the coastline was analyzed [46].

14.3.3 Drought

Drought is a natural phenomenon that occurs when there is insufficient rainfall below the minimum average [47]. High rates of mortality, hunger, and dehydration are frequently caused by prolonged droughts. Recent droughts in the US and Africa have led to a decline in food security and a significant reduction in food production [48]. Africa is one of the most vulnerable regions in the world to droughts because of its geographical position [49, 50]. Prediction and monitoring of drought conditions are necessary to help people and prevent further damage in the current situation [51]. Drought does not occur uniformly throughout the region but is instead spread across different areas. The World Meteorological Organization (WMO) has recommended different drought indices (DI). Among them, MODIS-derived DI, VCI, and SPI have been widely applied [52, 53]. The GEE provides users with indices for time-series analysis. The GEE facilitates researchers' access to remote sensing techniques for assessing conditions in combination with the indices mentioned above [54]. Table 14.2 lists the indices used for different types of droughts.

Table 14.2 Drought indices and their corresponding types and measures.

Index	Type of drought	Measure
SPI	Meteorological Drought Indicator	Precipitation deficit across the period of study
VCI	Agriculture Drought Indicator	Measure of vegetation situation based on water stress impact.
TCI	Temperature Drought Indicator	Stress on vegetation caused by temperatures and excessive wetness
VHI	Vegetation Drought Indicator	Composite index that integrates information about vegetation health (VCI) and temperature (TCI)

A study of drought indicators—SPI, VCI, TCI, and rates in Ethiopia was conducted based on the data obtained from MODIS using the GEE platform. Cloud noise pixels were eliminated using MOD13Q1 flags, which are available in GEE [55].

14.3.4 Flood

Floods occur when water bodies are unable to handle water flow, leading to water flowing into land areas. This affects the livelihoods of residents and destroys infrastructure [56]. Floods are among the most frequent and severe disasters in riverbanks and coastal areas. Several studies have been conducted on flood mapping using remote sensing tools, and they have been successful in predicting or assessing the damage caused by floods in the study area. Satellite remote sensing data are essential for such methodologies [57]. Studying flood risks across large areas facilitates proper planning for future development and guides further steps to prevent flooding in high-risk areas [58]. The GEE provides data coverage of flood events in different parts of the world. GEE enables the study of changes in flooded areas and assesses the damage inflicted even after restoration. Detecting water indices in an area over time helps to predict the possible occurrence of a disaster [59]. Sentinel-1 and Sentinel-2 satellite data were used to study flood damage in Mozambique [60]. The data cover labels the VH and VV imagery, which can be processed using ML to map floods.

14.3.5 Earthquake and Consequent Landslides and Flow Slides

Mortality due to this disaster has increased in recent years. Despite extensive investigations, no precursor has been found that reliably precedes every significant earthquake in various locations and time periods [61]. A study of different possible earthquake precursors based on MODIS-AOD (Aerosol Optical Depth) anomalies explains the role of GEE in the detection of seismic anomalies [62, 63]. GEE provides time-series data on ozone and chlorophyll, which were investigated using the median and LSTM methodologies. The boundaries of earthquake-affected regions can be determined using pre- and post-disaster satellite images [64–66]. A comprehensive study of the application of GEE in land monitoring after floods and earthquakes was conducted. Their study utilized data from the Sentinel-1 SAR sensor, processed through GEE, to examine land-use changes following flooding [67]. Landslides caused by earthquakes in hilly terrains and on islands were also analyzed using GEE. Hills are the most challenging to observe manually; hence, GEE with satellite images provides real-time data to monitor changes in hill forms and assess destruction in remote areas. Earthquake-induced landslides cause soil to travel long distances, changing the composition of regional soil distribution and behaving like a fluid [68–70].

14.4 Conclusion

The GEE offers a wide range of applications for monitoring and assessing various natural hazards that are not limited to those mentioned in this article. Damage caused by other disasters, such as avalanches and tsunamis, can also be studied using ice cover and NDWI land cover parameters. GEE provides data from remote areas where human observation is not feasible and helps appropriate organizations take preventive and protective measures. Therefore, GEE is very useful for monitoring hazards alone and in combination with machine learning algorithms. Although the GEE provides users with access to trustworthy data sources and accurate analyses, decision-making in an emergency still relies on the quality of the data and an understanding of its limitations. In particular, during significant occurrences, the GEE must be scalable to handle the growing amount and complexity of disaster data. There are also ethical concerns regarding privacy, data security, and possible algorithmic biases when using geographic data for disaster management. Although GEE provides effective disaster management solutions, issues with data handling, knowledge, accessibility, integration, and cost must be addressed for its successful deployment in the future.

AI disclosure statement: The author acknowledges the use of AI-based tools for minor language editing or grammatical correction. However, all intellectual contributions, analyses, and conclusions presented in this chapter are entirely the author's own.

References

1. Abdalzaher, M., Krichen, M., Yiltas-Kaplan, D., Ben Dhaou, I., Adoni, W., Early Detection of Earthquakes Using IoT and Cloud Infrastructure: A Survey. *Sustainability*, 15, 15, 11713, July 2023, doi:10.3390/su151511713.
2. Han, C. and Zang, S., A comprehensive review of disruptive technologies in disaster risk management of smart cities. *Clim. Risk Manage.*, 48, 100703, 2025, doi:10.1016/j.crm.2025.100703.
3. Botzen, W.J.W., Bouwer, L.M., van den Bergh, J.C.J.M., Climate change and hailstorm damage: Empirical evidence and implications for agriculture and insurance. *Resour. Energy Econ.*, 32, 3, 341–362, Aug. 2010, doi:10.1016/j.reseneeco.2009.10.004.
4. Raupach, T.H., *et al.*, The effects of climate change on hailstorms. *Nat. Rev. Earth Environ.*, 2, 3, 213–226, Feb. 2021, doi:10.1038/s43017-020-00133-9.
5. Krichen, M., Abdalzaher, M.S., Elwekeil, M., Fouda, M.M., Managing natural disasters: An analysis of technological advancements, opportunities, and challenges. *Internet Things Cyber-Phys. Syst.*, 4, 99–109, 2024, doi:10.1016/j.iotcps.2023.09.002.
6. K. Sundar, N. Vyas, and N. Bhati, Real-Time Earthquake Monitoring with Remote Sensing and AI Technology, in: Artificial Intelligence in Remote Sensing for Disaster Management, N. Dahiya, G. Singh, S. Singh, and A. Sharma (Eds.), pp. 303-322, Wiley-Scrivener, Hoboken, NJ, 2025, doi:10.1002/9781394287222.ch17.
7. Teh, D. and Khan, T., Types, Definition and Classification of Natural Disasters and Threat Level, in: *Handbook of Disaster Risk Reduction for Resilience*, pp. 27–56, Springer International Publishing, Cham, Switzerland, 2021, doi:10.1007/978-3-030-61278-8_2.
8. McBean, G. and Ajibade, I., Climate change, related hazards and human settlements. *Curr. Opin. Environ. Sustain.*, 1, 2, 179–186, Dec. 2009, doi:10.1016/j.cosust.2009.10.006.

9. Sun, R., Gao, G., Gong, Z., Wu, J., A review of risk analysis methods for natural disasters. *Nat. Hazard.*, 100, 2, 571–593, Dec. 2019, doi:10.1007/s11069-019-03826-7.
10. Criado, M., Martínez-Graña, A., San Román, J.S., Santos-Francés, F., Flood Risk Evaluation in Urban Spaces: The Study Case of Tormes River (Salamanca, Spain). *Int. J. Environ. Res. Public Health*, 16, 1, 5, Dec. 2018, doi:10.3390/ijerph16010005.
11. Tomaszewski, B., Judex, M., Szarzynski, J., Radestock, C., Wirkus, L., Geographic Information Systems for Disaster Response: A Review. *J. Homeland Secur. Emerg. Manage.*, 12, 3, Jan. 2015, doi:10.1515/jhsem-2014-0082.
12. Thach, N.N., *et al.*, Spatial pattern assessment of tropical forest fire danger at Thuan Chau area (Vietnam) using GIS-based advanced machine learning algorithms: A comparative study. *Ecol. Inf.*, 46, 74–85, July 2018, doi:10.1016/j.ecoinf.2018.05.009.
13. Scaioni, M., Longoni, L., Melillo, V., Papini, M., Remote Sensing for Landslide Investigations: An Overview of Recent Achievements and Perspectives. *Remote Sens.*, 6, 10, 9600–9652, Oct. 2014, doi:10.3390/rs6109600.
14. Voigt, S., *et al.*, Global trends in satellite-based emergency mapping. *Science*, 353, 6296, 247–252, July 2016, doi:10.1126/science.aad8728.
15. Singh, K.K. and Frazier, A.E., A meta-analysis and review of unmanned aircraft system (UAS) imagery for terrestrial applications. *Int. J. Remote Sens.*, 39, 15–16, 5078–5098, Jan. 2018, doi:10.1080/01431161.2017.1420941.
16. Bello, O.M. and Aina, Y.A., Satellite Remote Sensing as a Tool in Disaster Management and Sustainable Development: Towards a Synergistic Approach. *Procedia - Soc. Behav. Sci.*, 120, 365–373, Mar. 2014, doi:10.1016/j.sbspro.2014.02.114.
17. Tatem, A., Goetz, S., Hay, S., Fifty Years of Earth-observation Satellites. *Am. Sci.*, 96, 5, 390, 2008, doi:10.1511/2008.74.390.
18. Joyce, K.E., Dellow, G.D., Glassey, P.J., Assessing Image Processing Techniques for Mapping Landslides, in: *IGARSS 2008–2008 IEEE International Geoscience and Remote Sensing Symposium*, IEEE, II-1231–II-1234, 2008, doi:10.1109/igarss.2008.4779224.
19. Pradhan, B., Tehrany, M.S., Jebur, M.N., A New Semiautomated Detection Mapping of Flood Extent From TerraSAR-X Satellite Image Using Rule-Based Classification and Taguchi Optimization Techniques. *IEEE Trans. Geosci. Remote Sens.*, 54, 7, 4331–4342, July 2016, doi:10.1109/tgrs.2016.2539957.
20. Amani, M., *et al.*, Google Earth Engine Cloud Computing Platform for Remote Sensing Big Data Applications: A Comprehensive Review. *IEEE J. Sel. Top. Appl. Earth Observ. Remote Sens.*, 13, 5326–5350, 2020, doi:10.1109/jstars.2020.3021052.

21. Pham-Duc, B., Nguyen, H., Phan, H., Tran-Anh, Q., Trends and applications of google earth engine in remote sensing and earth science research: a bibliometric analysis using scopus database. *Earth Sci. Inform.*, 16, 3, 2355–2371, June 2023, doi:10.1007/s12145-023-01035-2.
22. Bly, J., Francescutti, L.H., Weiss, D., Disaster Management: A State-of-the-Art Review, in: *Natural Hazards - Impacts, Adjustments and Resilience*, E. Noroozinejad Farsangi (Ed.), IntechOpen, London, UK, June 30, 2021, doi:10.5772/intechopen.94489.
23. Chakraborty, S., Sasmal, S., Basak, T., Chakrabarti, S.K., Comparative study of charged particle precipitation from Van Allen radiation belts as observed by NOAA satellites during a land earthquake and an ocean earthquake. *Adv. Space Res.*, 64, 3, 719–732, Aug. 2019, doi:10.1016/j.asr.2019.05.020.
24. Jia, J. and Ye, W., Deep Learning for Earthquake Disaster Assessment: Objects, Data, Models, Stages, Challenges, and Opportunities. *Remote Sens.*, 15, 16, 4098, Basel, Switzerland, Aug. 2023, doi:10.3390/rs15164098.
25. Huayong, N., *et al.*, Comparison of Earthquake Environmental Effects and ESI intensities for recent seismic events in different tectonic settings: Sichuan (SW China) and Central Apennines (Italy). *Eng. Geol.*, 258, 105149, Aug. 2019, doi:10.1016/j.enggeo.2019.105149.
26. O'Mara, T., Meador, A.S., Colavito, M., Waltz, A., Barton, E., Navigating the evolving landscape of wildfire management: A systematic review of decision support tools. *Trees For. People*, 16, 100575, June 2024, doi:10.1016/j.tfp.2024.100575.
27. I, Y., D, M.G., A, A.D., Review of methods for modelling forest fire risk and hazard. *Afr. J. Environ. Sci. Technol.*, 9, 3, 155–165, Mar. 2015, doi:10.5897/ajest2014.1820.
28. Giordan, D., Hayakawa, Y., Nex, F., Remondino, F., Tarolli, P., Review article: the use of remotely piloted aircraft systems (RPASs) for natural hazards monitoring and management. *Nat. Hazards Earth Syst. Sci.*, 18, 4, 1079–1096, Apr. 2018, doi:10.5194/nhess-18-1079-2018.
29. Novellino, A., *et al.*, Mapping landslides from space: A review. *Landslides*, 21, 5, 1041–1052, Feb. 2024, doi:10.1007/s10346-024-02215-x.
30. Snehil and Goel, R., Flood Damage Analysis Using Machine Learning Techniques. *Procedia Comput. Sci.*, 173, 78–85, 2020, doi:10.1016/j.procs.2020.06.011.
31. Munawar, H.S., Hammad, A.W.A., Waller, S.T., A review on flood management technologies related to image processing and machine learning. *Autom. Constr.*, 132, 103916, Dec. 2021, doi:10.1016/j.autcon.2021.103916.
32. Arslan, M., Roxin, A.-M., Cruz, C., Ginhac, D., A Review on Applications of Big Data for Disaster Management, in: *2017 13th International Conference on Signal-Image Technology & Internet-Based Systems (SITIS)*, IEEE, pp. 370–375, Dec. 2017, doi:10.1109/sitis.2017.67.

33. Arachchige, S.M., Pradhan, B., Park, H.-J., A critical review of hurricane risk assessment models and predictive frameworks. *Geosci. Front.*, 16, 3, 102012, May 2025, doi:10.1016/j.gsf.2025.102012.
34. Vömel, H. and Dunion, J., Aircraft dropsonde campaigns, in: *Field Measurements for Passive Environmental Remote Sensing*, pp. 185–194, Elsevier, 2023, doi:10.1016/b978-0-12-823953-7.00021-6.
35. Romano, P., Di Lieto, B., Scarpetta, S., Benítez, C., Aoki, Y., Editorial: Applications of machine learning in volcanology. *Front. Earth Sci.*, 13, Apr. 2025, doi:10.3389/feart.2025.1604299.
36. Jahromi, M.N., Jahromi, M.N., Zolghadr-Asli, B., Pourghasemi, H.R., Alavipanah, S.K., Google Earth Engine and Its Application in Forest Sciences, in: *Environmental Science and Engineering*, pp. 629–649, Springer International Publishing, Cham, Switzerland, Oct. 09, 2020, doi:10.1007/978-3-030-56542-8_27.
37. Gorelick, N., Hancher, M., Dixon, M., Ilyushchenko, S., Thau, D., Moore, R., Google Earth Engine: Planetary-scale geospatial analysis for everyone. *Remote Sens. Environ.*, 202, 18–27, Dec. 2017, doi:10.1016/j.rse.2017.06.031.
38. Drusch, M., *et al.*, Sentinel-2: ESA's Optical High-Resolution Mission for GMES Operational Services. *Remote Sens. Environ.*, 120, 25–36, May 2012, doi:10.1016/j.rse.2011.11.026.
39. Kulinan, A.S., Cho, Y., Park, M., Park, S., Rapid wildfire damage estimation using integrated object-based classification with auto-generated training samples from Sentinel-2 imagery on Google Earth Engine. *Int. J. Appl. Earth Obs. Geoinf.*, 126, 103628, Feb. 2024, doi:10.1016/j.jag.2023.103628.
40. Gessesse, A.A. and Melesse, A.M., Temporal relationships between time series CHIRPS-rainfall estimation and eMODIS-NDVI satellite images in Amhara Region, Ethiopia, in: *Extreme Hydrology and Climate Variability*, pp. 81–92, Elsevier, Amsterdam, Netherlands, 2019, doi:10.1016/b978-0-12-815998-9.00008-7.
41. Hu, Y., Dong, Y., Batunacun, An automatic approach for land-change detection and land updates based on integrated NDVI timing analysis and the CVAPS method with GEE support. *ISPRS J. Photogramm. Remote Sens.*, 146, 347–359, Dec. 2018, doi:10.1016/j.isprsjprs.2018.10.008.
42. Guirado, E., Alcaraz-Segura, D., Cabello, J., Puertas-Ruíz, S., Herrera, F., Tabik, S., Tree Cover Estimation in Global Drylands from Space Using Deep Learning. *Remote Sens.*, 12, 3, 343, Jan. 2020, doi:10.3390/rs12030343.
43. Lan, W.-H., Lee, C.-M., Kuo, C.-Y., Lin, L.-C., Handoko, E.Y., Regional sea level budget around Taiwan and Philippines over 2002-2021 inferred from GRACE, altimetry, and *in-situ* hydrographic data. *J Geod.*, 99, 1, Dec. 2024, doi:10.1007/s00190-024-01928-0.
44. Gao, B., NDWI—A normalized difference water index for remote sensing of vegetation liquid water from space. *Remote Sens. Environ.*, 58, 3, 257–266, Dec. 1996, doi:10.1016/s0034-4257(96)00067-3.

45. Apostolopoulos, D.N. and Nikolakopoulos, K.G., SPOT vs Landsat satellite images for the evolution of the north Peloponnese coastline, Greece. *Reg. Stud. Mar. Sci.*, 56, 102691, Nov. 2022, doi:10.1016/j.rsma.2022.102691.
46. Boukhennaf, A. and Mezouar, K., Long and short-term evolution of the Algerian coastline using remote sensing and GIS technology. *Reg. Stud. Mar. Sci.*, 61, 102893, July 2023, doi:10.1016/j.rsma.2023.102893.
47. Patel, N.R. and Yadav, K., Monitoring spatio-temporal pattern of drought stress using integrated drought index over Bundelkhand region, India. *Nat Hazards*, 77, 2, 663–677, Jan. 2015, doi:10.1007/s11069-015-1614-0.
48. Kogan, F. and Guo, W., Early twenty-first-century droughts during the warmest climate. *Geomatics Nat. Hazards Risk*, 7, 1, 127–137, Jan. 2014, doi:10.1080/19475705.2013.878399.
49. Hermans, K. and Garbe, L., Droughts, livelihoods, and human migration in northern Ethiopia. *Reg. Environ Change*, 19, 4, 1101–1111, Feb. 2019, doi:10.1007/s10113-019-01473-z.
50. Nooni, I.K., *et al.*, Spatiotemporal Characteristics and Trend Analysis of Two Evapotranspiration-Based Drought Products and Their Mechanisms in Sub-Saharan Africa. *Remote Sens.*, 13, 3, 533, Feb. 2021, doi:10.3390/rs13030533.
51. Shiferaw, B., Tesfaye, K., Kassie, M., Abate, T., Prasanna, B.M., Menkir, A., Managing vulnerability to drought and enhancing livelihood resilience in sub-Saharan Africa: Technological, institutional and policy options. *Weather Clim. Extremes*, 3, 67–79, June 2014, doi:10.1016/j.wace.2014.04.004.
52. Bento, V.A., Gouveia, C.M., DaCamara, C.C., Libonati, R., Trigo, I.F., The roles of NDVI and Land Surface Temperature when using the Vegetation Health Index over dry regions. *Global Planet. Change*, 190, 103198, July 2020, doi:10.1016/j.gloplacha.2020.103198.
53. Xie, F. and Fan, H., Deriving drought indices from MODIS vegetation indices (NDVI/EVI) and Land Surface Temperature (LST): Is data reconstruction necessary? *Int. J. Appl. Earth Obs. Geoinf.*, 101, 102352, Sept. 2021, doi:10.1016/j.jag.2021.102352.
54. Wassie, S.B., Mengistu, D.A., Birlie, A.B., Agricultural drought assessment and monitoring using MODIS-based multiple indices: the case of North Wollo, Ethiopia. *Environ. Monit. Assess.*, 194, 11, Sept. 2022, doi:10.1007/s10661-022-10455-4.
55. Fentaw, A.E., Yimer, A.A., Zeleke, G.A., Monitoring spatio-temporal drought dynamics using multiple indices in the dry land of the upper Tekeze Basin, Ethiopia. *Environ. Challenges*, 13, 100781, Dec. 2023, doi:10.1016/j.envc.2023.100781.
56. Hoque, M.A.-A., Phinn, S., Roelfsema, C., Childs, I., Assessing tropical cyclone risks using geospatial techniques. *Appl. Geogr.*, 98, 22–33, Sept. 2018, doi:10.1016/j.apgeog.2018.07.004.

57. Antoine, R., *et al.*, Geoscientists in the Sky: Unmanned Aerial Vehicles Responding to Geohazards. *Surv. Geophys.*, 41, 6, 1285–1321, Sept. 2020, doi:10.1007/s10712-020-09611-7.
58. Van Alphen, J., Martini, F., Loat, R., Slomp, R., Passchier, R., Flood risk mapping in Europe, experiences and best practices. *J. Flood Risk Manage.*, 2, 4, 285–292, Nov. 2009, doi:10.1111/j.1753-318x.2009.01045.x.
59. Phiri, D., Simwanda, M., Nyirenda, V., Mapping the impacts of cyclone Idai in Mozambique using Sentinel-2 and OBIA approach. *S. Afr. Geogr. J.*, 103, 2, 237–258, Apr. 2020, doi:10.1080/03736245.2020.1740104.
60. Nhangumbe, M., Nascetti, A., Georganos, S., Ban, Y., Supervised and unsupervised machine learning approaches using Sentinel data for flood mapping and damage assessment in Mozambique. *Remote Sens. Appl. Soc. Environ.*, 32, 101015, Nov. 2023, doi:10.1016/j.rsase.2023.101015.
61. Geller, R.J., Earthquake prediction: a critical review. *Geophys. J. Int.*, 131, 3, 425–450, Dec. 1997, doi:10.1111/j.1365-246x.1997.tb06588.x.
62. Akhoondzadeh, M., Advances in Seismo-LAI anomalies detection within Google Earth Engine (GEE) cloud platform. *Adv. Space Res.*, 69, 12, 4351–4357, June 2022, doi:10.1016/j.asr.2022.03.033.
63. Kutoglu, S.H., Ghasempour, F., Sekertekin, A., Investigation of Possible MODIS AOD Anomalies as Earthquake Precursors for Global Earthquakes. *Adv. Space Res.*, 68, 9, 3531–3545, Nov. 2021, doi:10.1016/j.asr.2021.06.036.
64. Abad, L., Hölbling, D., Spiekermann, R., Prasicek, G., Dabiri, Z., Argentin, A.-L., Detecting landslide-dammed lakes on Sentinel-2 imagery and monitoring their spatio-temporal evolution following the Kaikōura earthquake in New Zealand. *Sci. Total Environ.*, 820, 153335, May 2022, doi:10.1016/j.scitotenv.2022.153335.
65. Lemnitzer, A., *et al.*, The September 19, 2017 MW 7.1 CENTRAL-Mexico earthquake: Immediate observations on selected infrastructure systems. *Soil Dyn. Earthquake Eng.*, 141, 106430, Feb. 2021, doi:10.1016/j.soildyn.2020.106430.
66. Montgomery, J., Candia, G., Lemnitzer, A., Martinez, A., The September 19, 2017 Mw 7.1 Puebla-Mexico City earthquake: Observed rockfall and landslide activity. *Soil Dyn. Earthquake Eng.*, 130, 105972, Mar. 2020, doi:10.1016/j.soildyn.2019.105972.
67. Ghosh, S., Kumar, D., Kumari, R., Cloud-based large-scale data retrieval, mapping, and analysis for land monitoring applications with Google Earth Engine (GEE). *Environ. Challenges*, 9, 100605, Dec. 2022, doi:10.1016/j.envc.2022.100605.
68. Salocchi, A.C., Minarelli, L., Lugli, S., Amoroso, S., Rollins, K.M., Fontana, D., Liquefaction source layer for sand blows induced by the 2016 megathrust

earthquake (Mw 7.8) in Ecuador (Boca de Briceño). *J. South Am. Earth Sci.*, 103, 102737, Nov. 2020, doi:10.1016/j.jsames.2020.102737.

69. Mason, H.B., *et al.*, East Palu Valley flowslides induced by the 2018 M 7.5 Palu-Donggala earthquake. *Geomorphology*, 373, 107482, Jan. 2021, doi:10.1016/j.geomorph.2020.107482.
70. Okamura, M., Ono, K., Arsyad, A., Minaka, U.S., Nurdin, S., Large-scale flowslide in Sibalaya caused by the 2018 Sulawesi earthquake. *Soils Found.*, 60, 4, 1050–1063, Aug. 2020, doi:10.1016/j.sandf.2020.03.016.

15

Innovative Business Models for Geospatial-Enabled Agricultural Solutions

Amit Sharma[1]*, Shubhi Sharma[2], Ruchi Sharma[3] and Harshita Doad[4]

[1]*Department of Computer Science and Engineering, Vivekananda Global University, Jaipur, Rajasthan, India*
[2]*Amity Business School, Amity University, Jaipur, Rajasthan, India*
[3]*School of Commerce and Management, Nirwan University, Jaipur, Rajasthan, India*
[4]*School of Engineering & Technology (Computer Science), Career Point University, Kota, Rajasthan, India*

Abstract

The incorporation of geospatial technologies, such as Geographic Information Systems (GIS), remote sensing, unmanned aerial vehicles (UAVs), and the Internet of Things (IoT), has transformed contemporary agriculture through data-driven decision-making, precision farming, and real-time crop monitoring. This requires rethinking how agricultural value is created, delivered, and monetized through tailored business models. Effective business models that are adaptable, accessible, and tailored to local needs can leverage geospatial technologies. In this chapter, the emerging new typologies of geospatial-enabled agribusiness models, namely, Direct-to-Farmer (D2F) platforms, Business-to-Business (B2B) services for agri-enterprises, Business-to-Government (B2G) analytics for government decision-making, and Platform-as-a-Service (PaaS) ecosystems, are outlined. Through case studies in India and Africa, we also demonstrate how design decisions, such as pricing strategies, modular design, and localization, affect adoption and scale in practice. Other challenges discussed in this chapter include the digital divide, policies that do not match the context, and a trust deficit with farmers. Focusing on principles such as offline-first architecture, adaptive pricing, and climate-smart advisory integration, this chapter suggests the tenets of sustainable

**Corresponding author*: dr.amittech@gmail.com; ORCID: 0000-0001-9825-3978
Shubhi Sharma: ORCID: 0009-0005-6856-3189
Ruchi Sharma: ORCID: 0009-0005-7193-7188
Harshita Doad: ORCID: 0009-0003-7959-5819

Gurwinder Singh, Vishakha Sood, Narayan Vyas, Surendra Yadav and Dankan Gowda V. (eds.)
Geospatial Intelligence for Precision Agriculture, (295–312) © 2026 Scrivener Publishing LLC

and scalable agri-tech solutions. The study ends with a call for participatory design and equitable value chains to democratize access to geospatial data. This in-depth analysis is intended for policymakers, entrepreneurs, and development agencies as they design innovative business models that link technological innovation to local impacts.

Keywords: Geospatial agriculture, precision farming, business models, remote sensing, technological innovations

15.1 Introduction

Agricultural production has undergone a significant transformation driven by technological advancements, particularly in geospatial tools [1]. Geographic Information Systems (GIS), remote sensing, unmanned aerial vehicles (UAVs), and the Internet of Things (IoT) are essential tools in the agronomic workflow, facilitating the adaptation of traditional agricultural systems to data-driven, precise, and flexible systems. For instance, GIS helps create comprehensive maps related to farms, soil properties, and crop spread. Satellite or drone remote sensing can help regularly monitor vegetation health, water stress, and pest infestation patterns [2]. In such cases, their ability to provide high-resolution imagery, particularly at sites with limited or seasonal satellite coverage, enhances real-time decision-making. In-field sensors and other IoT devices provide detailed metrics regarding soil moisture, nutrient levels, and microclimatic conditions, enabling timely interventions at the site [3]. In a sense, they empower farmers to evolve from instinct-driven operations to precision agriculture, minimizing waste, maximizing the use of inputs, and improving yield in both volume and quantity per unit area [4]. These insights also provide policymakers, agribusinesses, and researchers with the information needed to make evidence-based decisions on land use planning, resource allocation, and strategies for climate adaptation [5].

However, despite strong evidence of their effectiveness, geospatial tools have yet to be widely adopted and remain inaccessible to many small and marginal farmers in developing countries. The hurdles are not only technical but also deeply rooted in obsolete or antiquated business models. Traditional agri-service structures are often input-focused, such as seeds or fertilizers, and are applied across various agricultural geographies with unique requirements [6]. However, climate uncertainty, land fragmentation, market variability, and educational diversity call for more local, tailored, and scalable solutions. This requires rethinking how agricultural value is created, delivered, and monetized through innovative geospatial-enabled business models tailored to local needs [7]. They must be sustainable, economically viable over time, and inclusive, prioritizing access for smallholders, women farmers, and underserved regions. This chapter aims to discuss these revenue models and highlight their utilization in geospatial technology systems. This chapter provides a practical, future-focused roadmap for entrepreneurs, policymakers, and researchers seeking to create more equitable and technology-enabled agricultural futures, drawing on lessons learned from real-world case studies, revenue models, growth strategies, and institutional partnerships [8].

15.2 Understanding the Geospatial-Agriculture Nexus

Geospatial agriculture integrates geospatial technologies, including GIS, remote sensing, UAV, and IoT, for the detailed monitoring, assessment, and management of agricultural resources. These technologies provide critical data on environmental conditions, crop health, and farm-level factors. GIS aids in mapping and analyzing spatial information across various industries and sectors. Using satellite or drone images, remote sensing provides current information on vegetation conditions and land cover, tracking phenological stages [9]. UAVs provide low-altitude, high-resolution monitoring, which is often required in cloudy and difficult-to-access areas. These technologies are aided by IoT, which provides much of the real-time data about sensors in the ground that measure soil moisture, temperature, or nutrient content. This combination of technologies enables smarter and more localized decisions in agriculture. Through on-farm interventions, geospatial agriculture further complements decision-making at the regional and national levels for food security, water management, and climate resilience planning [10].

Geospatial agriculture has extensive practical applications and profound implications. By examining the Normalized Difference Vegetation Index (NDVI) or other vegetation indices calculated from satellite or drone imagery, early detection of crop stress or disease is possible, revealing crop health much earlier than the naked eye can. IoT-based soil moisture sensors and geospatial mapping enable precise water application, ensuring that it is applied at the right time and in the right proportions, thereby improving efficiency, conserving resources, and yielding better results [11]. Similarly, yield forecasting models leverage historical yield data, land cover classification, climate projections, and real-time vegetation indices to predict future outputs, empowering farmers to organize harvest logistics, complementary storage, and marketing. Such tools enhance productivity at the farm level and aid governments and agribusinesses in supply chain forecasting, insurance modeling, and providing early warning systems for drought or crop failure [12]. The granularity and accuracy achieved with geospatial applications can provide new insights into agricultural planning and operations [13].

GIS-based analysis can provide new insights. Nevertheless, geospatial tools are often underused in local agribusiness ecosystems, especially in developing countries [14]. GIS is a crucial technology and should not be viewed in isolation but rather as part of a broader value chain that encompasses advisory services, access to finance, market linkages, and institutional support. If, in the example, a precision irrigation solution is scaled but not accompanied by local water policies, pricing models, or farmer training, its impact will be marginal [15]. More importantly, these tools also need to be delivered at low costs, in a language that smallholders understand, and in an offline mode by agribusinesses that adapt them to the smallholder context. Models with a strong collaborative element between Farmer Producer Organizations (FPOs) and Non-Governmental Organizations (NGO) [16]. Geospatial tools can transform local agribusiness environments from reactive and intuition-based to proactive and evidence-based when embedded in decision-making at each stage, from input selection to planting, irrigation, pest management, and post-harvest logistics [17].

15.3 The Shift in Business Thinking: From Inputs to Intelligence

Most agribusinesses operate under a traditional supply chain model, focusing on selling physical inputs (such as seeds, fertilizers, and pesticides) with minimal interaction with farmers beyond the sale of these inputs [18]. However, the new paradigm revolves around information and intelligence, usually delivered through one or more digital platforms. This change is driven by the understanding that inputs alone cannot determine productivity; instead, data-driven decisions can optimize the outcomes. Based on field-level data, intelligence-as-a-service models provide farmers with insights on what, when, and how to plant, as well as which inputs to apply [19]. Weather alerts, satellite-based advisories, personalized agronomic recommendations, and early pest detection alerts can be provided as part of these services. Unlike product-oriented businesses, these models focus on maintaining long-term relationships with farmers, creating feedback loops, and progressively enhancing services.

The increasing availability of geospatial and farm-level data has led to the emergence of a new agribusiness model that emphasizes data monetization over traditional input-based monetization. Using satellite imagery, drone data, and Internet-of-Things sensors, companies are creating precision farming platforms that can deliver hyper-localized recommendations [20]. These platforms typically serve not only farmers but also agribusinesses, insurance companies, and government entities. For example, a geospatial platform could sell soil health analytics to fertilizer companies or yield-prediction APIs to banks [21]. This also creates multi-sided revenue models in which stakeholders pay for distinct insights from the same dataset. These types of business models drive alliances, interoperability, and innovation. However, they also create the need to question data ownership, privacy, and ethical issues surrounding smallholder-generated data. Solving these problems while building scalable models is essential for the long-term viability of precision-agriculture businesses [22].

15.4 Business Model Typologies in Geospatial AgTech

The Direct to Farmer (D2F) model is designed to put geospatial intelligence directly into the hands of farmers, typically through mobile phone-based platforms, SMS alerts, or voice advisories. They offer location- and crop-specific real-time recommendations on sowing dates, irrigation, pest and disease risks, and weather alerts [23]. The D2F model relies on accessibility, ease of use, and local content, which are offered under freemium or tiered subscription models. In India, notable examples include DeHaat and AgroStar, which provide a comprehensive package comprising geospatial advisories, doorstep delivery of agricultural inputs, and a buy-back option for produce. By addressing information asymmetry and offering timely Interventions, the D2F model addresses the key challenges faced by smallholders [24]. However, smartphone penetration remains limited. Digital literacy and willingness to adopt geospatial technologies vary among farmers, thereby impacting the effectiveness of D2F platforms. Language localization, offline work, and building trust with village-level agents are often important to a good D2F strategy.

In the Business-to-Business (B2B) paradigm, geospatial intelligence is sold to agribusinesses that require large-scale analytics to enhance operational efficiency, target their market, or improve product offerings. A fertilizer business may utilize soil maps acquired from satellites to customize nutrient products for geographical areas. A food processor can track crop maturity trends over thousands of hectares to procure them at the ideal time for the best yield. Leveraging dashboards, APIs, and analytics reports, these services are commonly offered as part of a subscription model or on a project basis. Simultaneously, the B2B model is more predictable in terms of revenue and customer acquisition costs, with higher-value contracts than the D2F model. For instance, SatSure provides satellite analytics to banks for making agricultural and livestock lending decisions, as well as to insurers for assessing crop losses [25]. Another impressive aspect of the B2B model is its scalability; one platform can cater to multiple enterprises across various geographical locations. However, there are typical risks of data confidentiality issues, integration complexity, and reliance on commercial cycles [26].

Above all, the B2G model focuses on geospatial data and data analytics provided to government agencies for planning, policy execution, and monitoring. They require accurate and timely data on cultivation areas, yield predictions, droughts, and pest outbreaks to effectively manage crop insurance, disaster relief, and subsidy schemes [27]. Government departments buy geospatial dashboards, policy briefs, and geospatial portals from start-ups or research organizations [28]. These are valuable time-bound (TB) projects, which are time-consuming but long-term. However, procurement requires considerable time, and compliance involves substantial work. They also call for accuracy, accountability, and alignment with the national data frameworks. Start-ups willing to enter this space should prepare for institutional negotiations, HIPAA validation processes, and bureaucratic engagement [29].

Platform-as-a-service (PaaS) models provide one or more geospatial services in an integrated manner to different stakeholders using open APIs or plug-and-play modules [30]. They act as marketplaces, that integrate data sources, analytical tools, and user interfaces. Farmers, cooperatives, banks, and agribusinesses are all users who can subscribe to services tailored to their specific requirements [31]. An example is IBM's Weather Company, which is a blend of geospatial and climate data that can be consumed for apps, advisory services, and more. Other examples include Microsoft Farm Beats, which provides plug-in analytics on the Azure infrastructure. This enables third-party innovation, interoperability, and rapid scaling. Licenses, application programming interface (API) usage, and data subscriptions generate revenue. However, maintaining a flexible, safe, and stable platform always incurs a substantial initial cost, comprehensive documentation, and an active partner ecosystem. The primary benefit of PaaS is that it provides flexibility when scaling geospatial solutions across disparate regions [32].

15.5 Innovation through Collaboration and Open Models

Co-design, or end-user involvement in developing and testing solutions, is one of the most overlooked success factors for the deployment of agri-tech geospatial tools. Created in a vacuum, often by engineers or data scientists, these tools do not consider the practical challenges faced by farmers [33]. Co-design ensures that solutions are correct, acceptable, logical, and usable. This is where cooperatives and NGOs are essential as intermediaries, as they possess knowledge of both community context and technology. Consider a scenario in which you collaborate with a FPO to design a pest alert system. This may result in voice alerts in local dialects, alerts at ideal times based on farming schedules, or peer-to-peer troubleshooting through group chats [34]. Involving end-users in the co-design process also improves ownership and adoption. If farmers observe their feedback integrated into app updates or platform designs, they will be more inclined to trust and consistently use the system [35].

As geospatial solutions are not widely adopted in India, public-private partnerships (PPP) play a significant role in expanding and enhancing the reach of affordable geospatial solutions. Governments have massive datasets and institutional access, while private players have innovation and agility. Consequently, they can develop tools that support decision-making in areas such as climate resilience, insurance modeling, and land-use planning. Incubators and accelerators are also crucial, providing early stage geospatial start-ups with access to investment, mentors, datasets, and pilots. The open-source movement in geospatial technology, including tools such as QGIS, Sentinel Hub, and OpenStreetMap, has significantly lowered the barriers to entry, enabling local customization and innovation. These tools bring spatial analytics to the masses, enabling local customization and tailoring solutions to specific needs [36]. Start-ups can create custom applications and are not bound by exorbitantly priced proprietary software. PPPs and open platforms support a sustainable pipeline of innovation for ag-geospatial applications.

Collaborative models have been successful in Asia and Africa. Digital Green partnered with NGOs and FPOs in India to provide video-based agricultural recommendations, and later complemented this effort with geospatial inputs to focus on specific districts by utilizing crop and weather data [37]. The Plant Village Nuru app, developed in collaboration with farmers and researchers in Kenya, assists in diagnosing crop diseases using Artificial Intelligence (AI) and satellite data, and sends alerts in local languages within Kenya [38]. A second example is the FAO's eLocust3 system, which combines field data with the Global Positioning System (GPS) and satellite monitoring to enable rapid responses to locust outbreaks, representing a joint effort by the FAO and various African agricultural ministries. By connecting the dots among different stakeholders, such as governments, technology innovators, local organizations, and farmers, these collaborative initiatives succeed in making data contextually relevant, accurate, and usable in the real world. Scalability is based on shared ownership, shared trust, and user-centered iterative improvement through use.

15.6 Real-World Case Studies

CropIn is one of India's leading agri-tech firms, offering a SaaS platform that leverages geospatial data, AI, and remote sensing for comprehensive farm monitoring and advisory services [39]. The company provides real-time crop insights through its SmartFarm and Smart Risk platforms by collaborating with agribusinesses, FPOs, and input providers. CropIn utilizes satellite data to track activities at the field level, including weather, crop health, and growth stages, which are combined with field inputs to provide actionable recommendations [40]. It is typically priced per acre or on a subscription basis, making it accessible to large enterprises and cooperatives. Its multilingual interface, mobile accessibility, and robust backend analytical tools have contributed to its adoption in over 50 countries [41]. One significant lesson from CropIn's journey is the importance of combining geospatial intelligence with a human advisory layer, including field agents, to build trust and adoption by farmers. The platform is highly scalable; however, its quality depends on the quality of the field data it can access. The high-touch nature of the solution means that maintaining ongoing client engagement in regions with lower digital penetration is key [42].

SatSure is a Bengaluru-based agri-analytics company that offers geospatial intelligence to financial institutions, insurance companies, and government bodies. At its core, satellite imagery and AI underpin their value proposition, enabling the provision of macro-level insights into agricultural lending, crop insurance, and disaster monitoring [43]. SatSure's business model is primarily B2B and B2G, driven by long-term contracts, licensing fees, and analytical dashboards [44]. Governments sometimes use SatSure's tools to track PMFBY claims or manage droughts. A key factor contributing to SatSure's success is its capability to deliver scalable, data-driven geospatial intelligence solutions [45]. To convert satellite data into easily consumable insights for non-technical stakeholders, such as policymakers and bankers. Its modular architecture and cloud infrastructure enable the platform to scale easily by rapidly deploying it to other countries. That said, bureaucracies plague governments and must align with data policies; this amplifies the call for institutional trust and a robust legal framework surrounding their use [46].

Based in South Africa, Aerobotics uses drones and AI-powered analytics for high-value tree crops, including citrus, almonds, and apples. They capture high-resolution drone imagery and fuse it with satellite data to provide orchard-level information on canopy health, pest risks, and fruit size [47]. They offer yield forecasting, early disease detection, and irrigation optimization as their main products, delivered *via* a simple dashboard and mobile application. The model requires hardware (drone imagery services) and a tiered subscription [48] for the analytics. Because the bottom-line impact of precision insights is readily apparent, Aerobotics focuses strongly on the commercial farming segment [49]. One of the most significant realizations is that combining AI with high-precision drone data for niche, high-margin markets is highly effective. This focus enables deeper integration and greater price flexibility in the market. However, adoption is limited by regulatory drone clearances and the technical expertise required for drone operation, which limits scalability in smallholder-heavy landscapes unless bundled with service providers or cooperatives [50].

15.7 Challenges in Scaling Geospatial Business Models

In agriculture, one of the main barriers to scaling geospatial business models is the long-standing digital divide, which is notably prevalent in rural locations [51]. Smallholder farmers often lack access to smartphones, stable Internet connections, or electricity, which makes it challenging for them to utilize app-based solutions consistently. However, severe connectivity constraints hindered the timely transmission of real-time data and the delivery of advisories, particularly in remote locations where device accessibility was limited. Equally important is the fact that digital literacy is not uniform; many farmers have limited experience with apps, cloud-based dashboards, and online subscriptions [52]. This absence is one of the gaps preventing the adoption and authentic engagement of geospatial tools in the field. Solutions for these issues may require companies to build "offline-first" systems that can store data temporarily until synchronization can occur, provide a voice-based experience in regional languages, and conduct training through field agents or village-level centers. Likewise, public sector support through subsidies or infrastructure investment is key to closing this gap [53].

Trust represents a significant barrier; even when the necessary technological infrastructure is available, its adoption is often hindered by a lack of user trust. Farmers are cynical about digital advisory tools; some do not trust their quality, and others believe that they are too complex [54]. Confidence has also been undermined by previous experiences in which input companies have not delivered inconsistent results or resorted to aggressive sales tactics. There are also concerns about how personal or farm-level data could be used if collected without the consent of the individuals or farms [55]. To mitigate this, data collection, storage, and usage policies must be transparent and effectively communicated. Companies can gain the trust of concerned residents by engaging with the community, forming partnerships with well-respected local organizations, and maintaining an opt-in process that allows residents to know that their information is being shared and with whom it is being shared. Open data policies, dashboards controlled by farmers, and anonymized analytics can enhance accountability and fairness [56].

15.8 Designing a Scalable, Inclusive Future

Finally, the scalability of geospatial-enabled agri-business models depends on a modular architecture that enables individual platform components (data ingestion, analytics, delivery, and visualization) to work independently or together as required. This design flexibility allows the integration of new technologies (such as satellite APIs or local weather sensors) more easily, expansion into new geographies, and customization of services for different stakeholders [57]. A basic NDVI mapping module could be offered by one company, with premium predictive analytics features available as an add-on from another company. It reduces development time, allows for gradual scaling, and makes everything more maintainable [58]. More generally, modular systems facilitate partnerships, allowing third parties to plug their datasets or analytical engines into the system. This process is similar to the microservices approach of scalable software platforms, in which speed and flexibility are paramount. Modularity is essential for agri-tech, since regions experience different farming conditions, grow different crops, and have distinctly different data needs. Figure 15.1 illustrates a modular, scalable geospatial agri-tech workflow that integrates diverse data sources, analytics, and multi-channel delivery for actionable precision farming insights.

Despite being commonplace in many aspects of daily life, connectivity remains a persistent challenge in rural areas, particularly in developing countries. Therefore, these agri-tech platforms must adopt an offline-first approach to ensure that the platform remains usable even when online availability is unavailable. This means that the platform, particularly mobile apps, should function in low- or no-connectivity areas, caching critical data locally and uploading them to the cloud once connectivity is restored. These include offline-first features such as downloading maps, SMS/IVR-based advisories, or pre-loading crop-stage alerts [59]. This enhances reliability, builds trust with the user, and delivers critical insights

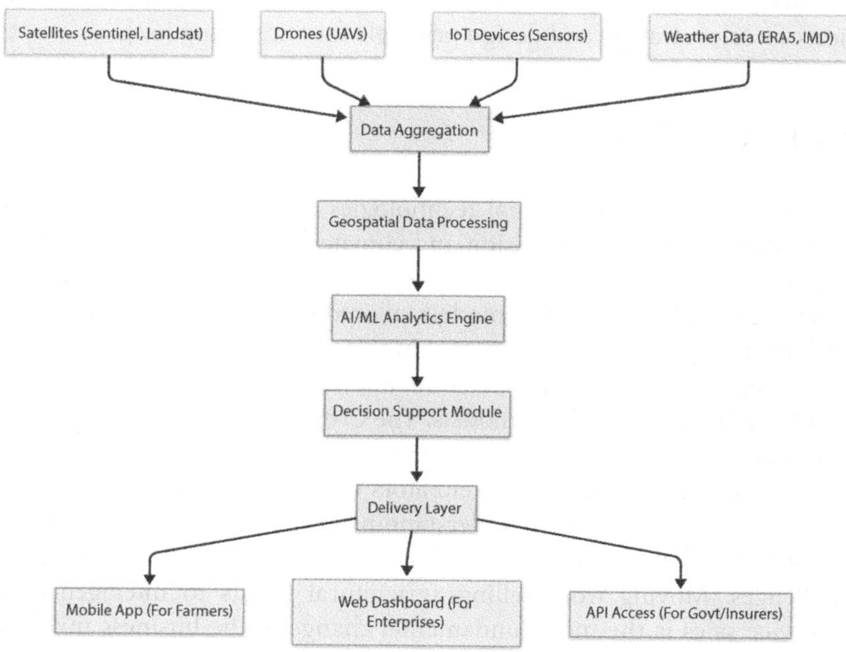

Figure 15.1 End-to-end workflow for geospatial-enabled precision agriculture.

to the farmer promptly during network outages. This aligns with inclusive innovation, providing farmers with basic phones or low-data-usage plans, as well as precision farming guidance. An offline-first approach requires thoughtful architecture, data must be well-compressed, and local storage should be optimized, but it pays dividends in terms of increased reach and retention [60].

Smallholder farmers often operate on such thin margins that affordability is a significant constraint. Therefore, scalable agri-business models should have more flexible pricing options, such as freemium or tiered pricing or a pay-as-you-use model. A freemium model offers free core services, such as climate alerts or general advisories, while monetizing more sophisticated analytics, specific soil wellness reports, or personalized insights. This can charge individual farmers, FPOs, and enterprise clients based on the scale of their benefits, offering tiered pricing [61]. This provides access to vital information for many people while allowing the business to operate with premium users. Moreover, the costs of geospatial services can be offset by bundling them with the supply of inputs, micro-insurance, or credit products, thus increasing the appeal of these platforms. Familiarity with the local context, including crop cycles, farmer income patterns, and existing subsidies, also enables the development of effective pricing strategies to increase adoption and retention. Ultimately, price flexibility is a combination of a sales strategy and design principle for fair scaling [62].

15.9 Conclusion

The introduction of geospatial technologies aims to transform modern agriculture by enabling precision and climate-resilient farming and data-driven decision-making. However, this change can only be truly effective if it is matched by new business models tailored to the needs of smallholders and diverse agricultural ecosystems. In this chapter, we delve deeper into the various pathways to monetization and the impact that can be created by D2F, B2B, B2G, and PaaS models. The CropIn, SatSure, and Aerobotics case studies found that modularity, trust-building mechanisms, and low-cost service delivery are key accelerators of scale. They excel not only in technical prowess but also in understanding local pain points, from pricing sensitivity to lower levels of digital outreach and regional infrastructure challenges. Moving from selling agricultural inputs to intelligence-as-a-service sales is the most fundamental change in the business mindset, driving the core of their value proposition toward making insights and risk mitigation. The future of geospatial-enabled agriculture must be inclusive, transparent, and collaborative. This journey of innovation cannot overlook smallholders, who comprise a large proportion of farmers worldwide.

Business models must incorporate offline-first approaches, local languages, flexible pricing, and co-design frameworks to resonate effectively with end users. Innovative geospatial governance, such as PPP, incubators, and open-source geospatial platforms, can expedite access and innovation by reducing barriers to access and providing collaborative solutions. Governments, start-ups, and development agencies must collaborate to design ecosystems that enable farmers to utilize geospatial insights and derive direct benefits, rather than merely being passive information consumers. The fundamental objective is to democratize spatial intelligence and ensure that technologies, often confined to analytical laboratories in developed nations, are translated into practical applications that enhance livelihoods, promote sustainability, and strengthen food security. This is both a business challenge and a social imperative for the future of the food industry.

AI disclosure statement: The author acknowledges the use of AI-based tools for minor language editing or grammatical correction. However, all intellectual contributions, analyses, and conclusions presented in this chapter are entirely the author's own.

References

1. Roberts, D.P., Short, N.M., Sill, J., Lakshman, D.K., Hu, X., Buser, M., Precision agriculture and geospatial techniques for sustainable disease control. *Indian Phytopathol.*, 74, 2, 287–305, Jun. 2021, doi:10.1007/s42360-021-00334-2.
2. Tarate, S.B., Patel, N.R., Danodia, A., Pokhariyal, S., Parida, B.R., Geospatial Technology for Sustainable Agricultural Water Management in India—A Systematic Review. *Geomatics*, 4, 2, Jun. 2024, doi:10.3390/geomatics4020006.
3. Ghosh, P., Kumpatla, S.P., Ghosh, P., Kumpatla, S.P., GIS Applications in Agriculture, in: *Geographic Information Systems and Applications in Coastal Studies*, IntechOpen, 2022, doi:10.5772/intechopen.104786.
4. Delgado, J.A., Short, N.M., Roberts, D.P., Vandenberg, B., Big Data Analysis for Sustainable Agriculture on a Geospatial Cloud Framework. *Front. Sustain. Food Syst.*, 3, Jul. 2019, doi:10.3389/fsufs.2019.00054.
5. Mathenge, M., Sonneveld, B.G.J.S., Broerse, J.E.W., Application of GIS in Agriculture in Promoting Evidence-Informed Decision Making for Improving Agriculture *Sustainability*: A Systematic Review. *Sustainability*, 14, 16, Jan. 2022, doi:10.3390/su14169974.

6. Bahn, R.A., Yehya, A.A.K., Zurayk, R., Digitalization for Sustainable Agri-Food Systems: Potential, Status, and Risks for the MENA Region. *Sustainability*, 13, 6, Jan. 2021, doi:10.3390/su13063223.
7. Eastwood, C., Ayre, M., Nettle, R., Dela Rue, B., Making sense in the cloud: Farm advisory services in a smart farming future. *NJAS Wagening. J. Life Sci.*, 90–91, 1, 1–10, Dec. 2019, doi:10.1016/j.njas.2019.04.004.
8. Cobben, D., Ooms, W., Roijakkers, N., Radziwon, A., Ecosystem types: A systematic review on boundaries and goals. *J. Bus. Res.*, 142, 138–164, Mar. 2022, doi:10.1016/j.jbusres.2021.12.046.
9. Kross, A., Kaur, G., Jaeger, J.A.G., A geospatial framework for the assessment and monitoring of environmental impacts of agriculture. *Environ. Impact Assess. Rev.*, 97, 106851, Nov. 2022, doi:10.1016/j.eiar.2022.106851.
10. Rodrigues, M.S., Castrignanò, A., Belmonte, A., da Silva, K.A., da T., Lessa, B.F., Geostatistics and its potential in Agriculture 4.0. *Rev. Ciênc. Agron.*, 51, e20207691, Aug. 2021, doi:https://doi.org/10.5935/1806-6690.20200095.
11. Alahmad, T., Neményi, M., Nyéki, A., Applying IoT Sensors and Big Data to Improve Precision Crop Production: A Review. *Agronomy*, 13, 10, Oct. 2023, doi:10.3390/agronomy13102603.
12. Vyas, N., Sethi, G.K., Singh, S., Comparative Analysis of Different Classifiers on Fusion of Sentinel Datasets for Agriculture Land, in: *2024 4th International Conference on Technological Advancements in Computational Sciences (ICTACS)*, pp. 1173–1177, Nov. 2024, doi:10.1109/ICTACS62700.2024.10840995.
13. Jabbari, A., Humayed, A., Reegu, F.A., Uddin, M., Gulzar, Y., Majid, M., Smart Farming Revolution: Farmer's Perception and Adoption of Smart IoT Technologies for Crop Health Monitoring and Yield Prediction in Jizan, Saudi Arabia. *Sustainability*, 15, 19, Jan. 2023, doi:10.3390/su151914541.
14. Manyise, T. and Dentoni, D., Value chain partnerships and farmer entrepreneurship as balancing ecosystem services: Implications for agri-food systems resilience. *Ecosyst. Serv.*, 49, 101279, Jun. 2021, doi:10.1016/j.ecoser.2021.101279.
15. Durán, A.P., *et al.*, Implementing ecosystem service assessments within agribusiness: Challenges and proposed solutions. *J. Appl. Ecol.*, 59, 10, 2468–2475, 2022, doi:10.1111/1365–2664.14250.
16. Quintas-Soriano, C., García-Llorente, M., Norström, A., Meacham, M., Peterson, G., Castro, A.J., Integrating supply and demand in ecosystem service bundles characterization across Mediterranean transformed landscapes. *Landscape Ecol.*, 34, 7, 1619–1633, Jul. 2019, doi:10.1007/s10980–019-00826–7.
17. Deguine, J.-P., Aubertot, J.-N., Flor, R.J., Lescourret, F., Wyckhuys, K.A.G., Ratnadass, A., Integrated pest management: good intentions, hard realities. A review. *Agron. Sustainable Dev.*, 41, 3, 38, May 2021, doi:10.1007/s13593–021-00689-w.

18. Reardon, T., *et al.*, Emerging outsource agricultural services enable farmer adaptation in agrifood value chains: A product cycle perspective. *Food Policy*, 127, 102711, Aug. 2024, doi:10.1016/j.foodpol.2024.102711.
19. Rashid, A.B. and Kausik, M.A.K., AI revolutionizing industries worldwide: A comprehensive overview of its diverse applications. *Hybrid Adv.*, 7, 100277, Dec. 2024, doi:10.1016/j.hybadv.2024.100277.
20. Miller, T., Mikiciuk, G., Durlik, I., Mikiciuk, M., Łobodzińska, A., Śnieg, M., The IoT and AI in Agriculture: The Time Is Now—A Systematic Review of Smart Sensing Technologies. *Sensors*, 25, 12, Jan. 2025, doi:10.3390/s25123583.
21. Gopal, S. and Pitts, J., Geospatial Finance: Foundations and Applications, in: *The FinTech Revolution: Bridging Geospatial Data Science, AI, and Sustainability*, S., Gopal and J., Pitts (Eds.), pp. 225–273, Springer Nature Switzerland, Cham, 2024, doi:10.1007/978-3-031-74418-1_6.
22. Zamlynskyi, V., *et al.*, Perspective Chapter: Exploring the Possibilities and Technologies of the Digital Agricultural Platform, in: *Agricultural Economics and Agri-Food Business*, IntechOpen, 2023, doi:10.5772/intechopen.112358.
23. Singh, N. and Dey, K., A typology of agricultural market information systems and its dimensions: Case studies of digital platforms. *Electron. Mark.*, 33, 1, 42, Aug. 2023, doi:10.1007/s12525-023-00665-0.
24. Longo, F., *et al.*, An overview of approaches and methodologies for supporting smallholders: ICT tools, blockchain, business models, sustainability indicators, simulation models. *Procedia Comput. Sci.*, 217, 1930–1939, Jan. 2023, doi:10.1016/j.procs.2022.12.393.
25. Vlachopoulou, M., Ziakis, C., Vergidis, K., Madas, M., Analyzing AgriFood-Tech e-Business Models. *Sustainability*, 13, 10, Jan. 2021, doi:10.3390/su13105516.
26. Anderson, E.G., Lopez, J., Parker, G.G., Leveraging value creation to drive the growth of B2B platforms. *Prod. Oper. Manage.*, 31, 12, 4501–4514, Dec. 2022, doi:10.1111/poms.13866.
27. Martens, B. and Duch Brown, N., The economics of business-to-government data sharing. JRC Digital Economy Working Paper, Working Paper 2020–04, 2020, Accessed: Jun. 29, 2025. [Online]. Available: https://www.econstor.eu/handle/10419/227699.
28. Sheoran, V., Rohtas, Punia, M., Kumari, A., Performance Evaluation of Pradhan Mantri Fasal Bima Yojana (PMFBY): A study on selected districts of Haryana. *Front. Health Inform.*, | EBSCOhost, 2024 Accessed: May 28, 2025, [Online] Available: https://openurl.ebsco.com/contentitem/gcd:184837773?sid=ebsco:plink:crawler&id=ebsco:gcd:184837773.
29. Murugesan, D. and Mathieu, R., Early season prevented and failed rice sowing detection using temporal Synthetic Aperture Radar data in coastal India for crop insurance applications. *Smart Agric. Technol.*, 10, 100868, Mar. 2025, doi:10.1016/j.atech.2025.100868.

30. Yangui, S., A Panorama of Cloud Platforms for IoT Applications Across Industries. *Sensors*, 20, 9, Art. no. 9, Jan. 2020, doi:10.3390/s20092701.
31. Giessmann, A. and Stanoevska-Slabeva, K., Business Models of Platform as a Service (PaaS) Providers: Current State and Future Directions. *J. Inf. Technol. Theory Appl. JITTA*, 13, 4, Apr. 2013, [Online]. Available: https://aisel.aisnet.org/jitta/vol13/iss4/4.
32. Borrero, J.D. and Mariscal, J., A Case Study of a Digital Data Platform for the Agricultural Sector: A Valuable Decision Support System for Small Farmers. *Agriculture*, 12, 6, Jun. 2022, doi:10.3390/agriculture12060767.
33. Abiri, R., N. Rizan, N., Balasundram, S.K., Shahbazi, A.B., Abdul-Hamid, H., Application of digital technologies for ensuring agricultural productivity, *Heliyon*, 9, 12, e22601, Dec. 2023, doi: 10.1016/j.heliyon.2023.e22601.
34. Steinke, J., Ortiz-Crespo, B., van Etten, J., Müller, A., Participatory design of digital innovation in agricultural research-for-development: insights from practice. *Agric. Syst.*, 195, 103313, Jan. 2022, doi:10.1016/j.agsy.2021.103313.
35. Kenny, U., Regan, Á., Hearne, D., O'Meara, C., Empathising, defining and ideating with the farming community to develop a geotagged photo app for smart devices: A design thinking approach. *Agric. Syst.*, 194, 103248, Dec. 2021, doi:10.1016/j.agsy.2021.103248.
36. Beamish, A.L., *et al.*, FERN.Lab: Bridging the gap between remote sensing academic research and society. *Remote Sens. Appl. Soc. Environ.*, 24, 100641, Nov. 2021, doi:10.1016/j.rsase.2021.100641.
37. Nagaraji, S., Kajale, D., Malik, S., Kadam, S., Digital needs & gaps assessment for farmer producer organizations in Maharashtra, India, Aug. 2023, Accessed: May 28, 2025. [Online]. Available: https://hdl.handle.net/10568/138469.
38. Silva, G., *et al.*, Plant pest surveillance: from satellites to molecules. *Emerg. Top. Life Sci.*, 5, 2, 275–287, Mar. 2021, doi:10.1042/ETLS20200300.
39. Verma, G. and Rawat, P.S., Cloud Computing-Based Service-Oriented Architecture for Agricultural Applications, in: *Soft Computing Principles and Integration for Real-Time Service-Oriented Computing*, Auerbach Publications, 2024.
40. Thilakarathne, N.N., Abu Bakar, M.S., Abas, P.E., Yassin, H., Internet of things enabled smart agriculture: Current status, latest advancements, challenges and countermeasures. *Heliyon*, 11, 3, e42136, Feb. 2025, doi:10.1016/j.heliyon.2025.e42136.
41. Johnraja, J.I., Leelipushpam, P.G.J., Shirley, C.P., Princess, P.J.B., Impact of Cloud Computing on the Future of Smart Farming, in: *Intelligent Robots and Drones for Precision Agriculture*, S., Balasubramanian, G., Natarajan, and P. R., Chelliah (Eds.), pp. 391–420, Springer Nature Switzerland, Cham, 2024, doi:10.1007/978-3-031-51195-0_18.
42. Arora, S., Thota, S.R., Gupta, S., Artificial Intelligence-Driven Big Data Analytics for Business Intelligence in SaaS Products, in: *2024 First International Conference on Pioneering Developments in Computer Science*

& Digital Technologies (IC2SDT), pp. 164–169, Aug. 2024, doi:10.1109/IC2SDT62152.2024.10696409.
43. Suresh, D., Choudhury, A., Zhang, Y., Zhao, Z., Shaw, R., The Role of Data-Driven Agritech Startups—The Case of India and Japan. *Sustainability*, 16, 11, Jan. 2024, doi:10.3390/su16114504.
44. Nagendra, N.P., Narayanamurthy, G., Moser, R., Satellite big data analytics for ethical decision making in farmer's insurance claim settlement: minimization of type-I and type-II errors. *Ann. Oper. Res.*, 315, 2, 1061–1082, Aug. 2022, doi:10.1007/s10479–020-03793–4.
45. Surendran, S. and B. N. M., Crop Insurance in India: An Analysis. *Indian J. Law Leg. Res.*, 5, 2, 1, 2023.
46. Satellite Big Data Analytics for Decision Intelligence - A Multi-Case Study Approach – ProQuest, Accessed: May 28, 2025, [Online] Available: https://www.proquest.com/openview/9c079542a590049ebb7fb7640667bc90/1?pq-origsite=gscholar&cbl=2026366&diss=y.
47. Kodors, S., *et al.*, Autonomous Yield Estimation System for Small Commercial Orchards Using UAV and AI. *Drones*, 8, 12, Dec. 2024, doi:10.3390/drones8120734.
48. Michailidis, E.T. and Vouyioukas, D., A Review on Software-Based and Hardware-Based Authentication Mechanisms for the Internet of *Drones. Drones*, 6, 2, Feb. 2022, doi:10.3390/drones6020041.
49. Talaviya, T., Shah, D., Patel, N., Yagnik, H., Shah, M., Implementation of artificial intelligence in agriculture for optimisation of irrigation and application of pesticides and herbicides. *Artif. Intell. Agric.*, 4, 58–73, Jan. 2020, doi:10.1016/j.aiia.2020.04.002.
50. Bazargani, K. and Deemyad, T., Automation's Impact on Agriculture: Opportunities, Challenges, and Economic Effects. *Robotics*, 13, 2, Feb. 2024, doi:10.3390/robotics13020033.
51. Al-Yadumi, S., Xion, T.E., Wei, S.G.W., Boursier, P., Review on Integrating Geospatial Big Datasets and Open Research Issues. *IEEE Access*, 9, 10604–10620, 2021, doi:10.1109/ACCESS.2021.3051084.
52. Breunig, M., *et al.*, Geospatial Data Management Research: Progress and Future Directions. *ISPRS Int. J. Geo-inf.*, 9, 2, Feb. 2020, doi:10.3390/ijgi9020095.
53. Aviv, I., Barger, A., Kofman, A., Weisfeld, R., References Architecture for Blockchain-Native Distributed Information System. *IEEE Access*, 11, 4838–4851, 2023, doi:10.1109/ACCESS.2023.3235838.
54. Gupta, M., Abdelsalam, M., Khorsandroo, S., Mittal, S., Security and Privacy in Smart Farming: Challenges and Opportunities. *IEEE Access*, 8, 34564–34584, 2020, doi:10.1109/ACCESS.2020.2975142.
55. Kaur, J., Hazrati Fard, S.M., Amiri-Zarandi, M., Dara, R., Protecting farmers' data privacy and confidentiality: Recommendations and considerations. *Front. Sustain. Food Syst.*, 6, Oct. 2022, doi: 10.3389/fsufs.2022.903230.

56. Nwaimo, C.S., Adegbola, A.E., Adegbola, M.D., Data-driven strategies for enhancing user engagement in digital platforms. *Int. J. Manag. Entrep. Res.*, 6, 6, Jun. 2024, doi:10.51594/ijmer.v6i6.1170.
57. Costa, L., Nunes, L., Ampatzidis, Y., A new visible band index (vNDVI) for estimating NDVI values on RGB images utilizing genetic algorithms. *Comput. Electron. Agric.*, 172, 105334, May 2020, doi:10.1016/j.compag.2020.105334.
58. Stamford, J.D., Vialet-Chabrand, S., Cameron, I., Lawson, T., Development of an accurate low cost NDVI imaging system for assessing plant health. *Plant Methods*, 19, 1, Jan. 2023, doi: 10.1186/s13007-023-00981-8.
59. Veesam, H., Nikam, V., Chand, P., Evaluating the effectiveness of extension and advisory services by Farmer Producer Organisations: an application of 'best fit' framework. *J. Agric. Educ. Ext.*, 32, 1, 103–127, Apr. 2025, doi: 10.1080/1389224x.2025.2496182.
60. Lamsal, R.R., Karthikeyan, P., Otero, P., Ariza, A., Design and Implementation of Internet of Things (IoT) Platform Targeted for Smallholder Farmers: From Nepal Perspective. *Agriculture*, 13, 10, Oct. 2023, doi:10.3390/agriculture13101900.
61. Garg, P.K., Next Generation Digital Technologies in Smart Society. *Technological Prospects and Social Applications of Society 5.0*, 251–259, Chapman and Hall/CRC, May 16, 2023. doi: 10.1201/9781003324720-22.
62. Adesiyan, T., Leveraging Agricultural Credit and Digital Finance for Enhancing Smallholder Productivity and Rural Economic Growth. *Int. J. Res. Publ. Rev.*, 6, 8, 642–667, Aug. 2025, doi: 10.55248/gengpi.6.0825.3090.

16

Economic Impacts of Geospatial Intelligence on Sustainable Farming

Gurwinder Singh*, **Aaskaran Bishnoi and Ranjit Singh**

Department of AIT-CSE, Chandigarh University, Mohali, Punjab, India

Abstract

Geospatial Intelligence (GI) has emerged as a transformative enabler in modern agriculture, providing powerful tools to enhance economic efficiency, optimize resources, and promote sustainability. This chapter examines the multifaceted economic impacts of GI technologies, including remote sensing, geographic information systems (GIS), Internet of Things (IoT) sensors, and artificial intelligence (AI)-driven analytics on farming practices across diverse scales and geographies. GI enables precision agriculture techniques that reduce input costs, enhance yield quality and quantity, and facilitate access to high-value markets such as organic or eco-certified sectors. Through detailed case studies and cost–benefit analyses, This chapter illustrates how the adoption of GI yields short-term benefits and long-term sustainability (e.g., improved soil health and increased carbon efficiency). This chapter further examines the benefits of risk mitigation, including resilience to climatic variability and enhanced supply chain efficiency. The chapter concludes with a roadmap of technological evolution, identification of policy gaps, and offers strategic recommendations for stakeholders. These include public-private partnerships (PPPs), targeted subsidies, training programs, and regulatory frameworks designed to democratize access to GI tools. Collectively, this analysis highlights GI as a cornerstone of data-driven, economically viable, and climate-resilient agriculture.

***Keywords*:** Geospatial intelligence (GI), precision agriculture, economic impact, remote sensing, sustainable farming

**Corresponding author*: singh1001maths@gmail.com

Gurwinder Singh, Vishakha Sood, Narayan Vyas, Surendra Yadav and Dankan Gowda V. (eds.) Geospatial Intelligence for Precision Agriculture, (313–338)

16.1 Introduction

Geospatial Intelligence (GI) refers to the integration of various technologies, such as Geographic Information Systems (GIS), remote sensing, the Internet of Things (IoT), and Artificial Intelligence (AI), to collect, process, analyze, and visualize spatial data for decision-making in agriculture. The role of GI in agriculture is rapidly growing, driven by the need to address the challenges of climate change, resource depletion, and the increasing global population that demands more food with fewer resources [1, 2]. In sustainable farming, GI enables farmers to make data-driven decisions regarding resource management, crop production, and environmental stewardship [3]. Spatially explicit data in GI enable optimized water, nutrient, and pesticide management, enhancing crop productivity while protecting the environment [4]. Satellite imagery is an essential tool in GI because it reveals comprehensive agricultural field views, soil moisture information, plant health, and crop development patterns [5]. These equipments enable farmers to observe specific areas within their fields that need irrigation, so they can apply water only in the necessary regions. Satellite-based remote sensing enables farmers to detect plant water stress, allowing for targeted irrigation to optimize water use [6]. An overview of the GIS is presented in Figure 16.1.

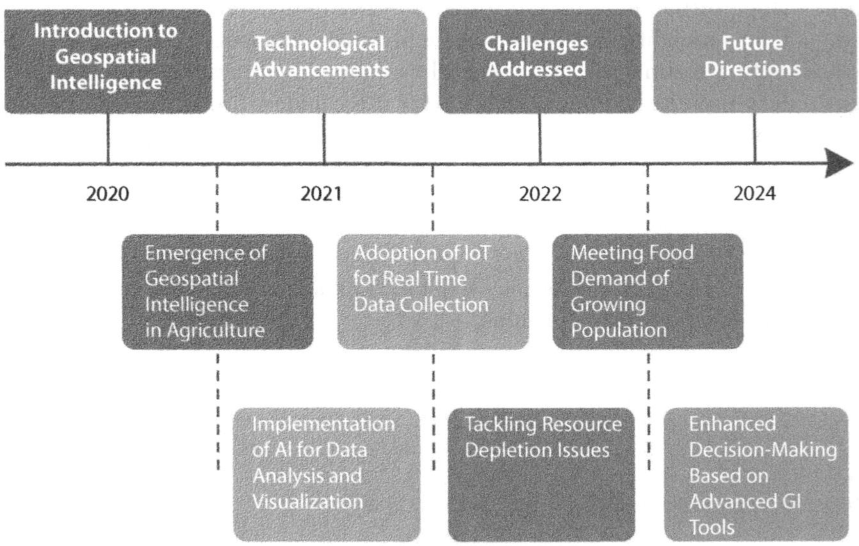

Figure 16.1 Overview of GIS.

In addition to water management, GI aids in the optimization of fertilizers and pesticides. Remote sensing technology and GIS systems enable farmers to locate areas in their fields that require fertilization or pest management. It can target nutrients and pesticides properly instead of using broad-field applications [7]. Precise chemical application using GIS creates a two-fold advantage by decreasing both operational spending and environmental damage from chemical pollution affecting surface waters [8]. GI frameworks integrate IoT devices for real-time environmental data monitoring, including the tracking of soil moisture levels, temperature data, and crop health status [9]. Qualitative field data obtained by sensors allow farmers to monitor conditions directly through GIS platforms and then take immediate corrective measures. The large amounts of data processed through AI and machine learning (ML) algorithms enable accurate prediction of crop yield and disease outbreaks, allowing decision-makers to develop better protection strategies [10]. The main goal of this chapter is to investigate the economic effects of GI technology on sustainable farming systems. GI technologies adopted for agriculture create diverse economic value while generating major effects on immediate profits and long-term farming viability [11].

GI adoption is the primary focus of this chapter, highlighting how it reduces operational expenses. The precise nature of GI technologies allows farmers to reduce waste and optimize their usage of water, fertilizers, pesticides and labor [12]. Operational expenses, which typically restrict farming system profitability, become lower when farmers use these technologies. GI provides a method for managing resources efficiently, as it reduces input consumption while achieving or surpassing production goals [13]. This section examines how GI implementation results in profitability gains. Increased crop yields are possible through GI because it provides optimized control of irrigation systems, fertilization practices, and pest management strategies [14]. The combination of enhanced crop yields and improved agricultural practices generates better product quality, leading to increased revenue. GI helps farmers enter premium markets for eco-friendly products through demonstrations of sustainable operations, which lead to higher selling prices [15]. Through their adoption, farmers can align with the rising requirements of sustainable farming standards, which consumers and governments are pushing forward [16]. A farm that adopts GI technologies can achieve improved market positioning and competitive advantage. The utilization of GI by farmers enables them to promote sustainable technology adoption, through which they secure advantages in marketing [17]. This section examines the GI adoption hurdles, which include high technology acquisition costs, complex system integration, and limited expansion capabilities for smallholder agriculturalists [18]. A thorough economic evaluation of GI potential for sustainable agriculture is presented in this chapter, examining its advantages and implementation obstacles.

This chapter provides a detailed exploration of the economic impact of GI technologies in agriculture. Case studies, together with cost–benefit analyses and future trend examinations, enable readers to understand both the immediate and unseen economic advantages of adopting GI according to [19]. This research aims to showcase various case studies that demonstrate the use of GI technology by both large- and small-scale farming operations, with the goal of reducing expenses and enhancing crop production and market penetration capabilities [20]. The practical uses of GI, alongside measurable farm economic effects, are illustrated through various study examples. The cost–benefit analysis presented in this chapter transforms the financial benefits of GI adoption into quantifiable numbers for ROI evaluation and break-even calculations. The analysis includes detailed information on technology expenditure, operational budget reduction, and revenue development, which together present a complete picture of the economic potential of sustainable farming with GI implementation. This chapter analyzes upcoming trends in GI alongside innovative AI applications and sensor systems, which demonstrate the potential to decrease prices and optimize farming operations to maximize agricultural profits in the coming years [22]. New, promising technological developments aim to provide GI solutions that are cheaper and more accessible to smallholder farmers. This chapter addresses a broad range of issues encompassing farmers, researchers, and policymakers. These insights help farmers implement GI technology for operational improvement, and researchers become informed about sustainable agricultural economic factors referred to in [23]. The information gained from this study enables policymakers to create incentives for the adoption of general integration and promote sustainable agricultural systems [24].

16.2 Background and Context

Sustainable farming is a viable solution for minimizing environmental degradation while maintaining long-term agricultural productivity. Despite its benefits, it poses several economic challenges that farmers must address. A primary issue is the escalating cost of essential inputs, such as fertilizers, water, and energy [25]. Farmers usually allocate major operational costs to these components of the farm. Farming operations in water-affected areas lead farmers to obtain advanced irrigation systems because water expenses increase, which drives them to reduce water consumption [27]. The unfavorable market environment from unpredictable supply-demand fluctuations, combined with weather changes and regulatory policy fluctuations, makes it impossible for farmers to accurately predict their revenue while maintaining spending control [28]. The sharp drop in commodity prices generates considerable profit margin reductions, which leads to potential financial instability. Transitioning to environmentally sustainable practices requires farmers to make substantial monetary investments. The adoption of organic farming requires substantial preliminary costs because farmers must purchase new equipment, secure certifications, and change pest and disease management practices [29]. The long time it takes to achieve returns from these investments makes it difficult for farmers to invest in such transitions immediately. Economic challenges in agriculture are becoming more severe due to climate change, as changing environmental conditions produce unstable agricultural harvests. The unpredictable weather patterns of droughts, together with untimely rainfall, create uncertainty regarding crop yields, which leads to problems with financial planning and prevents farmers from obtaining credit or engaging in investments [30].

16.3 Role of GI in Addressing Economic Challenges

GI technologies serve as strategic tools to combat economic challenges by implementing better business resource management, decreasing waste, and strengthening decision-making approaches [31]. Precision mapping made possible by GI enables farmers to strategically apply fertilizers and pesticides, resulting in reduced unnecessary use and reported cost reductions of up to 15% in large-scale farming scenarios, without affecting yields [26, 32]. Farmers use satellite imagery and remote sensors to track instantaneous measurements of soil moisture and crop health visibility [33]. The collected data direct the irrigation process to precise locations, thereby minimizing unnecessary water utilization. These optimization strategies reduce water and energy costs, thereby enhancing sustainability and profitability [34]. Farm operations benefit from enhanced management through data-driven insights, which GI delivers to streamline operational processes. Real-time tracking of essential environmental variables, including soil quality, humidity, and temperature, can be achieved by integrating GIS systems with IoT sensors, as described in [35]. Through real-time tracking, GI enables fast implementation, which increases output levels while diminishing the requirement for additional resources. Through its implementation, GI creates farming operations that require less expense while being more productive [36].

A small farm in India demonstrated success by using satellite data to check soil water content [20]. Over two years, the farm reduced its irrigation costs by 20% through accurate determination of water requirements across

Table 16.1 Economic benefits of GI in farming.

Farm type	Technology used	Economic benefits
Small-scale farm in India (5 ha, Maharashtra)	- Satellite imagery (ISRO Resourcesat-2) - NDVI and soil moisture mapping - GIS-based irrigation planning - Integrated with drip irrigation	- 25% reduction in water usage - $1,000/year irrigation cost savings - 10% increase in cotton yield - $500/year added revenue - Total economic benefit: $1,500/year
Large-scale farm in the USA (500 ha, Iowa)	- IoT soil sensors for moisture, temperature, and nutrients - -GIS platform with historical weather + satellite imagery - Variable-rate planting and fertilization	- 15% increase in crop yields - Corn yield: 180–207 bushels/acre - $50,000/year increased revenue - 10% less fertilizer, $15,000/year savings - ROI achieved within 1 year on a $30,000 investment
Mid-sized organic farm (Europe)	- Remote sensing for pesticide monitoring - GIS-supported organic compliance tracking	- Achieved organic certification - Up to 20% premium pricing for produce—Increased trust *via* traceability and labeling
Agri-cooperative in Africa (Multi-farm system)	- Shared cloud-based mobile GIS apps - Community-level sensor analytics - Affordable tools with NGO support	- Reduced collective irrigation input - Improved food security at the regional scale—Cost-sharing in training and maintenance

different field sections, supported by the improved water-use efficiency (see Table 16.1), thereby minimizing excessive irrigation and improving plant health. Through GI technologies, production farms of any resource capacity can achieve both financial savings and improved operational efficiency. The administrators of a major North American farm utilized precision agriculture tools to enhance their planting operations, as reported by [14]. The use of GIS and IoT systems enabled the farm to collect data on soil conditions, climate data, and crop information, resulting in a 10% increase in yield output, which generated substantial financial gains. The combination of real-time analytics and enhanced operational productivity, combined with improved resource management, demonstrated the financial advantages of GI applications on a large scale. Several economic benefits of GI farming are listed in Table 16.1.

These case studies and associated figures provide clear evidence that GI technologies are not only for large farms but can be used effectively in small operations, and a scalable solution can be offered for financial challenges in agriculture [14, 20]. By using GI to improve efficiency and stability, farms of all sizes can achieve better financial results.

16.4 Economic Benefits of GI

16.4.1 Cost Reduction

Targeted Resource Application: GI tools facilitate the efficient and site-specific application of essential agricultural inputs, such as fertilizers, pesticides, and irrigation water, by offering granular insights into field variability [32]. For instance, soil maps generated using GIS technologies help pinpoint areas with nutrient deficiencies. This enables farmers to apply fertilizers selectively, thereby avoiding blanket applications [13].

Operational Efficiency: The use of GI-driven precision agriculture and automated monitoring systems significantly boosts operational productivity [36]. Technologies such as drones and smart machinery, integrated with geospatial analytics, can scan and evaluate extensive agricultural fields by tracking parameters such as crop vigor, soil hydration, and growth stages. This reduces the dependency on manual field surveys, reduces labor expenses, and enhances overall farm management efficiency [7]. The application of GI technologies across various agricultural resources has shown considerable promise in reducing operational costs and enhancing sustainability, as presented in Table 16.2.

Table 16.2 Expanded cost reduction through GI.

Resource optimized	GI application	Economic impact/potential savings
Fertilizers	- GIS-based soil nutrient mapping - Variable-rate fertilizer application using IoT and GIS data	- Up to 10% reduction in fertilizer usage - Annual savings of $15,000 reported in large-scale farms - Reduced chemical runoff and environmental impact
Pesticides	- Remote sensing to detect early signs of pest infestation - Targeted spraying *via* precision drones	- Reduced pesticide application and waste - Cost-effective pest control - Improved crop health with fewer chemical residues
Water usage	- Satellite imagery and soil moisture indexing (e.g., NDVI) - Precision irrigation using GIS-drip integration	- Up to 25% reduction in water usage (small-scale Indian case) - $1,000/year irrigation cost savings—Lowered groundwater extraction and energy costs
Labor costs	- Use of automated drones for field inspections - Real-time monitoring with IoT sensors	- Reduced manual labour in large farms - Faster detection and response to crop stress - Enhanced operational efficiency
Fuel and machinery use	- GPS-guided tractors and route optimization - GIS-based logistics and transport planning	- Lower fuel consumption and machine wear–Reduced carbon emissions and maintenance costs
Time and planning efficiency	- Real-time alerts and predictive analytics for crop events - Automated data collection from multi-sensor networks	- Reduced decision-making delays - Streamlined operations and better resource allocation

16.4.2 Increased Productivity and Yield

Enhanced Crop Monitoring: A significant advantage of GI lies in its ability to improve crop monitoring through continuous, real-time data collection and analysis [35]. By utilizing IoT-activated sensors, farmers can track key variables, such as soil moisture levels, environmental temperatures, and nutrient availability. This uninterrupted data flow enables the early identification of problems, such as drought stress or nutrient imbalance, allowing for rapid corrective actions to support optimal crop growth [9].
Improved Crop Quality: In addition to increasing crop yield, GI provides significant enhancement to crop quality, offering remarkable economic benefits [15]. Through accurate input management, including calibrated watering, nutrient distribution, and insect control, farms can produce healthy plants with enhanced production qualities. High-quality crops command premium prices in the market, especially in specialized segments that value organic or sustainably cultivated products [17]. Table 16.3 highlights how the adoption of GI significantly enhances crop yield, quality, and resource efficiency across various agricultural practices.

16.5 Risk Mitigation

Early Detection of Issues: GI technologies allow farmers to access superior systems to detect crop health issues early on [5]. Remote sensing tools, such as satellite imagery and drone surveillance, enable the detection of the initial indicators of pest infestations, plant diseases, and nutrient imbalances. Early issue detection through advanced techniques provides farmers with the proper timing to act and stop problems from growing out of control, which reduces resource consumption and economic losses [7].

16.5.1 Predictive Analytics

The main benefit of GI stems from its ability to unite with predictive analytics for strategic organizational decision-making according to [10]. Evolved through the fusion of modern sensor data, existing agricultural records, and weather projection information, machine learning-based predictive models generate forecasts about potential pest life cycles in addition to seasonal weather forecasts and optimized farming schedules [22]. By forethinking operations, the planning process receives enhanced accuracy, resulting in improved crop performance and operational efficiency improvements. Table 16.4 demonstrates how geospatial intelligence (GI)

Table 16.3 Increased productivity and yield from GI adoption.

Benefit	GI application	Impact	Examples/case evidence
Crop monitoring	IoT sensors and remote sensing for real-time tracking of soil moisture, temperature, and nutrient levels	Increased yields by up to 12%	A large-scale farm in the USA improved yield by optimizing planting using soil data
Improved crop quality	Precision irrigation and site-specific pesticide application using GIS	Higher-quality produce with better market acceptance and premium pricing	Organic farms leveraging GI secured 20% higher prices in niche markets
Optimized planting strategies	GIS-based field variability analysis combined with weather prediction models	Enhanced crop establishment and reduced seedling mortality	Corn yield increased from 180 to 207 bushels/acre (USA case)
Stress detection and early intervention	Satellite imagery (e.g., NDVI) and drone surveillance for detecting plant stress early	Reduction in crop loss and better yield consistency	A small-scale farm in India identified water stress zones to improve irrigation
Nutrient management	GIS-guided variable-rate fertilizer application based on nutrient maps	Balanced growth, reduced over- or under-fertilization, and better biomass output	Large farms reduced fertilizer by 10% without affecting yield
Precision pest control	Remote sensing and AI to detect pest patterns before visible damage	Minimized pest-related yield loss and better crop uniformity	Preventive spraying in hotspots improved output and reduced pesticide use

Table 16.4 Risk mitigation *via* GI.

Risk mitigated	GI application	Impact	Examples/outcomes
Pests and diseases	Remote sensing, drone surveillance, and AI-based detection of crop stress or pest hotspots	Reduced crop loss by up to 15%; faster treatment response	Drones identified pest clusters early in maize crops, enabling targeted pesticide use
Weather events	Predictive analytics using historical weather, satellite, and sensor data	Optimized irrigation scheduling and planting windows; minimized crop damage	Farmers avoided flood-prone periods by adjusting sowing schedules
Market fluctuations	Market trend forecasting and integration of commodity price monitoring with GI dashboards	Improved financial planning and investment decisions	Farmers stored produce and delayed sales based on the forecasted price rise
Irrigation failures	Real-time soil moisture mapping *via* satellite and IoT sensors	Proactive irrigation intervention and prevention of water stress	Small farm in India reduced under-irrigation risks, saving 25% of water
Nutrient deficiencies	GIS-guided nutrient variability maps and NDVI analysis	Targeted intervention prevents yield loss due to imbalance	Large farm reduced nutrient deficiency zones by 30% with zonal fertilization
Regulatory and certification compliance	GI-based traceability and sustainability monitoring tools	Avoidance of non-compliance penalties; readiness for audits	Organic certification achieved through precise tracking of water/pesticide use

supports proactive risk mitigation in agriculture by enhancing resilience against pests, weather events, market volatility, and compliance challenges.

16.5.2 Sustainability and Long-Term Economic Gains

Soil Health Management: Process management of soil health is vital for sustained agricultural sustainability and economic value [8]. The use of GI technologies enables farmers to monitor land parameters, including nutrient concentrations, persistent pH levels, and moisture content. Detailed agricultural information enables farmers to optimize their tillage approach, thus preserving soil quality with low levels of fertilizers and pesticides [4].
Market Premiums: GI technologies enable farmers to improve sustainability by integrating optimized water management and reducing chemical applications, thus helping them fulfill organic and eco-certification requirements [15]. Pricing premiums for farms meeting the criteria enable certified sustainable products to fetch prices up to 20% higher because consumers choose green products [17]. Table 16.5 outlines how the long-term adoption of GI drives sustainable agricultural practices, leading to economic resilience through improved soil health, water conservation, premium market access, and reduced environmental impact.

16.6 Market Dynamics and Economic Opportunities

16.6.1 Access to New Markets

GI functions as a central tool that allows farmers to access developing markets because more consumers want sustainably produced agricultural products [16]. The increasing environmental consciousness of consumers has led to an increasing preference for agricultural products produced using sustainable and environmentally conscious cultivation methods. Modern agricultural technology enables farmers to boost production efficiency and prove their commitment to sustainable practices, which are rapidly becoming essential to customer buying behavior [17]. GI tools that enable soil quality monitoring as well as pesticide-low and optimal irrigation management create possibilities for farms to obtain premium markets through products that demonstrate environmentally friendly practices as per [15]. The organic certification process receives beneficial support through GI, as it provides precise data documentation of agricultural practices [29]. Documented scientific evidence shortens certification procedures while increasing transparency, which builds consumer trust and

Table 16.5 Long-term economic gains from GI adoption.

Sustainability benefit	GI application	Economic impact	Examples/ strategic outcomes
Soil health management	Real-time soil analysis tools (moisture, pH, nutrients) integrated with GIS	Reduced chemical input usage; improved soil fertility over time	Farmers reduced fertilizer input by 10%, enhancing long-term soil structure
Water resource conservation	Precision irrigation systems supported by satellite and IoT-based water mapping	Lower pumping and energy costs; reduced water stress risk	25% reduction in water use reported in Indian cotton farms
Market premiums	Traceability of sustainable inputs and eco-friendly practices *via* GI monitoring	Up to a 20% premium in organic and green-labeled markets	European farms accessed high-value organic markets through GI verification
Carbon footprint reduction	GIS tracking of field operations and input consumption	Access to carbon credits and compliance with emissions policies	GI-enabled farms qualified for climate-resilient farming schemes
Biodiversity and eco-balance	Mapping pesticide and fertilizer spread zones to minimize environmental runoff	Long-term ecosystem sustainability, regulatory compliance	Farmers avoided penalties and enhanced land-use certification eligibility
Future-proofing farm economics	Predictive models for rotation planning, degradation monitoring	Sustained productivity, lower future remediation costs	Farmers forecast soil depletion and adjusted rotations pre-emptively

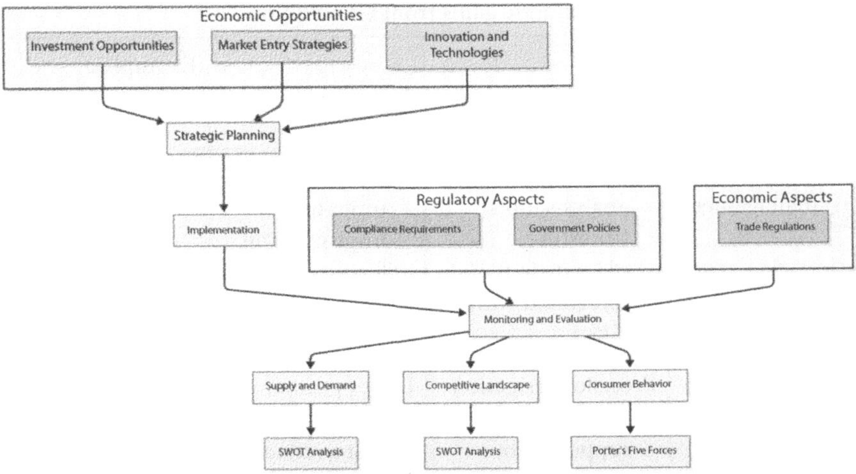

Figure 16.2 Market dynamics and economic opportunities.

raises market demand. An overview of the market dynamics and economic opportunities is presented in Figure 16.2.

GI enhances the ability to track products, which remains vital in the existing modern food supply network [16]. GIS tools help track source crops in detail, monitor environmental parameters, and record harvest times, creating complete transparency throughout the supply chain. The assurance of trackable origins plays a crucial role in global food sales, as it addresses safety concerns and enhances product value, leading to price increases for both final market customers and retail establishments [14]. Farmers who adopt GI practices and enable product tracking will attract environmentally minded consumers and gain entry into both specialized domestic and global market sectors [15].

16.6.2 Supply Chain Efficiency

Efficient supply chain management is essential for profitable farm operations, which GI technologies optimize agricultural logistics [16]. Real-time information on field conditions and crop maturity can be gathered through the combination of satellites and drones, together with IoT sensors under the GI system. By continuously tracking crops, farmers can receive alerts for the best harvest windows [9], thus avoiding spoilage and deterioration of crop quality [34]. Collecting crops when they reach their maximum maturity state results in a longer shelf life combined with superior market value. GI technologies help farmers conduct time-sensitive harvests,

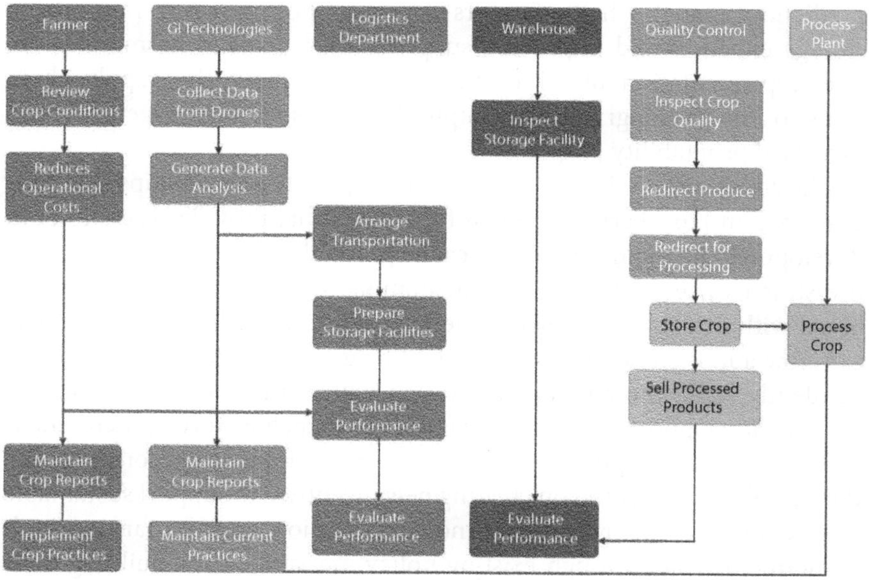

Figure 16.3 Flowchart illustrating the integration of GI technologies in the agricultural value chain.

which decreases post-harvest waste, thereby strengthening food security and boosting profits [36]. The agricultural value chain can be significantly enhanced by integrating geospatial intelligence and drone-based data analysis, as shown in Figure 16.3.

16.6.3 Policy and Incentive Alignment

GIS tools help farmers achieve maximum operational efficiency when determining how to move their harvests from the field to the market. Farmers use GIS-based mapping technology along with live traffic information to find the best delivery routes for their produce distribution. GIS tools contribute to lower fuel expenses and reduced travel duration, and enable quick delivery times for perishable products [16]. The implementation of efficient transportation methods helps maintain product excellence in retail outlets and control operational costs to build stronger farm financial performance [14]. The implementation of GI technologies leads to improved systems that control inventory and manage storage spaces. Farmer decision-making becomes more evident through crop yield prediction and harvest status monitoring, which enables them to organize storage areas and plan timing according to the results [35]. Advanced prediction analytics helps farmers decrease spoilage while minimizing storage expenses and improving supply-market demand harmonization. The implementation of GI technologies throughout the supply chain leads to improved agricultural output, better resource management, and increased profitability [16].

The adoption of GI benefits from public-private partnerships (PPPs) is described in [16]. Farmers benefit from PPPs that provide access to both development financing and technical help, coupled with the infrastructure investments needed for GI implementation. Through public funds combined with private sector involvement, small- and medium-sized farmers can now adopt GI solutions that would be expensive to implement independently [24]. The alignment between national and international policy agendas supports the United Nations Sustainable Development Goals (SDGs) related to responsible production and environmental conservation. Public policy has focused on creating new incentives to support sustainable practices in the long run. GI technology adoption provides farmers with economic benefits through existing policy measures while building their capability to satisfy planned governmental standards and upcoming market requirements regarding product transparency and sustainability [16].

16.7 Challenges and Economic Barriers

16.7.1 Initial Investment Costs

The first hurdle for farmers in adopting GI technologies is their high initial investment costs, which mainly affect smallholder farmers' access [18]. The required capital investments exceed what farmers with limited financial means can afford. Acquiring GI technologies involves significant financial investments in both system hardware and software, as well as the necessary infrastructure to effectively deploy and operate them. Farm operators who must manage limited profit space encounter substantial barriers due to demanding initial expense requirements. While the long-term advantages of GI, such as improved productivity and cost savings, are well documented, the delayed return on investment (ROI) makes the technology less attractive, particularly in low-income or resource-constrained farming communities. The absence of immediate financial gains can lead to skepticism or reluctance among small-scale farmers, who may already face income uncertainty due to unpredictable environmental or market factors.

16.7.2 Training and Skill Development

Beyond equipment costs, effective GI adoption requires adequate training and skill acquisition. Technologies such as GIS software, drone operation, and real-time sensor interpretation require a certain level of technical competence, which many farmers may not initially possess. Training costs, estimated at $500 per farmer in developed regions, add to the financial burden [24]. When training regarding GI system implementation is overlooked, the underutilization of these systems poses a significant economic risk by reducing the productivity and limiting its return on investment. Farmers who do not understand how to read GIS maps and set up sensor-based irrigation systems are unlikely to fully realize the benefits of the system, resulting in unsatisfactory returns on investment. Some situations reveal that technology exists for deployment, yet insufficient knowledge hinders the proper execution systems. GI technologies are continually evolving; therefore, users need to consistently learn new aspects and skills. Software development and tool innovation prompt periodic updates of user capabilities because new capabilities are added to the available features. Technical support and educational resources require constant maintenance owing to technological evolution; however, these resources often remain unavailable, primarily to farmers situated in rural and less-served areas.

16.7.3 Scalability Issues

The effectiveness of GI has been established in decision-making processes at the farm level, although difficulties are faced when expanding applications across different farm sizes and geographical areas. Sophisticated GI deployments find practical support among larger commercial farms, mainly because of their well-developed operational and financial resources. The systems created for GI appear difficult to reach for smallholder farmers on financial and technical levels. Achieving scalability is difficult because of the abundant variations in farming operations. Large-scale farms require processed data systems along with integrated platforms, which increases their operating expenses. Small farms encounter difficulties when trying to modify generic GI solutions to meet their agricultural circumstances regarding land types, field variations, and geographical landscape requirements. Tailored solutions must be implemented to address this accessibility problem. The accessibility of GI to small-scale operations has improved through inexpensive hardware, user-friendly software, and analytical tools developed for specific regions. Cloud-based platforms, together with mobile applications, create the possibility for farmers to lower their infrastructure requirements. Smallholder farmers in rural areas can access GI technologies using smartphones and tablets, which offer affordable digital solutions. Table 16.6 presents a SWOT analysis of geospatial intelligence (GI) adoption in agriculture, revealing its strengths in precision and efficiency, weaknesses in cost and skill gaps, opportunities in market expansion and policy support, and threats from digital inequality and cybersecurity risks.

Table 16.6 SWOT analysis of GI adoption in agriculture.

Factor	Key aspects	Examples/ implications
Strengths	- Enhances precision in resource application - Reduces input waste and operational costs - Improves yield forecasting and crop quality - Enables data-driven decisions and automation	- GI helps optimize water, fertilizer, and pesticide use - Precision agriculture enhances productivity and sustainability
Weaknesses	- High upfront investment in tech and training - Limited digital skills among smallholder farmers - Challenges in integrating GI with traditional practices - Data security and privacy risks	- Small farms may avoid GI due to cost barriers - Lack of training reduces ROI from GI tools
Opportunities	- Access to organic and sustainable markets - Availability of government incentives and subsidies - Rapid growth in smart aggrotech startups - Potential for low-cost mobile-based GI platforms	- GI adoption can attract 20% higher market premiums - National schemes (e.g., PM-KUSUM, EU Green Deal) promote adoption
Threats	- Unequal access to digital infrastructure in rural areas - Resistance to change among traditional farmers - Frequent updates and tech shifts increase costs - Exposure to cyber threats, data misuse	- The digital divide may widen income gaps in agriculture - Farmers may abandon GI due to data breach or policy confusion

16.8 Case Studies or Practical Applications

16.8.1 Case Study 1: Small-Scale Farm in India

A 5-hectare farm in Maharashtra's semi-arid region has struggled continuously due to a lack of water resources and rising water management costs. The farm dedicated its land to cotton and soybean production, which requires high amounts of water, using standard irrigation systems to uniformly spread moisture over the entire area. Water waste occurred together with crop performance decline, as this method resulted in both zones receiving excessive irrigation and other zones receiving insufficient moisture. The farm's large yearly irrigation expenditure of $5,000 represented its biggest operational cost; therefore, the need for cost reduction became crucial for maintaining financial stability. Table 16.7 shows how a small-scale Indian farm achieved a 25% reduction in water usage and a 10% increase in cotton yield through GI adoption, resulting in an annual economic benefit of $1,500 and improved environmental sustainability.

Adoption of GI: In 2018, the farm adopted a GI-based solution to remove irrigation inhibition. Through a government-supported agricultural extension

Table 16.7 Case study 1: Small-scale farm in India—key metrics.

Aspect	Value/result
Farm size	5 ha
Annual irrigation costs	$5,000
Water usage reduction	25% reduction in water usage
Cost savings	$1,000 annually
Increase in crop yield (cotton)	10% increase
Additional revenue	$500 annually
Total economic benefit	$1,500 annually (combined savings and revenue)
Environmental impact	Reduced water usage, conservation of groundwater
Subsidy	The government program subsidized GI adoption costs

initiative, the farm gained access to satellite images from the Indian Space Research Organization (ISRO), especially from the Resourcesat-2 satellite. Remote measurement data generally provide important vegetative indicators, such as the Normalized Difference Vegetation Index (NDVI) and soil moisture indices. With these insights, the farmer implemented a variable-rate irrigation strategy, targeting only those zones requiring water rather than applying it uniformly across the field. This approach was further supported by a low-cost drip irrigation system that significantly enhanced water distribution efficiency.

Outcomes and Economic Impact: Between 2018 and 2020, the farm recorded a 20% reduction in irrigation-related expenses, translating to annual savings of \$1,000. This was largely driven by a 25% decrease in overall water consumption, made possible by the precise application enabled by GI tools. Concurrently, crop health improved, leading to a 10% increase in cotton yield and generating an additional \$500 in revenue each year. The combined economic benefit, including cost savings and enhanced productivity, amounted to \$1,500 annually. On the environmental front, reduced water usage contributes to the conservation of local groundwater reserves, supporting broader sustainability objectives. Notably, the initial costs associated with satellite data access and GIS software were subsidized under a government scheme, ensuring affordability for smallholder farmers.

Relevance and Scalability: This case exemplifies how GI technologies can empower small-scale farmers in water-constrained areas to improve economic and environmental outcomes. By integrating satellite-derived insights with localized irrigation strategies, the farm achieved greater resource efficiency and profitability. The success of this model highlights its potential for replication among other smallholder farmers in India and in similar agroecological settings.

16.8.2 Case Study 2: Large-Scale Farm in the USA

A large-scale farm in Iowa, USA, covering 500 ha, specializes in the production of corn and soybeans. The farm operated in a highly competitive market where maximizing yield and operational efficiency was critical for profitability. Despite adopting conventional precision agriculture techniques, such as GPS-guided tractors, the farm faced challenges in optimizing planting decisions owing to variability in soil conditions and weather patterns across its expansive fields. In 2015, the farm's annual revenue was approximately \$1.2 million; however, rising input costs and fluctuating market prices necessitated further improvements in productivity to maintain profit margins. Table 16.8 highlights how a large-scale farm in the

USA leveraged GI and IoT technologies to achieve a 15% yield increase and $65,000 in annual economic gains, successfully recovering its $30,000 investment in the first year.

Implementation of GI: In 2016, the farm integrated a GI-based system to enhance precision planting by utilizing a combination of IoT sensors and GIS technology. The farm deployed a network of IoT soil sensors across the fields to collect real-time data on soil moisture, temperature, and nutrient content. These sensors were connected to a cloud-based GIS platform that integrated the sensor data with historical weather data and satellite imagery to generate detailed planting prescriptions. The prescriptions accounted for the spatial variability in soil conditions, recommending optimal planting dates, seed densities, and fertilizer applications for each section of the field. The farm used this data to guide its variable-rate planting equipment, ensuring that seeds were planted under optimal conditions for growth.
Results and Economic Impact: The adoption of GI for precision planting resulted in a 15% increase in crop yields over three years (2016–2019), adding $50,000 to the farm's annual revenue. The yield increase was attributed to the improved accuracy of planting decisions, which ensured that seeds were sown in areas with optimal soil conditions, thereby reducing seedling mortality and enhancing overall crop establishment. The precise

Table 16.8 Case study 2: Large-scale farm in the USA—key metrics.

Aspect	**Value/result**
Farm size	500 ha
Annual revenue	$1.2 million
Yield increase	15% increase in crop yields
Increase in corn yields	From 180 bushels/acre to 207 bushels/acre
Additional revenue	$50,000 annually
Fertilizer usage reduction	10% reduction in fertilizer use
Cost savings	$15,000 annually (fertilizer savings)
Total economic benefit	$65,000 annually (combined revenue and savings)
Initial investment	$30,000 for IoT sensors and GIS platform
Recoupment of investment	Within the first year

application of fertilizers based on sensor data resulted in a 10% reduction in fertilizer usage, saving approximately $15,000 annually in terms of input costs. The total economic benefit, combining increased revenue and cost savings, was $65,000 per year. The initial investment in the IoT sensors and GIS platform was $30,000, which was recouped within the first year owing to substantial economic gains.

Significance: This case study highlights the transformative potential of GI in large-scale farms in developed agricultural systems. The integration of IoT sensors with GIS enables data-driven decision-making that significantly enhances productivity and profitability while reducing input costs and environmental impact through efficient resource use. This approach serves as a model for other large-scale farms seeking to leverage advanced technologies to remain competitive in the global market.

16.9 Future Economic Implications

16.9.1 Emerging Trends

The future of GI in agriculture is being shaped by rapid advancements in technologies such as AI, IoT, edge computing, and machine learning. AI-driven analytics make it possible to analyze vast quantities of real-time data, helping farmers make predictive decisions regarding irrigation scheduling, disease prevention, and yield forecasting [22]. Sensor technologies are becoming significantly more affordable. While early generation soil and moisture sensors could cost upwards of $500, newer alternatives are available for under $100, thanks to economies of scale and open-source hardware platforms [9]. These low-cost sensors allow dense deployments across large farms or shared use in smallholder cooperatives, democratizing access to precision data. Similarly, drone technology, once cost-prohibitive, is becoming viable for mid-sized operations owing to modular and AI-assisted models. Moreover, the integration of GI with mobile platforms is driving their adoption in low-connectivity regions. Offline-capable apps that sync with cloud GIS platforms when connectivity is available allow real-time spatial insights even in remote areas, paving the way for inclusive smart farming ecosystems.

16.9.2 Scaling Economic Benefits

As GI technologies become more accessible, their large-scale adoption could lead to systemic economic benefits for the global agricultural sector [16], especially when adopted alongside complementary technologies, such as weather forecasting systems and sustainable water management infrastructure. In developing regions, where inefficiencies in resource use, market access, and crop management are more pronounced, the economic benefits could be even more significant. Smallholder farmers using GI tools have already demonstrated yield increases of up to 20% and cost reductions of 15% in pilot programs [21]. At the macroeconomic level, GI adoption can contribute to national food security strategies, reduce import dependencies, and stimulate rural economies. Integrating GI into climate-smart agriculture can mitigate the economic impact of climate variability and improve long-term resilience and productivity [24].

16.9.3 Policy Recommendations

Coordinated policy frameworks and incentive structures are essential to unlock the full economic potential of GI. Governments should consider offering capital subsidies covering up to 50% of the costs of GI hardware and software, especially for small and mid-sized farms. These subsidies can be modelled after successful programs, such as the European Union's Common Agricultural Policy (CAP) and India's PM-Kisan and Digital Agriculture Mission schemes [12].

Free or subsidized training programs, both online and in-person, should be established to build digital literacy among farmers. Partnerships with universities, agricultural extension services, and private agri-tech firms can facilitate the development of practical and scalable training modules. GI certification could become part of broader farm sustainability standards, enabling access to global markets. Furthermore, PPPs can play a vital role in piloting GI adoption and reducing the initial risks for farmers. Investment in digital infrastructure (e.g., broadband and cloud data hubs) is equally important for ensuring equitable access to GI platforms. Data privacy policies must be instituted to protect farmers' data rights and ensure trust in GI platforms. Such forward-thinking policy actions will not only ensure broader GI adoption but also align with the Sustainable Development Goals (SDGs), particularly those focused on zero hunger, responsible consumption and production, and climate action [24].

16.9.4 Conclusion

GI has emerged as a transformative tool for advancing the economics of sustainable agriculture. Through precision input management, enhanced crop monitoring, and data-informed decision-making, GI enables farmers to reduce operational costs, increase yield efficiency, and access higher value markets. Smallholder and large-scale farms have benefited from GI technologies, such as satellite-based irrigation planning, IoT-driven soil monitoring, and AI-powered pest detection. However, economic barriers, most notably high initial investments, skill gaps, and data infrastructure requirements, continue to hinder widespread adoption, particularly in low- and middle-income regions. These challenges must be systematically addressed to unlock the full potential of GI for inclusive climate-resilient agricultural development. Recognizing the long-term economic potential of GI in agriculture necessitates a collective commitment from stakeholders across the government, academia, private sector, and civil society. Researchers are encouraged to explore GI-enabled innovations tailored to smallholder contexts and develop scalable cost-benefit frameworks for adoption. Governments and international organizations must expand their roles by subsidizing GI infrastructure, supporting digital literacy programs, and incentivizing sustainable practices through favorable policies. Additionally, collaborations with agri-tech startups and knowledge institutions should be strengthened to drive research translation and capacity-building.

AI disclosure statement: The author acknowledges the use of AI-based tools for minor language editing or grammatical correction. However, all intellectual contributions, analyses, and conclusions presented in this chapter are entirely the author's own.

References

1. Thenkabail, P.S., Lyon, J.G., Huete, A., Hyperspectral remote sensing of vegetation: Knowledge gain and knowledge gap after 40 years of research. *IEEE J. Sel. Topics Appl. Earth Observ. Remote Sens.*, 11, 12, 4562–4582, Dec. 2018.
2. Liaghat, S. and Balasundram, S.K., A review: The role of remote sensing in precision agriculture. *Am. J. Agric. Biol. Sci.*, 5, 1, 50–55, 2010.
3. Zhang, C. and Kovacs, J.M., The application of small unmanned aerial systems for precision agriculture: A review. *Precis. Agric.*, 18, 4, 693–712, Aug. 2017.

4. Mulla, D.J., Twenty five years of remote sensing in precision agriculture: Key advances and remaining knowledge gaps. *Biosyst. Eng.*, 114, 4, 358–371, Apr. 2013.
5. Atzberger, C., Advances in remote sensing of agriculture: Context description, existing operational monitoring systems and major information needs. *Remote Sens.*, 5, 2, 949–981, Feb. 2013.
6. Jones, H.G. and Vaughan, A.L., Remote sensing of vegetation: Principles, techniques, and applications. *Remote Sens. Environ.*, 193, 1–12, May 2017.
7. Shanahan, J.F., Cassman, K.G., Roberts, D.C., Precision agriculture: Opportunities and challenges for the future. *Agron. J.*, 107, 6, 2351–2358, Nov. 2015.
8. Foley, J.A., *et al.*, Solutions for a cultivated planet. *Nature*, 478, 7369, 337–342, Oct. 2011.
9. Kamilaris, A. and Prenafeta-Boldu′, F.X., Deep learning in agriculture: A survey. *Comput. Electron. Agric.*, 147, 70–90, Apr. 2018.
10. Liakos, K.G., *et al.*, Machine learning in agriculture: A review. *Sensors*, 18, 8, 2674, Aug. 2018.
11. Pierpaoli, E., Carli, G., Pignatti, E., Canavari, M., Drivers of precision agriculture technologies adoption: A literature review. *Procedia Technol.*, 8, 61–69, 2013.
12. Schimmelpfennig, D., Farm profits and adoption of precision agriculture. *USDA Econ. Res. Serv.*, ERR-217, Oct. 2016.
13. Bongiovanni, R. and Lowenberg-DeBoer, J., Precision agriculture and sustainability. *Precis. Agric.*, 5, 4, 359–387, Aug. 2004.
14. King, A., Technology: The future of agriculture. *Nature*, 544, 7651, S21–S23, Apr. 2016.
15. Lowenberg-DeBoer, J. and Erickson, B., Setting the record straight on precision agriculture adoption. *Agron. J.*, 108, 4, 1552–1557, Jul. 2016.
16. Wolfert, S., Ge, L., Verdouw, C., Bogaardt, M.-J., Big data in smart farming – A review. *Agric. Syst.*, 153, 69–80, May 2017.
17. Aubert, B.A., Schroeder, A., Grimaudo, J., IT as enabler of sustainable farming: An empirical analysis of farmers' adoption decision of precision agriculture technology. *Decis. Support Syst.*, 54, 1, 510–520, Dec. 2012.
18. Kutter, T., Tiemann, E., Siebert, H., Fountas, S., The role of communication and co-operation in the adoption of precision farming. *Precis. Agric.*, 12, 1, 2–17, Feb. 2011.
19. McBratney, A., Whelan, B., Ancev, T., Bouma, J., Future directions of precision agriculture. *Precis. Agric.*, 6, 1, 7–23, Feb. 2005.
20. Raj, R., Jeong, S.J., Lee, B.-W., Use of satellite imagery in precision agriculture: A case study in India. *Remote Sens. Appl.: Soc. Environ.*, 19, 100343, Aug. 2020.
21. Low, S.A. and Isserman, A.M., Where are the innovative entrepreneurs? A look at the economic impact of precision agriculture. *Am. J. Agric. Econ.*, 97, 2, 725–738, Mar. 2015.

22. Basso, B., Cammarano, D., Carfagna, A., Review of crop yield forecasting methods and early warning systems. *Agric. Syst.*, 148, 1–12, Oct. 2016.
23. Stafford, J.V., Implementing precision agriculture in the 21st century. *J. Agric. Eng. Res.*, 76, 3, 267–275, Mar. 2007.
24. Eastwood, C., Dela Rue, B., Gray, D., Precision agriculture and the digital divide: The case of smallholder farmers. *Comput. Electron. Agric.*, 165, 104928, Oct. 2019.
25. Tilman, D., *et al.*, Global food demand and the sustainable intensification of agriculture. *Proc. Natl. Acad. Sci. USA*, 108, 50, 20260–20264, Dec. 2011.
26. Huang, W., Yang, Y., Wang, K., Trends in global fertilizer prices and their impact on agricultural productivity. *J. Agric. Econ.*, 68, 3, 693–710, Sep. 2017.
27. Elliott, J., *et al.*, Constraints and potentials of future irrigation water availability on agricultural production under climate change. *Proc. Natl. Acad. Sci. USA*, 111, 9, 3239–3244, Mar. 2014.
28. Hazell, P. and Wood, S., Drivers of change in global agriculture. *Philos. Trans. R. Soc. B*, 365, 1554, 495–507, Feb. 2010.
29. Crowder, D.W. and Reganold, J.P., Financial competitiveness of organic agriculture on a global scale. *Proc. Natl. Acad. Sci. USA*, 112, 24, 7611–7616, Jun. 2015.
30. Lobell, D.B., Schlenker, W., Costa-Roberts, J., Climate trends and global crop production since 1980. *Science*, 333, 6042, 616–620, Jul. 2011.
31. Blackmore, S., The role of precision farming in sustainable agriculture. *Agric. Eng. Int.: CIGR J.*, 5, 1–10, 2003.
32. Gebbers, R. and Adamchuk, V.I., Precision agriculture and food security. *Science*, 327, 5967, 828–831, Feb. 2010.
33. Calera, A., Campos, I., Osann, A., D'Urso, G., Remote sensing for crop water management: From spectral measurements to irrigation scheduling. *Remote Sens.*, 9, 5, 408, Apr. 2017.
34. Smith, P., *et al.*, Precision irrigation: A pathway to sustainable water use in agriculture. *Agric. Water Manage.*, 199, 1–10, Feb. 2018.
35. Tzounis, A., Katsoulas, N., Bartzanas, T., Kittas, C., Internet of Things in agriculture: Recent advances and future challenges. *Biosyst. Eng.*, 164, 31–48, Dec. 2017.
36. Whelan, B. and Taylor, J., *Precision Agriculture for Grain Production Systems*, CSIRO Publishing, 2015.

17

Future Directions and Challenges for Geospatial Technologies: Shaping India's Policy Landscape for Sustainable Renewable Energy

Deepali Singh[1]*, Harshita Sharma[2], R. B. Singh[3], Seema Ahmad[4] and Ronak Duggar[5]

[1]Research & Development Cell, Vivekananda Global University, Jaipur, Rajasthan, India
[2]Department of Development Studies, Vivekananda Global University, Jaipur, Rajasthan, India
[3]Department of English, NIMS University, Jaipur, Rajasthan, India
[4]Department of Fine Arts, Aligarh Muslim University, Aligarh, Uttar Pradesh, India
[5]School of Computer Science & Engineering, Geeta University, Panipat, Haryana, India

Abstract

India, the world's third-largest energy consumer, is undergoing a significant energy transformation by embracing renewable energy to meet growing demands while addressing environmental sustainability. This chapter explores India's journey toward a cleaner energy future, with a policy target of achieving 500 GW of renewable energy capacity by 2030. It outlines the strategic role of geospatial technologies, international collaborations, and national policy initiatives, such as the National Action Plan on Climate Change (NAPCC), PM-KUSUM, and the National Green Hydrogen Mission, in facilitating this transition. This chapter reviews the progress of solar, wind, hydro, and bioenergy, highlighting their applications and barriers, such as financial limitations, land acquisition issues, grid infrastructure gaps, and policy fragmentation. A particular focus is on the solar and small hydropower sectors, with an emphasis on their developmental patterns and challenges thereof. The influence of international partnerships, notably the US-India PACE program,

**Corresponding author*: deepalisingh1616@gmail.com

Gurwinder Singh, Vishakha Sood, Narayan Vyas, Surendra Yadav and Dankan Gowda V. (eds.) Geospatial Intelligence for Precision Agriculture, (339–352) © 2026 Scrivener Publishing LLC

is also discussed for their contribution to research, technology transfer, and capacity building. Additionally, the chapter evaluates the socio-economic benefits of renewable energy, including job creation and rural development. In conclusion, it recommends policy alignment, technology enhancement, and strengthened private-sector engagement to ensure India's continued leadership in the global renewable energy landscape.

***Keywords*:** Fossil fuel, renewable energy, greenhouse gas, global warming, geospatial

17.1 Introduction

India is the world's third-largest consumer of energy and has one of the fastest growing economies. The nation's ambitious goal of reaching 500 GW of renewable energy capacity by 2030 demonstrates its dedication to sustainability [1]. Economic reforms have drastically changed India's economic growth. On one hand, economic growth in India leads to high Gross Domestic Production (GDP), and, on the other hand, economic growth is very challenging. One of the primary challenges is the need for additional energy sources. Urbanization, industrialization, land use, population growth, demographic transformation, technology adoption, and the expanding transport sector all require more energy, and this demand for energy consumption is expected to increase in the near future. India faces the challenge of adopting alternative energy sources to ensure a reliable and adequate energy supply for economic growth and universal access to energy. Greenhouse gas (GHG) emissions from increasing fossil fuel use pose a significant environmental challenge that threatens India's sustainable economic development.

India adopted the National Action Plan on Climate Change (NAPCC), with a mission plan that focuses on power generation, renewable energy, and energy efficiency [2]. It tries to reduce GHG emissions through decreasing the intensity of industrial production using energy and encouraging the use of renewable energy sources such as solar power, hydropower, and wind power. Geospatial tools assist in mapping renewable energy resources, assessing their availability (e.g., water for hydropower or biomass), and suggesting sustainable management methods for them. Geospatial technology enhances renewable energy development by enabling the precise mapping of resources such as solar, wind, and hydropower potential [3]. International partnerships, such as the US-India PACE program, support India's renewable energy development through technology transfer and funding. Thus, international economic cooperation has made the world economy more interdependent. In addition, it increased the level of International Trade, Production increased, and International Capital Markets became larger and more active [4]. A significant portion of villages lacks access to the grid, hindering the impact of solar power. Even electrified villages may not have sufficient grid capacity to handle additional solar power generation. The high cost of solar panels compared to traditional sources, such as oil-fired generation, further hinders their widespread adoption. Although the government invests in research and development (R&D) for more efficient solar technologies, the initial cost remains a barrier. Land scarcity in India poses a challenge for large-scale solar installations. There is also a need to streamline the financing structure for solar projects. The current system, which involves multiple agencies such as the IREDA, MNRE, and SNA, can be complex and bureaucratic. Hydropower also faces uncertainties, as fluctuations in power supply and potential failures in generation, transmission, and distribution can disrupt the power system. The key applications of renewable energy are shown in Figure 17.1.

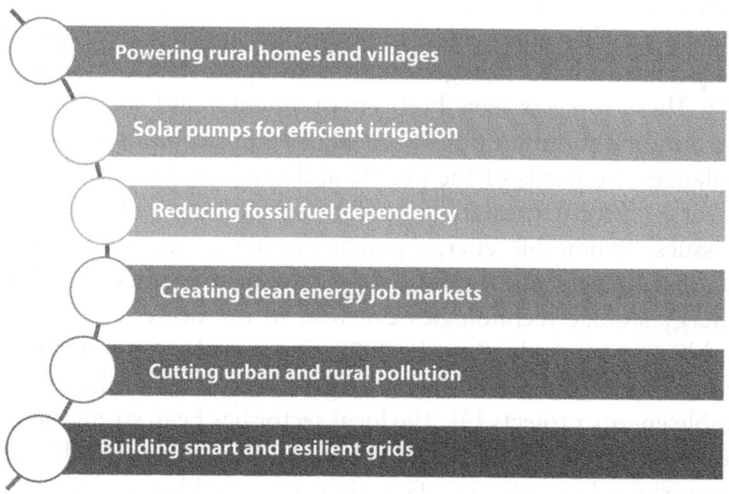

Figure 17.1 Applications of renewable energy.

17.2 The Policy Framework for Renewable Energy in India

India's renewable energy policy development has experienced substantial updates over the last decade. The government has been encouraging the introduction of solar power in the Indian energy mix through policies like the NAPCC and the National Solar Mission [5]. State-level policies since 2015, along with amendments to the Electricity Act (2003), have supported renewable energy deployment. The Ministry of New and Renewable Energy indicates that rural areas and communities will benefit from solar energy through initiatives such as the Pradhan Mantri Kusum Yojana and Atal Jyoti Yojana.

17.3 Obstacles in the Implementation of Policies

A detailed set of policies exists; however, several obstacles make it challenging to implement these laws. The implementation of renewable energy projects faces challenges such as financial constraints and land acquisition issues [6]. Expansion faces three main challenges: limited financial support, insufficient grid infrastructure, and difficulty in acquiring land. Integrating variable renewables, such as solar and wind energy into the grid requires advanced storage solutions and grid stabilization technologies [7]. These projects have both environmental and social impacts on the surrounding communities. We must develop a comprehensive strategy that balances the needs of the people and the planet within the existing frameworks. Effective results require comprehensive policies to address these issues. Renewable energy policies create numerous opportunities for the development and innovation of new technologies [8]. New solar and energy storage technologies can lead to the decrease of the cost of renewable energy production. In 2021, the growth of the public-private collaboration and the rise in the use of green bonds aided the financing of renewable energy projects [9]. The local sector has been strengthened, and its need for imports has decreased because of the Made in India initiative and support for domestic manufacturing of renewable energy components [10]. Additionally, the advancement of digital technology and smart grids provides prospective options for improving energy management and distribution [11, 12].

Although not without flaws, India's renewable energy regulations have significantly impacted the country's efforts to achieve a sustainable energy future. India has increased its renewable energy capacity [13]. However, innovation and ongoing government support are necessary to achieve long-term goals [14]. According to an upcoming study, a comprehensive approach to energy sustainability will need to integrate renewable energy with other sustainable practices, including demand response and energy efficiency. India's success in transitioning to renewable energy will depend on the creation of stronger policies that capitalize on new opportunities and address the existing challenges [15]. According to the literature, India has developed solid policies to boost renewable power; however, implementing these policies remains challenging. The actual practice of creating new energy policies presents several challenges. India pursues both renewable energy programs and future energy objectives. The path to a sustainable energy future must overcome technical, financial, and environmental challenges, while driving innovation. Their success depends on technical progress and new economic arrangements.

17.4 Impact of Policy and Future Directions

The evolution of India's energy solutions towards sustainability depends largely on its recent renewable energy regulations. The government has initiated significant energy programs in the past decade. Since 2010, initiatives such as the National Solar Mission and PM Kusum Yojana have driven renewable energy growth. Initial efforts like the National Solar Mission, the Renewable Energy Development Fund provided the backbone to the development of renewable energy even as later programs like PM-KUSUM have led to increased growth of renewable energy with India having a renewable capacity of 175 GW by 2023. The government offers development frameworks for solar, wind, hydro, and biomass power. The research indicates that successful policies require ongoing development and government support to achieve energy goals. Ongoing government investment in energy innovations will help achieve our distant objectives.

The latest study indicates that India's renewable energy sector must address financial issues, power system controls, and enhanced energy networks. Renewable power and other sustainable energy systems like energy efficiency and demand side management can be adopted to improve energy sustainability. India requires more effective policies to capitalize on emerging business opportunities and address enduring challenges in technology, finance, and the environment [16]. Renewable energy project funding improved when public-private partnerships entered the Green Bond market. New financial mechanisms, such as public-private partnerships and green bonds, create opportunities for private investors and reduce the financial burden on governments for renewable-energy projects. Under the "Make in India" program, India drives the domestic production of wind turbines and solar panels, helping its economy and reducing imports. These developments, along with the deployment of smart grids and Internet of Things-based energy management systems, are essential for guaranteeing India's energy security and sustainability.

17.5 Solar Energy Vision

Solar energy is a carbon-free renewable source. Solar energy is a promising renewable energy source though its extensive use is dependent on the efficiency of the technology and storage; however, practical limitations, such as efficiency and storage, require technological advancements. This highlights the significance of solar technology in harnessing renewable energy from the sun. The solar energy systems in India such as photovoltaic (PV) and concentrated solar power (C) rely on the connection to the grid to distribute electricity on large scale. Solar PV technologies consist of semiconductor or molecular devices called photovoltaic or solar cells, which convert sunlight into direct electricity. On the other hand, CSP is the technology that converts solar energy into heat energy and then into electrical energy. This system can store energy and provide electricity for several hours. Small-scale grid-connected and off-grid electricity generation are crucial to India's electricity supply. India possesses significant solar energy potential, receiving approximately 5,000 TWh of solar energy annually, which is ideal for large-scale harnessing [17].

17.6 Hydropower Position

However, the small hydropower (SHP) potential is 21,133 MW, and about 5,000 MW has been installed so far. Over 7,133 potential sites have been identified. This sector has been opened for private participation and investment [18–20]. The Ministry of Renewable Energy has set a target of 2% to come from minor hydro projects and a target of 3 GW for the 12th Plan [21]. The SHP program invited both the private and public sectors to set up hydropower projects. Previous estimates showed that there were approximately 228 private SHP projects with an existing capacity of 1,230 MW. India's hydropower remains low by global standards despite its large size. However, this sector faces numerous environmental and political challenges.

In addition, loss of biodiversity, rehabilitation, and sharing of water have become obstacles to investing in the hydropower sector. Fossil fuel consumption will persist until renewable energy sources, supported by robust policies and infrastructure, reliably meet energy demands.

17.7 National Renewable Energy Policy Measures

Over the last five years, there has been a remarkable shift in renewable energy in the Indian energy landscape. To achieve this target, the Government of India has taken major initiatives to promote ecologically sustainable growth. India-US cooperation, including the PACE program, has advanced renewable energy technologies, such as solar and wind. India's advanced policies, such as the National Solar Mission and PM-KUSUM, drive renewable energy growth, and nuclear energy complements broader energy security goals. In addition, both nations are deliberating nuclear energy research and development (R&D) cooperation in fields like advanced materials characterization and validation in light water reactor (LWR) applications, characterization of manufacturing defects in reactor components, constitutive modeling of austenitic and nickel-based alloys, and high-temperature reactor technologies [22].

17.8 The Influence of Partnership to Advance Clean Energy

The primary target of the US–India PACE collaboration is to accelerate the adoption of renewable power systems. The US–India PACE operates unique India-based research (PACE-R) and application (PACE-D) programs. PACE operates through two key programs, PACE-R (research) and PACE-D (deployment), supported by broader frameworks such as the US–India Energy Cooperation (2009).

17.9 Policy Interconnections and Energy Security

Through the 2005 civil nuclear accord, India enhanced its energy security by teaming up with the United States. US–India collaborations, such as the PACE program, support the growth of renewable energy sectors, such as solar and wind, through technology transfer and research. Advancements in solar power and India's aim to surpass its 2030 renewable targets motivate their success, according to official ministry reports. Although the development of these issues is great, they still exist.

Infrastructure Restrictions: The current power transmission system lacks proper capacity, and energy storage technology needs to advance further. Our electric grid must work reliably with energy storage to ensure the effective operation of renewable power systems, including wind and solar power sources.
Financial and Regulatory Challenges: Research into case studies and surveys shows that strict rules and limited money present the largest challenges to entrepreneurs. New research aligns with other findings to prove that better financial systems and faster business rules make a critical difference [23].
Inconsistencies in Policy Implementation: Technical differences between state policies show the need for a national system for policy enforcement. When policies are not standardized properly, the nation will struggle to reach renewable energy targets while using resources poorly. Scholars see this issue as proof of India's fragmented method of controlling its policies [24].

Currently, there are many ways to improve India's renewable energy initiatives:

> *Developments in Technology:* Current technological solutions in energy storage, smart grid management, and sustainable power production can solve renewable energy challenges today. New technologies help renewable energy systems to operate better and more effectively [25].
> *Financial Support and Policy Alignment:* Defining better incentives for renewable programs and collaboration between federal and state departments will help projects overcome practical challenges. Meeting our protection goals requires better financial policy systems that work together across all states.
> *Collaborative Ventures:* When governments work with private companies, they can promote investments that create new renewable energy solutions. Through partnerships, the public and private sectors share knowledge that helps solve problems and move progress forward [26].

India has joined research and technology efforts with the Indian Institute of Science (IISc) and the National Renewable Energy Laboratory (NREL) in the US for better renewable power development.

Engagement of the Private Sector: Many renewable energy cost reductions come from private companies, such as SunEdison. Collaborating with the government has brought significant benefits to India's renewable energy sector.

Alternative Energy Sources: The deployment of renewable power sources outside traditional infrastructure brings significant benefits to India's rural populations. Wind power and solar technology help improve everyday lives and generate electricity in distant locations [27].

Global Leadership in Clean Energy: Through successful US cooperation, India is becoming a global powerhouse in the field of renewable energy. India's leadership position in creating environmentally friendly power generation builds on its partnership with the United States.

17.10 India's Renewable Energy

India has made remarkable progress in its renewable energy sector, achieving a major milestone by crossing 200 GW of installed renewable energy capacity by October 2024. According to PIB data, renewable energy accounted for about 46% of India's installed electricity capacity as of October 2024, reflecting a strong shift toward sustainable energy. Solar energy leads the renewable mix with over 92 GW, followed by wind power at 47.72 GW, large hydro at 46.93 GW, and other sources such as small hydro and bioenergy, which contribute significantly. This growth is further supported by India's policy initiatives, such as the National Green Hydrogen Mission, PM-KUSUM, and production-linked incentives (PLI) for solar manufacturing. These efforts align with India's commitment under the Paris Agreement to derive 50% of its electricity capacity from non-fossil fuel sources by 2030 and to achieve net-zero emissions by 2070. Leading states, such as Rajasthan, Gujarat, and Tamil Nadu, are spearheading this transformation. Overall, India's renewable energy journey reflects a multi-dimensional approach toward enhancing energy security and supporting economic growth [28].

17.11 Conclusion

India's progress towards renewable energy depends on effective laws and ambitious targets. Geospatial technologies can support renewable energy development through improved resource mapping, site selection, and planning. Current policies serve as a stable foundation; however, we need to address further challenges and incorporate modern geospatial methods. Researchers should prioritize accurate geospatial data and cross-sector partnerships to enhance access to renewable energy. By adopting these approaches, India can achieve more reliable power generation while conserving energy. Geospatial technology can help India build a sustainable energy path when it works in conjunction with public and private partners. Government policies and renewable energy targets have contributed to the expansion of India's renewable energy capacity in recent years. Despite recent achievements in this sector, our path to renewable energy use faces substantial challenges. To seamlessly connect renewable energy sources to the power grid, the nation's energy systems require updates. Investments in energy storage and grid transmission systems are crucial for ensuring a reliable and renewable energy supply. One way to accelerate technology implementation is to develop consistent nationwide energy policies and improve regulations. Funding problems are a significant issue that requires attention. Private investors and renewable energy businesses will attract more funding when the government provides special financial benefits and develops new ways of financing projects.

AI disclosure statement: The author confirms that no artificial intelligence (AI)-generated content or AI-assisted writing tools were used in the preparation of this chapter (including summaries and the abstract). All text, analysis, and conclusions are the result of the author's original work.

References

1. Dwivedi, S., Singh, A.K., Gupta, A.K., Chand, S., Introduction to Sustainable Development and Geospatial Technologies, in: *Sustainable Development and Geospatial Technology*, pp. 1–16, Springer Nature, Switzerland, 2024, doi:10.1007/978-3-031-65683-5_1.
2. Chandel, S.S., Shrivastva, R., Sharma, V., Ramasamy, P., Overview of the initiatives in renewable energy sector under the national action plan on climate

change in India, in: *Renewable Sustainable Energy Rev.*, vol. 54, pp. 866–873, Elsevier BV, Feb. 2016, doi:10.1016/j.rser.2015.10.057.

3. Nagavi, J.C., Shukla, B.K., Bhati, A., Rai, A., Verma, S., Harnessing Geospatial Technology for Sustainable Development: A Multifaceted Analysis of Current Practices and Future Prospects, in: *Sustainable Development and Geospatial Technology*, pp. 147–170, Springer Nature, Switzerland, 2024, doi:10.1007/978-3-031-65683-5_8.
4. Singh, D., The Role of India towards Economic-Growth and Development of BRICS Member Countries. *Asian J. Educ. Soc. Stud.*, 40, 3, Science domain International, 1–9, Mar. 01, 2023, doi:10.9734/ajess/2023/v40i3873.
5. MoEFCC, *National Action Plan on Climate Change*, Ministry of Environment, Forest and Climate Change, Government of India, 2015, Retrieved from https://www.moef.gov.in/wp-content/uploads/2018/09/NAPCC.pdf.
6. Hassan, Q., Algburi, S., Sameen, A.Z., Salman, H.M., Jaszczur, M., A review of hybrid renewable energy systems: Solar and wind-powered solutions: Challenges, opportunities, and policy implications. *Results Eng.*, 20, 101621, Dec. 2023, doi:10.1016/j.rineng.2023.101621.
7. Thahir, R., Altway, A., Juliastuti, S.R., Susianto, Production of liquid fuel from plastic waste using integrated pyrolysis method with refinery distillation bubble cap plate column. *Energy Rep.*, 5, Elsevier BV, 70–77, Nov. 2019, doi:10.1016/j.egyr.2018.11.004.
8. Heredia-Fonseca, R., Kumar, S., Ghosh, S., Thakur, J., Bhattacharya, A., Modeling a 100% renewable energy pathway in developing Countries: A case study of State of Goa, India. *Energy Convers. Manage.*, 315, Elsevier BV, 118800, Sep. 2024, doi:10.1016/j.enconman.2024.118800.
9. Zhang, Y., Role of green finance, green bonds, public private partnership, and technology innovation in carbon neutrality and sustainable development. *Heliyon*, 10, 18, e37189, Sep. 2024, doi:10.1016/j.heliyon.2024.e37189.
10. Elavarasan, R.M., *et al.*, A Comprehensive Review on Renewable Energy Development, Challenges, and Policies of Leading Indian States With an International Perspective. *IEEE Access*, 8, Institute of Electrical and Electronics Engineers (IEEE), 74432–74457, 2020, doi:10.1109/access.2020.2988011.
11. Dawn, S., Tiwari, P.K., Goswami, A.K., Mishra, M.K., Recent developments of solar energy in India: Perspectives, strategies and future goals. *Renewable Sustainable Energy Rev.*, 62, Elsevier BV, 215–235, Sep. 2016, doi:10.1016/j.rser.2016.04.040.
12. Adhikary, P. and Kundu, S., Renovation modernization uprating & life extension: optimal solution for small hydropower development. *Int. J. Eng. Sci. Adv. Technol.*, 4, 3, 300–306, 2014.
13. Mishra, B.K., *et al.*, A Review of India's Renewable Energy Transition: Policies, Achievements, and Integration Challenges, in: *2024 10th International Conference on Electrical Energy Systems (ICEES)*, IEEE, pp. 1–7, Aug. 22, 2024, doi:10.1109/icees61253.2024.10776919.

14. Inampudi, S. and Gaurav, A.K., Transforming Institutions for Inclusive and Sustainable Governance: An Empirical Analysis of NITI Aayog's Role in India, in: *Governance and Sustainable Development in South Asia*, pp. 177–195, Springer Nature Switzerland, 2024, doi:10.1007/978-3-031-67616-1_10.
15. Kumar, A., Singh, E., Mishra, R., Lo, S.L., Kumar, S., Global trends in municipal solid waste treatment technologies through the lens of sustainable energy development opportunity. *Energy*, 275, Elsevier BV, 127471, Jul. 2023, doi:10.1016/j.energy.2023.127471.
16. Kumar, J.C.R. and Majid, M., Advances and development of wind–solar hybrid renewable energy technologies for energy transition and sustainable future in India. *Energy Environ.*, 35, 5, SAGE Publications, 2517–2565, Jan. 29, 2023, doi:10.1177/0958305x231152481.
17. Mousavi, R., *et al.*, Revolutionizing solar energy resources: The central role of generative AI in elevating system sustainability and efficiency. *Appl. Energy*, 382, 125296, Mar. 2025, doi:10.1016/j.apenergy.2025.125296.
18. Ul-Haq, A., Jalal, M., Sindi, H.F., Ahmad, S., Energy Scenario in South Asia: Analytical Assessment and Policy Implications. *IEEE Access*, 8, Institute of Electrical and Electronics Engineers (IEEE), 156190–156207, 2020, doi:10.1109/access.2020.3019648.
19. Government of India, *Renewable Energy in India: Growth and Targets*, Ministry of New and Renewable Energy, New Delhi, 13 May 2015.
20. Hunt, J.D., Stilpen, D., de Freitas, M.A.V., A review of the causes, impacts and solutions for electricity supply crises in Brazil. *Renewable Sustainable Energy Rev.*, 88, Elsevier BV, 208–222, May 2018, doi:10.1016/j.rser.2018.02.030.
21. Doepfert, M. and Castro, R., Techno-economic optimization of a 100% renewable energy system in 2050 for countries with high shares of hydropower: The case of Portugal. *Renewable Energy*, 165, Elsevier BV, 491–503, Mar. 2021, doi:10.1016/j.renene.2020.11.061.
22. Bhattacharyya, S., Ed., *Rural Electrification Through Decentralised Off-grid Systems in Developing Countries*, Springer, London, 2013, doi:10.1007/978-1-4471-4673-5.
23. Olabi, A.G., *et al.*, Renewable energy systems: Comparisons, challenges and barriers, sustainability indicators, and the contribution to UN sustainable development goals. *Int. J. Thermofluids*, 20, 100498, Nov. 2023, doi:10.1016/j.ijft.2023.100498.
24. Koontz, T.M. and Thomas, C.W., Use of science in collaborative environmental management: Evidence from local watershed partnerships in the Puget Sound. *Environ. Sci. Policy*, 88, Elsevier BV, 17–23, Oct. 2018, doi:10.1016/j.envsci.2018.06.007.
25. Weitzel, M., Saveyn, B., Vandyck, T., Including bottom-up emission abatement technologies in a large-scale global economic model for policy assessments. *Energy Econ.*, 83, Elsevier BV, 254–263, Sep. 2019, doi:10.1016/j.eneco.2019.07.004.

26. Niti Aayog, *Renewable Energy Integration: Policy Recommendations and Future Directions*. Niti Aayog, Government of India, 2022, Retrieved from https://www.niti.gov.in/sites/default/files/202201/RE_Integration_Policy_Recommendations.pdf.
27. Bose, D., Saini, D.K., Yadav, M., Shrivastava, S., Parashar, N., Review of Sustainable Grid-Independent Renewable Energy Access in Remote Communities of India. *Integr. Environ. Assess. Manage.*, 17, 2, Oxford University Press (OUP), 364–375, Jan. 01, 2021, doi:10.1002/ieam.4373.
28. Press Information Bureau, Government of India, India's renewable energy capacity hits new milestone, Nov. 13, 2024. [Online]. Available: https://www.pib.gov.in/PressReleasePage.aspx?PRID=2073038.

18

Future Directions and Challenges for Geospatial Intelligence-Based Precision Agriculture

Amit Sharma[1]*, Shubhi Sharma[2], Ruchi Sharma[3] and Balasani Raghupathi[4]

[1]*Department of Computer Science and Engineering, Vivekananda Global University, Jaipur, Rajasthan, India*
[2]*Amity Business School, Amity University, Jaipur, Rajasthan, India*
[3]*School of Commerce and Management, Nirwan University, Jaipur, Rajasthan, India*
[4]*School of Engineering & Technology (Computer Science), Career Point University, Kota, Rajasthan, India*

Abstract

This chapter examines the transformative impact of geospatial intelligence (GEOINT) on advancing precision agriculture by integrating remote sensing, artificial intelligence (AI), machine learning (ML), unmanned aerial vehicles (UAVs), Internet of Things (IoT) sensors, and cloud platforms. GEOINT supports real-time monitoring, data-driven predictions, and informed decision-making, enabling farmers to manage inputs more efficiently, plan crops effectively, and accurately forecast yields for sustainable agriculture. This chapter reviews key technologies, including optical and radar satellite imagery, hyperspectral sensors, and real-time analytics, highlighting their applications in crop health monitoring, yield prediction, and resource optimization. It also addresses critical challenges, such as limited digital infrastructure, lack of data interoperability, model generalization issues, and inadequate inclusion of smallholder farmers and indigenous knowledge. Future directions emphasize the development of region-specific models, multisensor data fusion, edge computing, and participatory approaches to bridge current gaps. GEOINT can advance sustainable agriculture by integrating advanced technologies and promoting collaboration among farmers, researchers,

**Corresponding author*: dr.amittech@gmail.com; ORCID: 0000-0001-9825-3978
Shubhi Sharma: 0009-0005-6856-3189
Ruchi Sharma: 0009-0005-7193-7188

Gurwinder Singh, Vishakha Sood, Narayan Vyas, Surendra Yadav and Dankan Gowda V. (eds.) Geospatial Intelligence for Precision Agriculture, (353–376) © 2026 Scrivener Publishing LLC

and policy makers. This chapter examines emerging technologies and provides policy recommendations to promote their widespread adoption and measurable benefits in the agricultural sector.

***Keywords*:** Geospatial intelligence, precision agriculture, remote sensing, artificial intelligence, data fusion

18.1 Introduction

The global agricultural sector is evolving rapidly to address the increasing demand for food security, environmental responsibility, and sustainable resource use. Precision agriculture (PA), a modern farming management concept, has emerged as a transformative approach that integrates information technologies to optimize agricultural productivity and input efficiency [1]. PA, defined as the science of using advanced technologies to observe, measure, and respond to inter- and intra-field variability in crops, facilitates site-specific management strategies for improved agricultural outcomes. At the center of this shift is geospatial intelligence (GEOINT), which uses location-based data and analytics to guide smarter and more precise decisions in farming. GEOINT integrates remote sensing, GIS, GPS, and AI/ML to provide precise data for agricultural decision making [2]. GEOINT supports critical functions, such as real-time crop monitoring, yield forecasting, land use analysis, and efficient resource management in precision agriculture. GEOINT integrates multi-temporal and multi-sensor satellite imagery with field data to model agricultural systems, providing clear insights into the factors affecting crop performance [3, 4]. Geographic information systems have evolved into powerful platforms for integrating, modeling, and visualizing spatial data. Furthermore, the rise of artificial intelligence (AI) and machine learning (ML) algorithms has significantly enhanced the capability to analyze complex geospatial datasets, enabling predictive analytics and automated decision support [5]. The rise of the Internet of Things (IoT) has strengthened agricultural insights by linking field sensors with real-time data systems (e.g., soil moisture and weather parameters) with geospatial datasets, facilitating near-real-time decision-making [6]. The evolution of several GEOINT-based technologies in agriculture is illustrated in Figure 18.1.

Despite substantial progress, the full potential of GEOINT in agriculture remains underexploited, particularly in smallholder-dominated and data-scarce areas. Infrastructural, economic, and institutional barriers often constrain the adoption of geospatial technologies in agriculture [7].

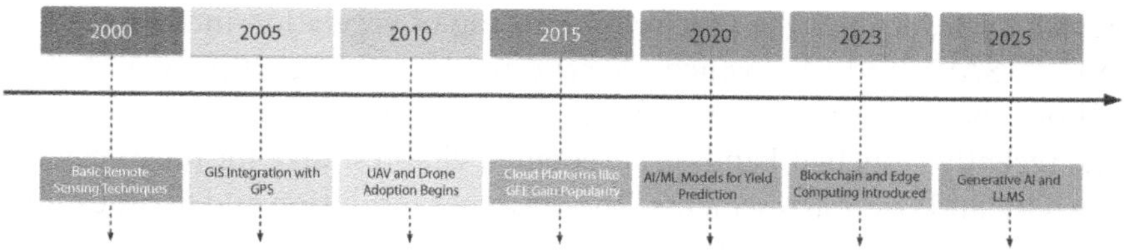

Figure 18.1 Evolution of GEOINT technologies in agriculture.

Moreover, future development must address critical challenges related to data interoperability, scalability, algorithm generalization across diverse geographies, and climate resilience. The convergence of multi-source data fusion, cloud computing, edge analytics, and participatory mapping holds promise for overcoming these limitations, paving the way for the next generation of intelligent, responsive, and sustainable agricultural systems. The objectives of this chapter are as follows: (a) to highlight the role of GEOINT in enhancing precision agriculture practices; (b) to explore the integration of technologies such as remote sensing, AI/ML, GIS, and IoT in smart farming; and (c) to identify future trends and key challenges for scalable, climate-resilient agricultural systems.

18.2 Current State of GEOINT in Precision Agriculture

The integration of GEOINT into precision agriculture has enabled data-driven decision-making at unprecedented scales and resolutions. The integration of geospatial data with agricultural knowledge enables targeted, efficient, and sustainable farming practices [8]. GEOINT in agriculture leverages high-resolution satellite imagery, unmanned aerial vehicles (UAVs), field sensors, and GPS for precise crop monitoring and management. These components operate synergistically to collect, analyze, and disseminate spatially explicit data to inform field-level decisions.

18.2.1 Satellite Imagery: Sentinel, Landsat, and MODIS

Satellite remote sensing plays a fundamental role in geospatial agricultural monitoring. The Sentinel series, particularly Sentinel-1 and Sentinel-2, under the Copernicus Program, offers free high-resolution multispectral and radar data [9]. Sentinel-2, with its 13 spectral bands and 5-day revisit cycle, enables the precise monitoring of vegetation indices, such as NDVI, SAVI, and EVI, which are crucial for assessing crop vigor, disease outbreaks, and water stress. Landsat missions, particularly Landsat-8 and the recently launched Landsat-9, remain a vital resource for historical and current land cover assessments. The 30-meter resolution and decades-long archive make Landsat a preferred choice for long-term crop pattern analysis and soil health assessment [10]. In contrast, the Moderate Resolution Imaging Spectroradiometer (MODIS) aboard NASA's Terra and Aqua satellites provides near-daily observations at moderate spatial resolutions (250 m–1 km) [11]. Its temporal frequency makes it ideal for monitoring crop phenology, drought dynamics, and evapotranspiration trends at a continental scale. Collectively, these satellite systems empower farmers and researchers to observe crop development patterns, detect anomalies, and make informed decisions regarding irrigation, fertilization, and harvesting schedules.

18.2.2 UAV and Drone-Based Monitoring Systems

While satellites provide macro-level views, UAVs and drones offer micro-level, high-resolution insights with great flexibility [12]. Drones equipped with RGB, multispectral, and thermal cameras are widely used for crop scouting, plant stress detection, and yield estimation. UAVs, typically offering spatial resolutions in the centimeter-to-meter range, can detect intra-field variability that is often not captured by satellite data. Drone imagery is particularly beneficial for precision pesticide spraying, soil moisture assessment, and early identification of pest infestations [13]. Moreover, drone missions can be automated and georeferenced using GPS to ensure consistency across data collection events. Recent innovations have integrated AI-based analytics directly into drone systems, thereby enabling real-time anomaly detection during flights.

18.2.3 Field Sensors and GPS Technologies

Ground-based sensors, including soil moisture probes, temperature loggers, pH meters, and nutrient analyzers, complement satellite and UAV data by providing continuous and real-time observations at the soil and canopy levels [14]. These sensors, often connected through IoT networks, feed data to centralized platforms for analysis and predictions. GPS and GNSS technologies enable georeferencing of agricultural activities, including planting, fertilization, and harvesting. Farmers can use water, fertilizers, and pesticides more efficiently by pinpointing exact field locations, thereby reducing costs, minimizing environmental harm, and minimizing manual labor. Modern tractors with precision GPS auto-steering systems can follow pre-programmed paths with sub-meter accuracy, ensuring consistent row spacing, reducing fuel consumption, and avoiding crop damage [15]. When combined with GIS, GPS systems also help generate prescription maps for seed density, irrigation scheduling, and variable-rate fertilization.

18.2.4 Case Studies Highlighting Current Implementations

18.2.4.1 Case Study 1: India—Rice Crop Monitoring with Sentinel-1 and GEE

In eastern India, researchers have used Sentinel-1 SAR imagery in the Google Earth Engine (GEE) to monitor rice crop phenology in flood-prone areas of Bihar. They successfully mapped crop planting dates and biomass accumulation through time-series analysis and backscatter signature modeling. This approach provides early yield estimates, enabling government agencies to plan food procurement and subsidies more efficiently [16].

18.2.4.2 Case Study 2: USA—Corn and Soybean Monitoring with UAVs

In Iowa, USA, farmers have adopted UAV-based scouting to monitor the health of corn and soybeans crops during the growing season. Analyzing vegetation indices from UAV imagery can detect nitrogen deficiencies and water stress up to 10 days earlier than traditional scouting methods. This approach resulted in increased crop yield along with a notable reduction in fertilizer usage [17].

18.2.4.3 Case Study 3: Africa—IoT-Based Sensor Networks for Smallholder Farmers

In Kenya, the "Digital Green" initiative has deployed soil moisture sensors and mobile GPS applications to assist smallholder farmers with precision irrigation. The integration of field sensors with a cloud-based analytics dashboard allowed local cooperatives to identify under-irrigated plots and optimize water distribution. The solution is scalable, cost-effective, and significantly improves water-use efficiency [18].

Table 18.1 systematically summarizes the recent advancements in the application of GEOINT tools in precision agriculture. Covering seven key studies from 2019 to 2025, the table illustrates the transition from basic

Table 18.1 Literature review of existing GEOINT tools in precision agriculture.

Tools used	Application area	Accuracy/ outcome	Data sources	Sources
AI, DL, IoT, Smart Sensors, Blockchain	Precision crop management (Ag5.0)	High accuracy in disease detection, efficient decision-making	UAV multispectral data, IoT sensors, big data	[19]
IoT, WSN, UAVs, NDVI	Crop health monitoring, intelligent automation	Accurate real-time monitoring with NDVI, WSN	Soil/temp data, UAV imagery	[20]
AI, computer vision, satellite imagery	Crop monitoring, pest detection, yield prediction	High prediction accuracy	Satellite, drone, IoT, historical crop data	[21]
Smart sensors, GIS, RS	Irrigation, greenhouse control, disease detection	Early issue detection, optimized input usage	WSN, IoT cloud platforms	[22]
UAVs, CNN, RF, YOLO, IoT	Crop health assessment, yield estimation	High accuracy in vegetation health and yield prediction	Drone + satellite imagery, environmental sensors	[23]
AI, GIS, spatiotemporal models	Climate resilience, pest monitoring	Improved pest outbreak forecasts and climate adaptation	Satellite, aerial imagery, IoT + spatial models	[24]
GIS, RS, GPS	Land use and cadastre mapping	Enhanced mapping and land management	Land use data, cadastral records, satellite data	[25]

sensor-based monitoring to advanced technology-driven precision agriculture systems. The authors have employed diverse technologies, such as IoT sensor networks, UAV-based remote sensing, AI, and advanced spatial models, to address challenges in crop monitoring, resource management, and agricultural sustainability. Studies have consistently shown that combining ML and deep learning (DL) with high-resolution remote sensing data leads to more accurate crop classification, faster analysis, and better yield forecasting. These studies span a range of application areas, including disease detection, irrigation optimization, land use mapping, and climate-resilient farming practices.

18.3 Emerging Technologies and Trends

Emerging technologies, such as AI, cloud computing, and sensor integration, are driving the advancement of GEOINT applications in precision agriculture. These include advanced AI models for geospatial interpretation, scalable cloud platforms for data storage and analysis, multimodal data fusion techniques, and secure data architectures, such as blockchain and edge computing [2]. These technologies enhance the accuracy of geospatial data, enabling automated field-level decisions for farmers. Together, they represent the next frontier in transforming agriculture into a fully digital and intelligent enterprise.

18.3.1 AI and ML for Geospatial Data Interpretation

One of the most transformative trends in GEOINT is the application of AI and ML to analyze large-scale heterogeneous geospatial datasets. Satellite imagery, UAV data, and sensor streams are inherently complex and contain high-dimensional spatiotemporal information that traditional analytical methods often fail to exploit fully. DL models, particularly Convolutional Neural Networks (CNNs), have effectively captured spatial features such as texture, edge patterns, and vegetation indices from high-resolution imagery. These models can be trained to classify crops, detect plant diseases, and accurately map land use. For instance, CNNs have been successfully employed to differentiate between wheat and maize using Sentinel-2 data, with classification accuracies of over 90% [26]. As training datasets grow, these AI systems are expected to achieve near-human interpretability of multisource geospatial data.

18.3.2 Cloud Computing Platforms: GEE, AWS, Microsoft Planetary Computer

With the exponential increase in geospatial data volume and velocity, cloud computing has become a cornerstone of the modern GEOINT infrastructure. Cloud-based platforms provide elastic storage, parallel processing capabilities, and access to open-source satellite datasets, thereby eliminating the need for extensive local computational resources. The GEE remains a widely adopted platform, particularly in academic and research settings. GEE provides access to petabytes of geospatial datasets, including Sentinel, Landsat, and MODIS, as well as powerful JavaScript and Python API for large-scale analysis. Applications range from crop monitoring to mapping deforestation and managing the water resources. Amazon Web Services (AWS) offers geospatial data lakes, such as the AWS Open Data Registry, and provides services such as Amazon SageMaker and Lambda to run ML models on large-scale geospatial data streams. The AWS also hosts satellite data cubes and provides processing pipelines for UAV and hyperspectral imagery [27]. The Microsoft Planetary Computer is an emerging platform that combines planetary-scale geospatial data with Jupiter Notebook-based computing environments. Its strength lies in environmental intelligence applications, with robust libraries for spatial analysis, cloud classification, and climate-resilience modeling. These platforms democratize access to geospatial tools and enable real-time and scalable analytics, which are critical in dynamic and heterogeneous agricultural environments [28].

18.3.3 Multisource and Multitemporal Data Fusion

The fusion of multisource data, especially optical, radar, and hyperspectral imagery, enables a more comprehensive and accurate understanding of agricultural systems. Optical sensors, such as Sentinel-2 and Landsat, capture visual and near-infrared data, which are valuable for analyzing chlorophyll content and canopy structure [29]. Synthetic Aperture Radar (SAR) systems, such as Sentinel-1, offer the advantage of cloud-penetrating capabilities, providing insights into crop structure and soil moisture even under adverse weather conditions. Hyperspectral sensors capture hundreds of narrow spectral bands, enabling the detailed identification of biochemical properties in plants, such as nitrogen levels and stress caused by disease. Together, these complementary data sources enhance decision-making in precision agriculture [30]. Multitemporal fusion, which combines imagery from different time steps, enables accurate phenology tracking, seasonal classification, and yield forecasting. Additionally, the temporal stacking of fused datasets is employed in DL models to train spatiotemporal classifiers for detecting seasonal variability and predicting climate-induced shifts in cropping patterns.

18.3.4 Blockchain and Edge Computing for Secure, Real-Time Intelligence

The future of precision agriculture lies in decentralization and real-time intelligence systems. With its inherent transparency and immutability, blockchain technology offers a secure way to manage and share sensitive agricultural data, such as land ownership, pesticide usage logs, and crop certifications. This ensures data integrity in collaborative farming ecosystems and enables traceability in agri-food supply chains. For example, blockchain-enabled systems can verify whether produce was grown with minimal pesticide use, appropriately stored, and transported under approved temperature conditions, thus boosting consumer confidence and market access for farmers. In contrast, edge computing pushes computation to the data source, whether a drone, tractor, or field sensor. This reduces latency, bandwidth usage, and dependency on unreliable Internet connectivity in rural areas. Edge-AI devices can process UAV images, detect crop anomalies, and send alerts to a farmer's mobile device with minimal delay. Projects such as IBM's PAIRS Geoscope and local startups in India and Africa are exploring hybrid models in which field devices conduct preprocessing, and only critical data are uploaded to the cloud for deeper analysis or blockchain logging [31].

18.4 Challenges in GEOINT-Driven Precision Agriculture

Although GEOINT shows strong promise for advancing agriculture, its widespread adoption still faces several key challenges. These challenges include technical issues with sensors and analytics, limited infrastructure, financial barriers, and ethical concerns. Understanding and addressing these barriers is critical to ensuring that GEOINT-driven precision agriculture evolves into an inclusive, sustainable, and resilient farming paradigm [32]. A mapping of GEOINT challenges to their respective solutions is presented in Figure 18.2.

18.4.1 Technical Challenges

18.4.1.1 Spatial and Spectral Resolution Trade-Offs

One of the most persistent challenges in remote sensing applications for agriculture is the trade-off between the spatial, spectral, and temporal

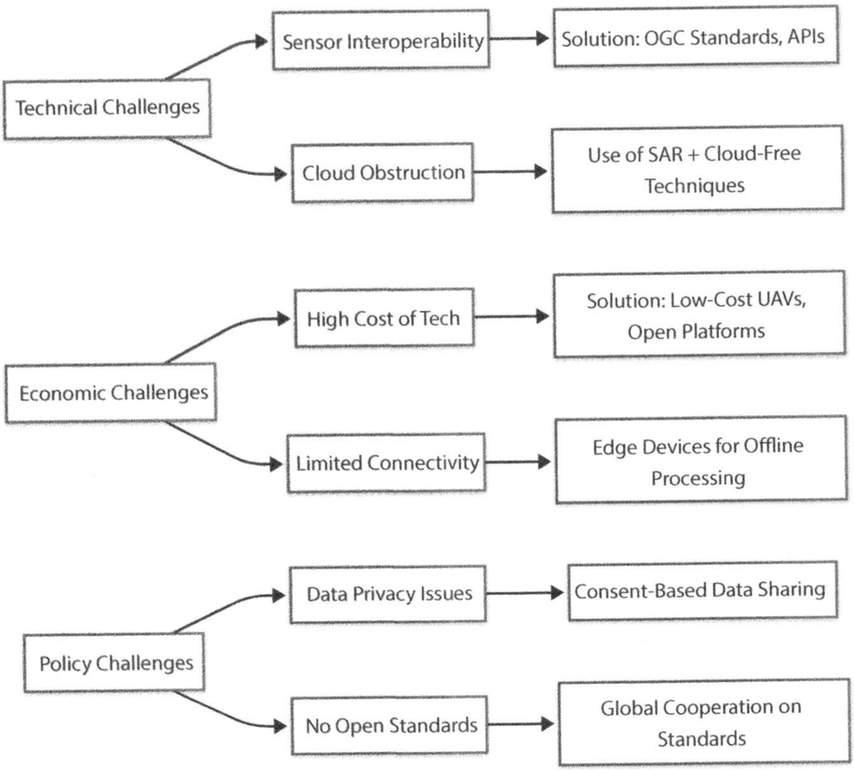

Figure 18.2 Mapping of GEOINT challenges to solutions.

resolutions of satellite data. High spatial resolution is necessary to monitor intra-field variability; however, such datasets often have limited spectral detail or a low revisit frequency. For instance, commercial satellites, such as Worldview or Planet Scope, offer <1-meter resolution but may not capture the rich spectral detail necessary for plant biochemistry or are prohibitively expensive for smallholders. Conversely, moderate-resolution sensors, such as Sentinel-2, provide wide-area coverage and adequate spectral diversity but may fail to capture sub-field variations that are critical for targeted input applications. Striking an optimal balance between these parameters or developing fusion-based approaches that preserve details without increasing processing complexity remains an ongoing technical hurdle [33].

18.4.1.2 Cloud Cover, Atmospheric Noise, and Time-Series Inconsistencies

Optical satellite systems are inherently affected by cloud cover, haze, and atmospheric scattering, which distort reflectance data and create gaps in time series observations. This is especially problematic in tropical regions and during monsoon seasons, when critical crop monitoring phases coincide with persistent cloudiness [34]. SAR systems, such as Sentinel-1, provide all-weather imaging but require complex backscatter analysis, posing challenges for interpretation without specialized training. Additionally, combining radar and optical imagery requires preprocessing steps such as speckle filtering, radiometric calibration, and co-registration, which increase computational requirements. Moreover, inconsistencies in acquisition dates, viewing angles, and radiometric responses among satellite missions often introduce artifacts in long-term monitoring, posing a significant challenge for change detection, yield modeling, and trend analysis [35].

18.4.1.3 Interoperability between Sensors, Platforms, and Algorithms

Modern precision agriculture involves the fusion of data from diverse sources, such as satellites, UAVs, field sensors, and weather stations, each with its own spatial resolution, data format, and metadata structures. The lack of standardization across sensors and software platforms hinders seamless integration, limiting the ability to synthesize multimodal datasets for a holistic analysis [36]. For example, hyperspectral data collected from UAVs may need to be georeferenced and resampled to align with the satellite imagery. Simultaneously, field sensors generate time-stamped point data in proprietary formats that cannot be directly fed into GIS models. Furthermore, differing coordinate systems, projection formats, and spatial resolutions often require complex geoprocessing workflows that demand significant technical expertise. Despite efforts such as Open Geospatial Consortium (OGC) standards and interoperable APIs by Google Earth Engine, the full integration of heterogeneous data streams remains elusive, especially in low-resource settings [37].

18.4.2 Economic and Infrastructural Challenges

18.4.2.1 High Cost of Technology Adoption

The initial investment required to set up GEOINT-enabled precision agriculture systems is a significant deterrent, particularly for smallholder farmers in developing economies. Expenses related to high-resolution satellite imagery, UAV procurement and maintenance, sensor networks, cloud subscriptions, and skilled labor can escalate quickly. Although platforms such as GEE offer free access to satellite archives, the real costs lie in the computational infrastructure, hardware deployment, training, and technical support. Moreover, although private-sector solutions are more robust, they often operate on subscription models that are financially inaccessible to most marginal farmers. This economic gap restricts access to digital agriculture tools, enabling large commercial farms to adopt innovations while excluding many smallholder farmers from these advancements [38].

18.4.2.2 Lack of Digital Literacy and Rural Connectivity

Another formidable barrier is the widespread lack of digital literacy in rural populations. The effective use of GEOINT tools requires familiarity with mobile applications, sensor interfaces, and basic data interpretation skills. Many farmers lack access to training resources or support systems to understand satellite imagery, GPS maps, and crop indices. Additionally, the digital infrastructure in several agricultural regions remains inadequate [39]. Unstable electricity supply, poor Internet connectivity, and limited access to smartphones or computing devices severely constrain the use of cloud-based analytics platforms and mobile advisory services. Without targeted capacity-building programs and last-mile connectivity solutions, the transformative potential of GEOINT cannot be fully realized at the grassroots level of society.

18.4.3 Policy and Ethical Challenges

18.4.3.1 Data Privacy, Ownership, and Consent

As agriculture increasingly relies on geospatial, biometric, and behavioral data, concerns regarding data privacy and ownership are growing increasingly pressing. Farmers often contribute data directly *via* sensors and mobile applications or through UAV imagery without a clear understanding or control over how their data will be used, stored, or monetized. Without stringent data governance frameworks, there is a risk of misuse, unauthorized surveillance, and exploitation by private entities [40]. Precise geolocation data on crop yields could potentially be used by agrochemical companies or insurers, highlighting the need for robust data governance policies. Ensuring informed consent, data anonymization, and transparent data-sharing agreements are crucial for fostering trust and ethical implementation of GEOINT technologies.

18.4.3.2 Lack of Open-Access Standards and Global Cooperation

Although several governments and space agencies have made remote sensing data freely available, a lack of uniform global data collection, preprocessing, and dissemination standards remains. The absence of open-source algorithms, common data schemas, and interoperable analytical models hampers cross-border collaboration and technology transfer. Moreover, developing nations often lack institutional capacity or access to cutting-edge research in geospatial analytics, resulting in a dependency on foreign solutions that may not align with local agroecological contexts. There is a pressing need for international cooperation, multi-stakeholder engagement, and open innovation ecosystems that promote the transparency, localization, and adaptability of GEOINT-based solutions. Policy frameworks, such as the FAO's Digital Agriculture Action Plan or Global Open Data for Agriculture and Nutrition (GODAN), represent commendable steps. However, much more must be done to institutionalize ethical data use and participatory development models.

18.5 Comparative Review of Global Research and Gaps

18.5.1 Developed vs. Developing Country Perspectives

In developed countries such as the United States, Germany, Australia, and Japan, GEOINT in precision agriculture has matured considerably. These countries have invested heavily in national satellite missions, AI-driven geospatial platforms, and advanced sensor networks [1]. Public-private collaborations, strong academic research ecosystems, and government subsidies have enabled the widespread adoption of technologies such as variable rate application (VRA) tools, autonomous tractors equipped with GPS-guided systems, UAVs for in-season monitoring, and AI-powered yield prediction systems that utilize multi-temporal remote sensing. The U.S. Department of Agriculture's (USDA) National Agricultural Statistics Service (NASS) integrates Landsat and MODIS data to monitor cropland patterns nationwide. Similarly, the European Union's Copernicus Programme supports precision farming by providing free access to Sentinel satellite data and thematic services tailored to agricultural needs. In contrast, developing countries, including India, Kenya, Nigeria, and Bangladesh, face foundational limitations that hinder the seamless adoption of GEOINT technology. Although platforms such as GEE have improved access to satellite data, there is a shortage of region-specific tools, customized models, and practical decision-making support for farmers. Many smallholder farmers operate in informal agricultural systems, often with limited financial resources, poor Internet connectivity, and limited exposure to digital technologies. Consequently, GEOINT initiatives in developing countries are usually donor-driven and pilot-scale and fail to scale beyond experimental or research boundaries. Moreover, technology is frequently introduced top-down, with limited participation from the communities it serves.

18.5.2 Gaps in Accessibility, Scalability, and Localized Model Training

One of the key gaps in global GEOINT research is the lack of scalable and context-specific models for diverse agroecological zones. Most ML or DL models trained on geospatial data are developed using datasets from high-income areas. These models are often calibrated to specific soil types, weather conditions, or cropping systems and fail to generalize effectively when applied to different geographic contexts. For example, a crop classification model trained on Sentinel-2 imagery from France may not perform well in sub-Saharan Africa because of spectral signature variability resulting from differences in crop varieties, irrigation practices, or soil reflectance. The lack of transfer learning frameworks and regionally fine-tuned models results in poor performance, which limits farmers' trust and adoption. Additionally, language and user interface barriers exist. Many tools assume digital literacy or provide dashboards in non-local languages, further alienating the smallholders. Another gap lies in the lack of ground-truth datasets for training and validating ML models. High-quality labeled datasets are vital for supervised learning models; however, collecting them requires time, money, and skilled personnel. This data scarcity limits innovation, as researchers in resource-limited regions are unable to develop or test new algorithms because of the unavailability of verified training data. Agricultural GEOINT systems are rarely integrated with other national spatial infrastructures (e.g., forestry, water, and health), lacking the opportunity to build synergistic platforms to support sustainable rural development [41].

18.5.3 Role of Indigenous Knowledge and Community Participation

A crucial yet often overlooked aspect of global GEOINT research is the integration of indigenous knowledge systems and community-based participatory approaches. Indigenous farmers and rural communities possess rich contextual knowledge of soil characteristics, crop varieties, microclimatic patterns, and ecological responses. However, most GEOINT models are built exclusively using satellite-derived or sensor-based data, with minimal input from local actors. For example, traditional crop calendars, water harvesting methods, and soil fertility indicators are rarely digitized or integrated into precision agriculture platforms. The failure to integrate indigenous knowledge with modern tools reduces their relevance and adoption among local farming communities. To address this, participatory mapping, crowdsourced data collection, and tool co-development are being explored as viable solutions. Citizen science platforms and farmer-led GIS training programs have shown promise in bridging this knowledge gap. One successful case is the Digital Green project in India, where local extension workers collect agro-advisory videos in regional dialects guided by satellite forecasts and community feedback [42]. There is also a growing recognition of the need to translate insights from GEOINT platforms into actionable formats, such as voice-based advisories, visual alerts, or simple rule-based recommendations, so that local communities can make timely decisions without understanding complex satellite maps or indices. Additionally, fostering community data ownership is essential. Projects that enable farmers to access, control, and benefit from their data are more likely to succeed than those that are imposed externally. Community co-ownership builds trust and ensures the long-term relevance and utility of geospatial technologies.

Table 18.2 presents an in-depth review of seven contemporary research papers that applied GEOINT tools in sustainable agriculture, environmental monitoring, and thoughtful land use planning. Each study was examined for its core technologies, regional context, application area, and practical outcomes. The table highlights various tools, including GIS, remote sensing, IoT, AI, and Web GIS-based platforms. These technologies have been applied to smallholder GHG mitigation, smart farming, illegal

Table 18.2 Comparative review of GEOINT-based agricultural practices.

Tools/ technologies	Use case	Key outcome/ impact	Region/ context	Sources
GIS, RS, field measurements	GIS, RS, field measurements	Standardized, low-cost methods adaptable to diverse farm systems	Africa, Asia, Latin America	[43]
GIS, RS	Agricultural sustainability decision-making	Strengthened evidence-based policy; exposed regional policy gaps	Sub-Saharan Africa (focus)	[44]
IoT, AI, drones, GIS	Soil health and biodiversity in smart farming	Green Infrastructure improved resilience and ecosystem services	Global (USA, Africa cases)	[45]
ML Models, big data	Agricultural big data systems	Identified scaling issues; emphasized need for robust architectures	Global	[46]
GEOINT, GIS, RS, Benford's Law	Timber laundering and deforestation monitoring	Detected 70% anomalies; supported forest law enforcement	Brazilian Amazon	[47]
GIS, RS, GPS, AI, big data	Smart land use planning	Improved zoning precision *via* AI-integrated geospatial tools	Global	[48]
WebGIS, RS, IoT, AI, big data	Precision agriculture and environment (SPAE)	Proposed SPAE model with Digital Twin for site-specific management	Global	[49]

deforestation detection, and data-driven land-use planning. The reviewed literature spans global case studies, with particular attention to developing regions where agricultural and climate resilience challenges are more severe. The limitations cited in the studies include policy gaps, infrastructure challenges, data interoperability, and scalability issues.

18.6 Future Research Directions

Advancements in GEOINT are expected to make the next phase of precision agriculture more efficient, responsive to local conditions, and accessible to diverse farming communities. As climate change, resource scarcity, and food insecurity intensify, future research must focus on building systems that are not only technologically advanced but also context-aware and people centered. This section highlights five transformative research directions that can shape the future of GEOINT-based precision agriculture [50].

18.6.1 Real-Time Analytics Integrated with AI-Based Prediction

Timeliness is critical in agricultural decisions related to irrigation, pest control, and harvesting, which must often be made within narrow timeframes. The fusion of real-time data analytics with AI-based prediction models represents a key frontier for future research [51]. Current remote sensing systems often suffer from data latency owing to limited satellite revisit cycles, cloud interference, or processing bottlenecks. Addressing this requires exploring edge AI systems that process sensor and UAV data locally, data pipelines for near-real-time ingestion, preprocessing, and inference, and event-based triggers that initiate advisory alerts as soon as certain thresholds (e.g., NDVI drop or pest infestation risk) are reached. AI models trained on historical time-series data can be made predictive by integrating near-real-time sensor streams (soil moisture, temperature, and humidity) to simulate future crop growth or forecast risks. For example, LSTM-based crop yield models, when enhanced with real-time environmental inputs, can support proactive decision-making at the farm level. This direction demands robust research on streamlined ML architectures, temporal data fusion, and scalable cloud-edge hybrid models optimized for low-bandwidth and high-variability environments [52].

18.6.2 Generative AI and Large Language Models for Agricultural Advisories

If emerging Generative AI and LLMs are adapted for agriculture, they could potentially transform extension services by providing tailored advisories based on geospatial data. These models can synthesize complex geospatial datasets, weather forecasts, and agronomic practices into contextualized and conversational advisories tailored to individual farmers [53]. Future research should focus on fine-tuning LLMs using agricultural datasets in local languages, training models to interpret geospatial indices such as NDVI, soil maps, or yield forecasts, generating audio or voice outputs for illiterate farmers, and embedding models within offline-capable mobile applications or rural kiosks.

18.6.3 Multi-Modal Sensor Integration: Thermal, Microwave, Spectral, and Soil Sensors

While current GEOINT systems primarily rely on optical or radar imagery, the future lies in multimodal sensor integration, which combines various data types to generate more holistic and robust agricultural insights.

Key sensor modalities for future integration include the following:

- Thermal sensors: Detect plant and soil temperatures for irrigation management and heat stress detection.
- Microwave sensors (e.g., SAR): Penetrate cloud cover and soil to estimate moisture content even in adverse weather conditions.
- Hyperspectral and multispectral sensors: Identify biochemical markers such as chlorophyll, nitrogen, and plant stress.
- *In situ* soil sensors: Provide ground-truth data on pH, EC, nitrogen, and soil moisture.

The fusion of these modalities with AI-driven models enables multidimensional crop monitoring. For example, combining hyperspectral UAV imagery with soil sensor data can improve precision fertilization models by considering aboveground and belowground dynamics. Research must address data alignment, temporal synchronization, and dimensionality reduction techniques to process heterogeneous sensor inputs into unified predictive models [54].

18.7 Conclusion

GEOINT is revolutionizing the agricultural sector by enabling precise, data-driven, and responsive farming practices. By integrating remote sensing technologies, AI/ML algorithms, UAVs, IoT-based field sensors, and scalable cloud platforms, GEOINT facilitates accurate crop monitoring, predictive yield modeling, and site-specific resource management. These advancements lay the foundation for a more sustainable and efficient agricultural future [2]. However, the transformative potential of GEOINT has not been uniformly realized. Regions with limited infrastructure, digital literacy, or economic support, particularly smallholder-dominated areas, face persistent barriers to adoption and scaling. Although progress in satellite imagery resolution, real-time analytics, and multisensor data fusion has expanded the analytical capabilities of GEOINT, gaps remain in data interoperability, access to high-performance computing, and the availability of localized, context-aware models. Additionally, excluding indigenous knowledge and limited participatory frameworks reduces the relevance and acceptance of these technologies by grassroots communities. A collaborative approach is required to fully unlock the potential of GEOINT in precision agriculture. Policymakers must invest in digital infrastructure and inclusive data policies, researchers should focus on region-specific explainable AI models, and technology providers must prioritize low-cost, interoperable platforms. Equally important is empowering farmers through co-development, vernacular tools, and user ownership of the data. By fostering cross-sectoral partnerships and embracing inclusive innovation, GEOINT can evolve into a transformative tool that drives climate-resilient, equitable, and intelligent agriculture worldwide.

AI disclosure statement: The author acknowledges the use of AI-based tools for minor language editing or grammatical correction. However, all intellectual contributions, analyses, and conclusions presented in this chapter are entirely the author's own.

References

1. S.s., V.C., S., A.H., Albaaji, G.F., Precision farming for sustainability: An agricultural intelligence model. *Comput. Electron. Agric.*, 226, 109386, Nov. 2024, doi:10.1016/j.compag.2024.109386.
2. Dritsas, E. and Trigka, M., Remote Sensing and Geospatial Analysis in the Big Data Era: A Survey. *Remote Sens.*, 17, 3, Jan. 2025, doi:10.3390/rs17030550.

3. Heningtiyas, H. and Supriyadi, A.A., Study of the implementation of geoint and remote sensing in climate change. *Remote Sens. Technol. Def. Environ.*, 1, 2, Aug. 2024.
4. Zhang, Z. and Zhu, L., A Review on Unmanned Aerial Vehicle Remote Sensing: Platforms, Sensors, Data Processing Methods, and Applications. *Drones*, 7, 6, Jun. 2023, doi:10.3390/drones7060398.
5. Goel, A., Goel, A.K., Kumar, A., The role of artificial neural network and machine learning in utilizing spatial information. *Spat. Inf. Res.*, 31, 3, 275–285, Jun. 2023, doi:10.1007/s41324–022-00494-x.
6. Bilotta, G., Genovese, E., Citroni, R., Cotroneo, F., Meduri, G.M., Barrile, V., Integration of an Innovative Atmospheric Forecasting Simulator and Remote Sensing Data into a Geographical Information System in the Frame of Agriculture 4.0 Concept. *Agriengineering*, 5, 3, Sep. 2023, doi:10.3390/agriengineering5030081.
7. Mathenge, M., The Spatial Dimension of Agriculture and Food Security: A GIS-based Spatially Explicit Approach for Integration of Smallholder Agriculture into Agribusiness, PhD-Thesis - Research and graduation internal, Proefschrift Maken, S.n., 2022.
8. Radočaj, D., Plaščak, I., Jurišić, M., Global Navigation Satellite Systems as State-of-the-Art Solutions in Precision *Agriculture*: A Review of Studies Indexed in the Web of Science. *Agriculture*, 13, 7, Jul. 2023, doi:10.3390/agriculture13071417.
9. Song, X.-P., Huang, W., Hansen, M.C., Potapov, P., An evaluation of Landsat, Sentinel-2, Sentinel-1 and MODIS data for crop type mapping. *Sci. Remote Sens.*, 3, 100018, Jun. 2021, doi:10.1016/j.srs.2021.100018.
10. Dhillon, M.S., Dahms, T., Kübert-Flock, C., Steffan-Dewenter, I., Zhang, J., Ullmann, T., Spatiotemporal Fusion Modelling Using STARFM: Examples of Landsat 8 and Sentinel-2 NDVI in Bavaria. *Remote Sens.*, 14, 3, Jan. 2022, doi:10.3390/rs14030677.
11. Dhillon, M.S., *et al.*, Evaluation of MODIS, Landsat 8 and Sentinel-2 Data for Accurate Crop Yield Predictions: A Case Study Using STARFM NDVI in Bavaria, Germany. *Remote Sens.*, 15, 7, Jan. 2023, doi:10.3390/rs15071830.
12. Duffy, J.P., Anderson, K., Fawcett, D., Curtis, R.J., Maclean, I.M.D., Drones provide spatial and volumetric data to deliver new insights into microclimate modelling. *Landsc. Ecol.*, 36, 3, 685–702, Mar. 2021, doi:10.1007/s10980–020-01180–9.
13. Sharma, H., Sidhu, H., Bhowmik, A., Remote Sensing Using Unmanned Aerial Vehicles for Water Stress Detection: A Review Focusing on Specialty Crops. *Drones*, 9, 4, Apr. 2025, doi:10.3390/drones9040241.
14. Zhang, X., Feng, G., Sun, X., Advanced technologies of soil moisture monitoring in precision agriculture: A Review. *J. Agric. Food Res.*, 18, 101473, Dec. 2024, doi:10.1016/j.jafr.2024.101473.
15. Obi Reddy, G.P., Dwivedi, B.S., Ravindra Chary, G., Applications of Geospatial and Big Data Technologies in Smart Farming, in: *Smart*

Agriculture for Developing Nations: Status, Perspectives and Challenges, K., Pakeerathan (Ed.), pp. 15–31, Springer Nature, Singapore, 2023, doi:10.1007/978-981-19-8738-0_2.
16. Singha, C. and Swain, K.C., Rice crop growth monitoring with sentinel 1 SAR data using machine learning models in google earth engine cloud. *Remote Sens. Appl. Soc. Environ.*, 32, 101029, Nov. 2023, doi:10.1016/j.rsase.2023.101029.
17. Xiao, J., *et al.*, Enhancing assessment of corn growth performance using unmanned aerial vehicles (UAVs) and deep learning. *Measurement*, 214, 112764, Jun. 2023, doi:10.1016/j.measurement.2023.112764.
18. Bayih, A.Z., Morales, J., Assabie, Y., de By, R.A., Utilization of Internet of Things and Wireless Sensor Networks for Sustainable Smallholder Agriculture. *Sensors*, 22, 9, Jan. 2022, doi:10.3390/s22093273.
19. Taha, M.F., *et al.*, Emerging Technologies for Precision Crop Management Towards *Agriculture* 5.0: A Comprehensive Overview. *Agriculture*, 15, 6, 582, Mar. 2025, doi:10.3390/agriculture15060582.
20. Shafi, U., Mumtaz, R., García-Nieto, J., Hassan, S.A., Zaidi, S.A.R., Iqbal, N., Precision Agriculture Techniques and Practices: From Considerations to Applications. *Sensors*, 19, 17, 3796, Sep. 2019, doi:10.3390/s19173796.
21. Gupta, G. and Kumar Pal, S., Applications of AI in precision agriculture. *Discov. Agric.*, 3, 1, 61, Apr. 2025, doi:10.1007/s44279-025-00220-9.
22. Soussi, A., Zero, E., Sacile, R., Trinchero, D., Fossa, M., Smart Sensors and Smart Data for Precision Agriculture: A Review. *Sensors*, 24, 8, 2647, Apr. 2024, doi:10.3390/s24082647.
23. Niedbała, G., Piekutowska, M., Hara, P., New Trends and Challenges in Precision and Digital Agriculture. *Agronomy*, 13, 8, 2136, Aug. 2023, doi:10.3390/agronomy13082136.
24. López-Quílez, A., AI, IoT and Remote Sensing in Precision Agriculture. *Appl. Sci.*, 15, 6, 2890, Mar. 2025, doi:10.3390/app15062890.
25. Escandón-Panchana, P., Herrera-Franco, G., Cuevas, S.M., Geomatic Tools in Agricultural Management, in: *CIGEO 2023*, MDPI, pp. 2, Dec. 2023, doi:10.3390/environsciproc2023028002.
26. Choi, Y., GeoAI: Integration of Artificial Intelligence, Machine Learning, and Deep Learning with GIS. *Appl. Sci.*, 13, 6, Jan. 2023, doi:10.3390/app13063895.
27. Kussul, N., Shelestov, A., Yailymov, B., Cloud Platforms and Technologies for Big Satellite Data Processing, in: *Information and Communication Technologies and Sustainable Development*, S., Dovgyi, O., Trofymchuk, V., Ustimenko, and L., Globa (Eds.), pp. 303–321, Springer Nature Switzerland, Cham, 2023, doi:10.1007/978-3-031-46880-3_19.
28. Wilkinson, R., *et al.*, Environmental impacts of earth observation data in the constellation and cloud computing era. *Sci. Total Environ.*, 909, 168584, Jan. 2024, doi:10.1016/j.scitotenv.2023.168584.

29. Dehghan-Shoar, M.H., Pullanagari, R.R., Kereszturi, G., Orsi, A.A., Yule, I.J., Hanly, J., A Unified Physically Based Method for Monitoring Grassland Nitrogen Concentration with Landsat 7, Landsat 8, and Sentinel-2 Satellite Data. *Remote Sens.*, 15, 10, Jan. 2023, doi:10.3390/rs15102491.
30. Wan, L., Li, H., Li, C., Wang, A., Yang, Y., Wang, P., Hyperspectral Sensing of Plant Diseases: Principle and Methods. *Agronomy*, 12, 6, Jun. 2022, doi:10.3390/agronomy12061451.
31. Ting, L., Khan, M., Sharma, A., Ansari, M.D., A secure framework for IoT-based smart climate agriculture system: Toward blockchain and edge computing. *J. Intell. Syst.*, 31, 1, 221–236, Jan. 2022, doi:10.1515/jisys-2022-0012.
32. Yambal, S. and Waykar, Y.A., Benefits and Opportunities of Geospatial AI Adoption, in: *Recent Trends in Geospatial AI*, pp. 271–296, IGI Global Scientific Publishing, 2025, doi:10.4018/979-8-3693-8054-3.ch010.
33. Segarra, J., Buchaillot, M.L., Araus, J.L., Kefauver, S.C., Remote Sensing for Precision Agriculture: Sentinel-2 Improved Features and Applications. *Agronomy*, 10, 5, May 2020, doi:10.3390/agronomy10050641.
34. Vyas, N., Sethi, G.K., Singh, S., Comparative Analysis of Different Classifiers on Fusion of Sentinel Datasets for Agriculture Land, in: *2024 4th International Conference on Technological Advancements in Computational Sciences (ICTACS)*, pp. 1173–1177, Nov. 2024, doi:10.1109/ICTACS62700.2024.10840995.
35. Caballero, G., *et al.*, Synergy of Sentinel-1 and Sentinel-2 Time Series for Cloud-Free Vegetation Water Content Mapping with Multi-Output Gaussian Processes. *Remote Sens.*, 15, 7, Jan. 2023, doi:10.3390/rs15071822.
36. Al-Yadumi, S., Xion, T.E., Wei, S.G.W., Boursier, P., Review on Integrating Geospatial Big Datasets and Open Research Issues. *IEEE Access*, 9, 10604–10620, 2021, doi:10.1109/ACCESS.2021.3051084.
37. Chen, D., Zhang, X., Chen, N., Yang, J., Gong, J., Geospatial sensor web adaptor for integrating diverse internet of things protocols within smart city. *ISPRS Ann. Photogramm. Remote Sens. Spat. Inf. Sci.*, V-4–2020, 115–121, Aug. 2020, doi:10.5194/isprs-annals-V-4-2020-115–2020.
38. Nhamo, L., *et al.*, Prospects of Improving Agricultural and Water Productivity through Unmanned Aerial Vehicles. *Agriculture*, 10, 7, Jul. 2020, doi:10.3390/agriculture10070256.
39. Shaurub, E.-S.H., Remote sensing and geographic information system applications as early-warning tools in monitoring fall armyworm, Spodoptera frugiperda: a review. *Int. J. Trop. Insect Sci.*, 44, 5, 2241–2258, Oct. 2024, doi:10.1007/s42690–024-01327–0.
40. Joshi, R. and Pandey, K., IoT-Enabled UAV, in: *Unmanned Aircraft Systems*, pp. 93–135, John Wiley & Sons, Ltd, 2024, doi:10.1002/9781394230648.ch3.
41. Grieco, R., *et al.*, The role of geospatial technologies for sustainable livestock manure management: A systematic review. *Sci. Total Environ.*, 954, 176687, Dec. 2024, doi:10.1016/j.scitotenv.2024.176687.

42. Sarkar, N.C., *et al.*, Enhancing livelihoods in farming communities through super-resolution agromet advisories using advanced digital agriculture technologies. *J. Agrometeorol.*, 25, 1, Feb. 2023, doi:10.54386/jam.v25i1.2080.
43. Cucho-Padin, G., Loayza, H., Palacios, S., Balcazar, M., Carbajal, M., Quiroz, R., Development of low-cost remote sensing tools and methods for supporting smallholder agriculture. *Appl. Geomat.*, 12, 3, 247–263, Sep. 2020, doi:10.1007/s12518-019-00292-5.
44. Mathenge, M., Sonneveld, B.G.J.S., Broerse, J.E.W., Application of GIS in Agriculture in Promoting Evidence-Informed Decision Making for Improving Agriculture *Sustainability*: A Systematic Review. *Sustainability*, 14, 16, Jan. 2022, doi:10.3390/su14169974.
45. Ogwu, M.C. and Kosoe, E.A., Integrating Green Infrastructure into Sustainable Agriculture to Enhance Soil Health, Biodiversity, and Microclimate Resilience. *Sustainability*, 17, 9, Jan. 2025, doi:10.3390/su17093838.
46. Cravero, A., Pardo, S., Sepúlveda, S., Muñoz, L., Challenges to Use Machine Learning in Agricultural Big Data: A Systematic Literature Review. *Agronomy*, 12, 3, Mar. 2022, doi:10.3390/agronomy12030748.
47. Perazzoni, F., Bacelar-Nicolau, P., Painho, M., Geointelligence against Illegal Deforestation and Timber Laundering in the Brazilian Amazon. *ISPRS Int. J. Geo-inf.*, 9, 6, Jun. 2020, doi:10.3390/ijgi9060398.
48. Batistatos, M.C., de Cola, T., Kourtis, M.A., Apostolopoulou, V., Xilouris, G.K., Sagias, N.C., AGRARIAN: A Hybrid AI-Driven Architecture for Smart *Agriculture. Agriculture*, 15, 8, Jan. 2025, doi:10.3390/agriculture15080904.
49. Rodrigues, H., *et al.*, Remote and Proximal Sensors Data Fusion: Digital Twins in Irrigation Management Zoning. *Sensors*, 24, 17, Jan. 2024, doi:10.3390/s24175742.
50. Kolluru, V., Mungara, S., Chintakunta, A.N., Kolluru, L., Telaganeni, C.S., Geospatial Intelligence Enhancement Using Advanced Data Science And Machine Learning: A Systematic Literature Review, Social Science Research Network, Rochester, NY, 4929468, Mar. 13, 2024, doi:10.2139/ssrn.4929468.
51. Villegas-Ch, W., García-Ortiz, J., Sánchez-Viteri, S., Toward Intelligent Monitoring in IoT: AI Applications for Real-Time Analysis and Prediction. *IEEE Access*, 12, 40368–40386, 2024, doi:10.1109/ACCESS.2024.3376707.
52. Althati, C., Malaiyappan, J.N.A., Shanmugam, L., AI-Driven Analytics: Transforming Data Platforms for Real-Time Decision Making. *J. Artif. Intell. Gen. Sci. (JAIGS)* ISSN3006–4023, 3, 1, Mar. 2024, doi:10.60087/jaigs.v3i1.155.
53. De Clercq, D., Nehring, E., Mayne, H., Mahdi, A., Large language models can help boost food production, but be mindful of their risks. *Front. Artif. Intell.*, 7, 1326153, Oct. 2024, doi:10.3389/frai.2024.1326153.
54. Maimaitijiang, M., Sagan, V., Sidike, P., Hartling, S., Esposito, F., Fritschi, F.B., Soybean yield prediction from UAV using multimodal data fusion and deep learning. *Remote Sens. Environ.*, 237, 111599, Feb. 2020, doi:10.1016/j.rse.2019.111599.

Index

Also of Interest

Check out these other related titles from Scrivener Publishing

Sustainable Approach to Land Use, Edited by Kundan Meshram and Chatrabhuj, ISBN: 9781394303250. This comprehensive guide transforms abstract sustainability theory into practical, real-world solutions, offering readers the proven policy frameworks, case studies, and green infrastructure strategies needed to balance urban economic growth with environmental stewardship.

Agricultural Supply Chain Optimization Using Federated Learning, Edited by Abhishek Kumar, Pooja Dixit, J.P. Ananth, S. Oswalt Manoj, and S. Panneerselvam, ISBN: 9781394461264. Master the next evolution of agricultural intelligence with this definitive guide to federated learning, providing the decentralized, privacy-preserving strategies needed to optimize global supply chains without compromising data sovereignty.

Climate Smart Agriculture, Edited by Anitha Velu, Prasanth Aruchamy, Raghu Ramamoorthy, Rajesh Kumar Dhanaraj, and Seifdine Kadry, ISBN: 9781394336333. Transform the future of sustainable farming with this guide to mastering the deep reinforcement learning architectures and algorithms that turn complex environmental data into precise, high-yield decisions for climate-smart agriculture.

Sustainable Agriculture Production Using Blockchain Technology, Edited by Rekh Ram Janghel, Rajesh Doriya, Jaykumar Lachure, and Yogesh Kumar Rathore, ISBN: 9781394248674. Revolutionize the agricultural supply chain with this essential guide, which provides the practical knowledge to leverage blockchain technology for transparency, traceability, and trust, alongside AI for overcoming modern farming challenges.

Mathematical Modeling in Agriculture, Edited by Sabyasachi Pramanik, Niranjanamurthy M., Ankur Gupta, and Ahmed J. Obaid, ISBN: 9781394233717. The main goal of the book is to explore the idea behind data modeling in smart agriculture using information and communication technologies and tools to make agricultural practices more functional, fruitful and profitable.

Agroforestry, Edited by Abhishek Raj, Manoj Kumar Jhariya, Arnab Banerjee, Ramesh Kumar Jha, and Krishan Pal Singh, ISBN: 9781394231133. This book offers comprehensive insights into the management of agroforestry for livelihood security and sustainable development in the tropics, addressing ecological interactions, productivity, and the monetization of carbon credits, while also outlining a future roadmap and policy challenges.

Smart Agritech: Robotics, AI, and Internet of Things (IoT) in Agriculture, Edited by Santosh Kumar Srivastava, Durgesh Srivastava, Korhan Cengiz, and Pramod Gaur, ISBN: 9781394302959. The main goal of *Smart Agritech: Robotics, AI, and the Internet of Things (IoT) in Agriculture* is to explore how emerging technologies such as robotics, artificial intelligence (AI), and IoT can be leveraged to improve efficiency, sustainability, and productivity in agriculture.

DIGITAL AGRICULTURAL ECOSYSTEM: Revolutionary Advancements in Agriculture, Edited by Kuldeep Singh and Prasanna Kolar, ISBN: 9781394242931. The book comprehensively explores the dynamic synergy between modern technology and agriculture, showcasing how advancements such as artificial intelligence, data analytics, and smart farming practices are reshaping the landscape to ensure food security in the era of climate change, as well as bridging the gap between cutting-edge research and practical implementation.

THE DIGITAL AGRICULTURAL REVOLUTION: Innovations and Challenges in Agriculture through Technology Disruptions, Edited by Roheet Bhatnagar, Nitin Kumar Tripathi, Nitu Bhatnagar and Chandan Kumar Panda, ISBN: 9781119823339. The book integrates computational intelligence, applied artificial intelligence, and modern agricultural practices and will appeal to scientists, agriculturists, and those in plant and crop science and management.

HANDBOOK OF AGRICULTURAL BIOTECHNOLOGY: Volume I Nanopesticides, Edited by Charles Oluwaseun Adetunji and Julius Kola Oloke, ISBN: 9781119836148. The book provides detailed information on the application of nanopesticides for the management of numerous agricultural pests and pathogens, as well as to foster an innovative pathway toward future sustainable agriculture and food systems.

HANDBOOK OF AGRICULTURAL BIOTECHNOLOGY: Volume II Nanobioherbicides, Edited by Charles Oluwaseun Adetunji, Julius Kola Oloke, ISBN: 9781119836155. The book provides a detailed examination of the application of nanobioherbicides that come from plants including information on the different metabolites derived from numerous plants that could become bioherbicides.

HANDBOOK OF AGRICULTURAL BIOTECHNOLOGY: Volume III Nanofungicides, Edited by Charles Oluwaseun Adetunji and Julius Kola Oloke, ISBN: 9781119836162. This book provides state-of-the-art information about recent advancements in the application of nanobiofungicides for effective management of post-harvest diseases and agricultural fungal diseases, including blights, mildews, molds, and rusts.

HANDBOOK OF AGRICULTURAL BIOTECHNOLOGY: Volume IV Nanoinsecticides, Edited by Charles Oluwaseun Adetunji and Julius Kola Oloke, ISBN: 9781119836179. The book provides detailed information about the application of repellent products that contain plant-based ingredients known as nanobioinsecticides, including the pesticide evaluation scheme guidelines for repellent testing, relevant information about the procedures to evaluate several repellent compounds, the development of new products that offer high repellency, and guidelines for consumer safety.

ECORESTORATION FOR SUSTAINABILITY, Edited by Arnab Banerjee, Manoj Kumar Jhariya, Surendra Singh Bargali, and Debnath Palit, ISBN: 9781119879718. With climate change and land degradation advancing at an alarming rate, the state of the global ecosystem and how it can be restored in our march toward sustainability is incredibly important, and this new, groundbreaking volume tackles these important issues head on.

FUNCTIONAL POLYMERS IN FOOD SCIENCE, Part 1: Food Packaging, Edited by Giuseppe Cirillo, Umile Gianfranco Spizzirri, and Francesca Iemma, ISBN: 9781118594896. Provides an extensive overview of the link between polymer properties and food quality. *NOW AVAILABLE!*

FUNCTIONAL POLYMERS IN FOOD SCIENCE, Part 2: Food Processing, Edited by Giuseppe Cirillo, Umile Gianfranco Spizzirri, and Francesca Iemma, ISBN: 9781118595183. Provides an extensive overview of the link between polymer properties and food quality. *NOW AVAILABLE!*

ADVANCES IN FOOD SCIENCE AND TECHNOLOGY Volume 1, Edited by Visakh, P.M., Sabu Thomas, Laura B. Iturriaga, and Pablo Daniel Ribotta, ISBN: 9781118121023. Written in a systematic and comprehensive manner the book reports recent advances in the development of food science and food technology areas. *NOW AVAILABLE!*

ADVANCES IN FOOD SCIENCE AND NUTRITION Volume 2, Edited by Visakh P.M., Laura B. Iturriaga, and Pablo Daniel Ribotta, ISBN: 9781118137093. This important book comprehensively reviews research on new developments in all areas of food chemistry/science and nutrition. *NOW AVAILABLE!*

Applied Water Science Volume 1: Fundamentals and Applications, Edited by Inamuddin, Mohd Imran Ahamed, Rajender Boddula and Tauseef Ahmad Rangreez, ISBN: 9781119724766. Edited by one of the most well-respected and prolific engineers in the world and his team, this is the first volume in a two-volume set that is the most thorough, up-to-date, and comprehensive volume on applied water science available today. *COMING IN SUMMER 2021*

Sustainable Water Purification, by Safiur Rahman and M. R. Islam, ISBN: 9781119650997. This is the only book that takes a zero-waste approach to propose 100% sustainable water purification techniques. *NOW AVAILABLE!*

Books in the series, "Bioprocessing in Food Science"

NUTRACEUTICALS FROM FRUIT AND VEGETABLE WASTE, Edited by Vidisha Tomer, Navnidhi Chhikara, Ashwani Kumar, and Anil Panghal, ISBN: 9781119803508. Nutraceuticals from Fruit and Vegetable Waste, the latest volume in the series, "Bioprocessing in Food Science," is an essential reference on the valorization of fruit and vegetable waste and the state of the art for fruit and vegetable processing.

NUTRITIONAL SCIENCE AND TECHNOLOGY: Concept to Application, Edited by Tejpal Dhewa, Anil Kumar Puniya, and Anil Panghal, ISBN: 97811198089. This new volume in the groundbreaking series, "Bioprocessing in Food Science," Nutritional Science and Technology, is a comprehensive resource covering topics such as human nutrition, food safety concerns, and emerging technologies like nanotechnology, and it also addresses the need for innovative food products, contamination, and regulatory challenges in the global food market.

HARVESTING FOOD FROM WEEDS, Edited by Prerna Gupta, Navnidhi Chhikara, and Anil Panghal, ISBN: 9781119791973. An exciting new volume in the groundbreaking new series, "Bioprocessing in Food Science," Harvesting Food from Weeds focuses on the latest processes, industrial applications, and leading research on this important, evolving new area, for engineers, scientists, students, and other industry professionals.

MICROBES IN THE FOOD INDUSTRY, Edited by Navnidhi Chhikara, Anil Panghal, and Gaurav Chaudhary, ISBN: 9781119775584. This newest volume in the groundbreaking new series, "Bioprocessing in Food Science," focuses on the latest processes, industrial applications, and leading research on microbes in the food industry, for engineers, scientists, students, and other industry professionals.

NOVEL TECHNOLOGIES IN FOOD SCIENCE, Edited by Navnidhi Chhikara, Anil Panghal, and Gaurav Chaudhary, ISBN: 9781119775584. The third book in the series, "Bioprocessing in Food Science," covers the latest and greatest new technologies in food science.

FUNCTIONAL FOODS, Edited by Navnidhi Chhikara, Anil Panghal, and Gaurav Chaudhary, ISBN: 9781119775560. Presenting cutting-edge information on new and emerging food engineering processes, *Functional Foods*, the second volume in the groundbreaking new series, "Bioprocessing in Food Science," is an essential reference on the modeling, quality, safety, and technologies associated with food processing operations today.

THERMAL FOOD ENGINEERING OPERATIONS, Edited by Nitin Kumar, Anil Panghal, and M. K. Garg, ISBN: 9781119775591. Presenting cutting-edge information on new and emerging food engineering processes, Thermal Food Engineering Operations, the first volume in the new series, "Bioprocessing in Food Science," is an essential reference on the modeling, quality, safety, and technologies associated with food processing operations today.

Printed and bound by CPI Group (UK) Ltd, Croydon, CR0 4YY
17/09/2026
14964265-0003